The Emerald Handbook of Tourism Economics and Sustainable Development

BUILDING THE FUTURE OF TOURISM

Series editor: Anukrati Sharma

The world is entering the Third Millennium in which great changes are expected in all areas of human interest, life, and activity. These changes have been brought on by past and present man-made events, which have had both positive and negative consequences. The coming millennium will be marked by significant social, political, demographic, and technological changes and will definitely differ from the last century. The future will bring more leisure time, a higher standard of living, and a better quality of life for us all. This series examines recent and the most probable changes and gives a wide range of visionary insights, as well as operational takeaways.

Forthcoming Volumes

Strategic Tourism Planning for Communities: Restructuring and Rebranding
Anukrati Sharma and Shruti Arora

Dark Tourism: Theory, Interpretation and Attraction
Anukrati Sharma, Shruti Arora, and Parag Shukla

Meaningful Tourism: Strategies and Futuristic Development
Pankaj Kumar Tyagi, Vipin Nadda, and Ajit Kumar Singh

Value Proposition for Tourism 'Co-opetition': Cases and tools
Adriana Fumi Chim Miki and Rui Augusto da Cost

The Emerald Handbook of Tourism Economics and Sustainable Development

EDITED BY

AHMED IMRAN HUNJRA

International University of Rabat, Morocco

AND

ANUKRATI SHARMA

University of Kota, India

emerald
PUBLISHING

United Kingdom – North America – Japan – India – Malaysia – China

Emerald Publishing Limited
Emerald Publishing, Floor 5, Northspring, 21-23 Wellington Street, Leeds LS1 4DL

First edition 2024

British Library Cataloguing in Publication Data
A catalogue record for this book is available from the British Library

ISBN: 978-1-83753-709-9 (Print)
ISBN: 978-1-83753-708-2 (Online)
ISBN: 978-1-83753-710-5 (Epub)

Printed and bound by CPI Group (UK) Ltd, Croydon, CR0 4YY

INVESTOR IN PEOPLE

Contents

List of Figures and Tables

Figures

Tables

About the Editors

Ahmed Imran Hunjra, Rabat Business School, International University of Rabat, Morocco. He has over 12 years of teaching experience in Finance and Quantitative Methods across a large number of universities internationally. His current area of research interest is in the domain of corporate finance, with a strong passion for topics related to the integrity and reputation of corporations, e.g., transparency, accountability, responsibility, fairness, and sustainability. He is a prolific author and has published in a broad range of highly cited international research journals, including *International Review of Financial Analysis, Journal of International Financial Markets, Institution & Money, Pacific Basin Finance Journal, Global Finance Journal, Finance Research Letters, International Review of Economics & Finance, Technological Forecasting & Social Change, Journal of Environmental Management, Economic Modelling, International Journal of Finance & Economics, Research in International Business & Finance, Quarterly Review of Economics & Finance*, among others. He is a senior area editor of the *International Journal of Emerging Markets* and sits on the editorial board of several ABS & ABDC indexed journals. Recently, he won the 'Best Researcher Award-2022' from Rabat Business School, 'Outstanding Reviewer Award-2021 & 2023' by Emerald Literati, the 'Distinguished Reviewer Award-2021' by Virtus Interpress, 'Best Paper Award-2022 & 2023' by the Institute of Chartered Accountants of India (ICAI), International Research Award and the 'Best University Teacher Award-2020' from PMAS-Arid Agriculture University Rawalpindi, Pakistan. He has supervised 6 PhDs and over 50 MS/MPhil scholars to successful completion.

Anukrati Sharma, Head and Associate Professor of the Department of Commerce and Management at the University of Kota (A State Govt. University) in Kota, Rajasthan, India. She is the Director of the Skill Development Centre of the same University. In 2015, she received a Research Award from the University Grants Commission (UGC), New Delhi, for her project 'Analysis of the Status of Tourism in Hadoti and Shekhawati Region/Circuit (Rajasthan): Opportunities, Challenges, and Future Prospects'. She has been appointed as the Official Partner of the BRICS Funded Project for the years 2022–2024. Dr Sharma is a Routledge featured author. Dr Sharma is the Book Series Editor of 'Building the Future of Tourism' published by Emerald Publishing. UK She is also the Book Series Editor of 'Routledge Insights in Tourism Series', Routledge, UK. She also serves as Book Series Editor of 'Perspectives and Anthropology in Tourism and Hospitality (PATH)', Apple Academic Press (CRC Press a Taylor and Francis Group).

About the Contributors

Muhammad Adnan, a skilled Network Engineer at Ovex Technologies, Pakistan, boasts expertise in routing protocols like BGP and OSPF, switching with knowledge in VLANs and STP, and device familiarity with Huawei routers and Cisco Catalyst series. He is adept in both Windows and Linux operating systems and displays proficiency in network security, covering AAA levels and URL filtering. Adnan's educational background includes an MCS from Lincoln University College, Malaysia. He has garnered numerous certifications, such as CCNA and HCIA. Notably, he coauthored a paper in 2023 on digital banking.

Naema Mohammed Al Bimani holds a PhD in Business Administration with concentration in Management from the University of South Alabama, USA, in 2022. Her Master's degree is in Media Management from Curtin University of Technology, Australia, 2008. She also has a Post Diploma in Project Management Professional from Maastricht School of Management, Netherlands, 2010. She currently serves as an Assistant Professor at the University of Nizwa. She has over 20 years of experience in both industrial and academic sectors in Oman, and she served as a graduate assistant at the University of South Alabama, USA. Her research interests include entrepreneurship, strategic management, dynamic capabilities, and leadership.

Manpreet Arora, a Senior Assistant Professor of Management at the Central University of Himachal Pradesh, Dharamshala, India, brings over 22 years of rich teaching experience. She holds academic accolades including a PhD in International Trade, an MPhil, a gold medalist, and several other academic distinctions from Himachal Pradesh University, Shimla. Dr Arora's diverse research interests encompass accounting, finance, strategic management, entrepreneurship, qualitative research, and microfinance. She works on mixed methods research. Noteworthy for guiding doctoral research and delving into microfinance, entrepreneurship, behavioral finance, and corporate reporting, she has presented at numerous seminars, delivering talks on various academic subjects across multiple universities and colleges. An accomplished academic, she has an impressive publication record, having authored over 30 papers in esteemed national and international journals listed in Scopus, WOS, and Category journals, alongside contributing to 55 book chapters in publications by reputed publishers like Emerald, Routledge, CABI, Springer Nature, AAP, and more. Her commitment to management research is evident through the editing of four books. Dr Arora is not only a dedicated researcher but also an ardent advocate for women's rights, combining academia with active social work in that domain. A motivational speaker and workshop facilitator, she inspires through communication and motivation workshops. Her impactful contributions showcase a multifaceted professional excelling in academia, research, and social advocacy.

Shruti Arora is currently working as a Guest Faculty in the Department of Commerce and Management, University of Kota, Rajasthan, India. She has an experience of 14 years in the education industry. She has attended various international conferences, and her publication includes various chapters in edited books like Routledge, Emerald, Springer and research papers in national and international UGC approved and peer-reviewed journals. Dr Arora has authored one book in and two edited under Apple Academic Press, CRC and the other two edited books under Emerald. She is appointed as the BRICS Community of Practice (CoP) Researcher for promoting cross-BRICS nations' collaborations in co-authorship, co-supervision, co-examining, co-publishing in DHET and SCOPUS accredited journals, co-conferencing participation and hosting rotating BRICS workshops, seminars and guest lectures in the academic years 2023 and 2024. The term 'Food Walk Tourism' is included in *Encyclopedia of Tourism Management and Marketing*, Editor-in-Chief, Professor Dimitrios Buhalis: Bournemouth University, Edward Elgar Publishing Limited, ISBN: 9781800377479, e-ISBN: 9781800377486 in 2022.

Saira Arsh is a PhD scholar in Economics at Lahore College for Women University, Lahore. Currently, she is working as a Visiting Faculty at Lahore College for Women University, Lahore. Her research interests are energy economics and environmental science.

Jahanvi Bansal is currently working as an Assistant Professor, School of Management, GSFC University, Vadodara, Gujarat (India). Her research papers have been published in reputable national and international journals. She was honored with 'Emerald Literati Award- Highly Commendable Paper 2020' by Emerald Publishing House, United Kingdom. Her research interests include workforce diversity, business ethics, corporate social responsibility, and SDGs.

Truong Thi Xuan Dao is the Head of the Tourism Department and a researcher at Hong Bang International University of Viet Nam. She has an MBA degree from the University of Economics Ho Chi Minh City and a PhD degree in Tourism at VNU University of Social Sciences and Humanities – VNU-USSH. She has more than 10 researches in business and tourism. Her field research includes marketing, business, management, and organizational commitment.

Antony Davis is an Assistant Professor at Christ academy Institute for Advanced Studies and is currently pursuing PhD from the school of Business and Management at Christ University, Bangalore. He holds an MBA in Finance specialization from CHRIST University, Bangalore and a degree in Philosophy from Dharmaram Vidya Kshetram, Bangalore. His research interests include social innovations, financial inclusion, social entrepreneurship, and their profound impact on well-being of rural artisans and people at the bottom of the pyramid.

He has twelve years of extensive experience in teaching and administrative and leadership roles in various educational institutions, he specializes in courses on Entrepreneurship and Skill Development. And actively engages with artisans by conducting orientation sessions on skill development in collaboration with the Export Promotion Council for Handicrafts in India, contributing significantly to empowering and uplifting these communities. Presented impactful papers on Social Entrepreneurship Models at National Law University, Gujarat, and Humanistic Management at IIM, Nagpur. He also holds a patent in product design in the domain of psychological wellbeing.

M. Bina Celine Dorathy is a Professor at Alliance University, Bangalore. Holding a PhD in Commerce from the University of Madras and an MCom from Madras Christian College, her expertise lies in foreign trade. Additionally, she has earned an MBA in Shipping and Logistics, enriching her insights into maritime affairs. With a teaching tenure exceeding 25 years, including a significant stint at a Maritime University, she has mentored students in international business and shipping intricacies. Dr Dorathy is an avid researcher, contributing prolifically to shipping and logistics, finance and international business literature. Her commitment extends to supervising PhD scholars and spearheading entrepreneurial initiatives, fostering a holistic educational experience for her students.

David T. Easow is a Professor and the Deputy Director – International Affairs, in Hindustan Institute of Technology and Science Chennai. He has Cross-Functional Experience and Cross-Industry Expertise in Cross-Cultural Environment Senior Cross-functional management executive with strong domestic and international experience in diverse industries and business management institutions. Ambitious, innovative, result-oriented, strategically focused, influential strategist with exceptional team-based leadership skills. Proven record of accomplishment in planning, building, and managing high-profile businesses and institutions from concept to completion.

Yuzine Esa is affiliated with the Department of Aquaculture, Universiti Putra Malaysia. He is currently providing services as Deputy Dean of Office [Development, Industry, and Community Relations] and has authored and coauthored multiple peer-reviewed scientific papers and presented works at many national and international conferences. His contributions have acclaimed recognition from honorable subject experts worldwide. He is actively associated with different societies and academies. His academic career is decorated with several reputed awards and funding. Her research interests include genetic and fish reproduction.

Sanjeev Gupta is presently the Dean of School of Commerce and Management. He had more than 60 publications in his credit in journals of repute. He has expertise in Data Science, Applied Forecasting, FinTech, and Financial Econometrics. He works on mixed methods research and currently deals with decentralized finance. He has widely traveled in India and abroad. He is a life member of various academic and professional bodies. He has strong administrative and career guidance skills. He has contributed a lot in terms of guest lectures and discussions in institutes of higher education on research and academics.

Suziana Hassan is a Senior Lecturer in the Center of Sustainable Inclusive and Development at the Faculty of Economics and Business, Universiti Kebangsaan Malaysia. She obtained her PhD from Faculty of Sciences in Copenhagen University, Denmark. Her areas of expertise are Nature Resource and Applied Economics which focus on the implementation of Choice Experiment Method for Economic Modelling. Her current fields of interest are economics valuation of ecotourism, natural resources, and biodiversity conservation. Her contributions include scientific publications, supervision, and teaching of the same discipline. She has supervised PhD and Master Science students of related fields.

Mohamad Syahrul Nizam Ibrahim is a Senior Lecturer from the Institute for Tropical Biology and Conservation, Universiti Malaysia Sabah. He is an expert in biodiversity and natural resource conservation. He has authored articles on Langkawi UNESCO Global Geopark and UNESCO World Heritage Sites of Gunung Mulu National Park, Malaysia. His research interests span environmental sociology, economics, community psychology, biodiversity, tourism, and protected areas. Notably, he serves as a peer reviewer for several reputable journals, including the *Journal for Nature Conservation*, *Humanities and Social Sciences Communications*, *Cogent Food and Agriculture*, and more. He is an active member of esteemed organizations like the Asian Association of Environmental Economics, Southeast Asia Research Academy, and others. Furthermore, he has actively participated in research collaborations with Sarawak Forestry Corporation and Sarawak Tourism Board, focusing on tourism and forestry. In addition to his early career, he was trained by the Academy Sciences of Malaysia through series of workshops via Chrysalis Award 2022.

Shazali Johari, a distinguished Senior Lecturer in the Faculty of Forestry and Environment at Universiti Putra Malaysia, holds a PhD from the same institution in the field of nature tourism. His areas of expertise encompass recreation programming, sustainable indigenous tourism, community-based tourism, outdoor recreation, and social entrepreneurship. Beyond academia, he has also made a significant impact as the principal assistant secretary of the Ministry of Tourism, Arts, and Culture in Sarawak. Dr Johari's multifaceted career extends to mentoring students and guiding research projects in these domains, nurturing future professionals with a profound understanding of tourism. His contributions bridge the gap between theory and practice, making him a respected figure in the field, with a lasting influence on the development of sustainable and responsible tourism practices.

Narentheren Kaliappen is an Associate Professor of International Business at School of International Studies, Universiti Utara Malaysia. He has published in WoS/Scopus journals from Emerald, Inderscience, and Sage, focusing on strategic management and business administration. He is the author of 'Match Your Strategies for Brilliant Performance' and 'Strategy Execution for Performance Excellence'. He has earned accolades, including the Distinguished Doctoral Dissertation Award, Best Paper Awards, and the 2016 Emerald Literati Award for Excellence.

K.Lakshmypriya holds a doctoral degree in Management on the research area work–life balance of women entrepreneurs. 20 years of teaching experience in UG and PG programs. Certified entrepreneurship educator from Stanford technological ventures and NEN. Underwent Faculty training in entrepreneurship from EDII, Ahmadabad. MDP in Design Thinking from XIME, MOOC course on Agile meets design thinking. Curated courses on Innovation and Entrepreneurship for students. Presented a paper on social entrepreneurship at IIM Trichy, INDAM conference, 2020 Consultant at SHG and women empowerment program at KRRA, Trivandrum, Kerala, 2016–2017. She has 31 publications in the domain of entrepreneurship and HRM. Ms Lakshmipriya is currently doing a university funded project on understanding succession planning and financial models and its impact on organizational sustainability of nonprofit organizations and has been involved in consultancy projects with start-ups.

Mustafa Malik received his PhD degree in Commerce (Tourism Management) from Aligarh Muslim University (India) in 2004 and his Master's degree in Tourism Administration (MTA) from Aligarh Muslim University in 2001. He also Qualified UGC-NET examination in Tourism Management in 2004. He is two times recipient of the University Medal from Aligarh Muslim University (AMU) for securing first position in BSC (Hon's) 1999 and in MTA 2001. Dr Malik is presently an Assistant Professor of Management, Assistant Dean for Undergraduate Studies, and Quality Management Officer in the College of Economics, Management and Information Systems, University of Nizwa, Oman. He has over 15 years of experience in teaching, research, and academic administration. His research interests include tourism and community development, sustainability, organizational leadership, and quality management in higher education. He has published in several international journals and completed three TRC-funded research projects.

Diana Oliveira Marques holds a Master in Marketing from the Polytechnic Institute of Cávado and Ave (IPCA), Barcelos, Portugal. Diana holds a Bachelor's degree in Tourism from Polytechnic Institute of Viana do Castelo (IPVC, Portugal). She strives to unravel the intricacies of marketing strategies in the context of the tourism industry and is committed to exploring innovative approaches that bridge the gap between marketing theory and the practical nuances of the tourism sector. Research areas: relational marketing, strategy, management, and health marketing.

Amit Mittal has over two decades of domestic and international experience in academic leadership, teaching, research, consulting, training, and mentorship. At Chitkara University, his current mandate is to manage and coordinate the PhD programs, research publications, and international/domestic research collaborations. Seventeen scholars have been awarded PhD degrees under his guidance, and he has published over 100 Scopus/SSCI indexed papers with a number of these included in the ABDC/ABS journal list. He was the recipient of the Career 360 'Outstanding Faculty Researcher Award' 2023, the Chitkara University Excellence award 2023 (Publications in Business School Category), 2021 (February) for highest cited author and publications with highest h-index (Business School category). He is an active resource person for FDPs, MDPs, and Corporate Trainings. He is a member of the thesis review board of a number of universities. He presently reviews for reputable journals such as *Technological Forecasting and Social Change* (Elsevier), *Public Health* (Elsevier), *Benchmarking* (Emerald), *Technology Analysis and Strategic Management* (Taylor and Francis), *International Journal of Consumer Studies* (Wiley), *Journal of Public Affairs* (Wiley), *Routledge Studies in Global Student Mobility*, *International Journal of Emerging Markets* (Emerald), *Sustainability* (MDPI), *Management Decision* (Emerald), *IIM KSMR* (Sage), etc. He serves on the editorial boards of Frontiers in Psychology (SSCI/Scopus indexed), Open Psychology Journal (Scopus indexed), Proceedings on Engineering Sciences (Scopus indexed), and Research on Enterprise in Modern Economy (Gdańsk Univeristy of Technology, Poland). He is currently guest editing a special issue in the *International Journal of Information Management Data Insights* (Elsevier).

Ravi Kant Modi is presently working as a Professor and Dean, Nirwan University Jaipur. He is a distinguished faculty member bringing with him an extensive 13 years of teaching experience. His contribution in research solidifies his position as a highly accomplished researcher and scholar. Dr Modi is an active member of several professional bodies, staying connected with the latest developments in his field. He has organized numerous online and offline conferences, seminars, FDP, workshops as Conference Secretary, showcasing his commitment to advancing knowledge and fostering intellectual exchange. As an author, Dr Modi has contributed to the academic community through his textbooks and edited books. Moreover, he has published over 40 papers in renowned national and international journals, including notable Scopus journals. His expertise in research is further exemplified by his position as an editorial board member for reputable journals. Dr Modi is an avid participant in academic conferences and seminars, having attended more than 50 events where he has presented his research. His comprehensive knowledge and rich academic background make him an invaluable and a dedicated mentor to the students.

Rozihan Mohamed is affiliated with the Department of Aquaculture, University Putra Malaysia. Dr Rozihan Mohamed has received a BSc degree from UPM in Fisheries Sciences, an MS from UPM in Biotechnology, and a PhD from UKM in Zoology. Dr Rozihan Mohamed is currently providing services as a Lecturer. Dr Rozihan Mohamed has published numerous publications in national and international peer-reviewed journals and presented scientific papers worldwide. Because of his active association with different societies and academies and his contributions, Dr Rozihan Mohamed has been recognized by subject experts around the world. Various reputed awards appreciate Dr Rozihan Mohamed's contributions. Dr Rozihan Mohamed's clinical and scientific research interests include fish population genetics.

Mohd Hairul Mohd Salleh, a Graduate Research Fellowship holder, started his PhD in Genetics in 2019 and graduated in 2024 at Universiti Putra Malaysia. He was born in Johor Bahru, Johor, on May 14, 1983. He received a Bachelor's degree in environmental biology from Universiti Sains Malaysia and a Master's degree in conservation biology from Universiti Kebangsaan Malaysia. He has about 19 years of working experience. Recently, he has published 20 articles with 121 citations, and his h-index is 5 (Google Scholar). From a biological standpoint, he is interested in turtle, tortoise, and penguin studies. Keep in touch with Hairul via the web: ResearchGate: https://www.researchgate.net/profile/Mohd-Hairul-Mohd-Salleh

Samia Nasreen is an Associate Professor at Lahore College for Women University, Lahore. Her research interests focus on various issues concerning macroeconomy, environment, growth, and development. Her research has featured in journals such as *International Journal of Finance and Economics*, *International Journal of Health Planning and Management*, *Energy Economics*, *Empirical Economics*, *Computational Economics*, and many others.

Xuan-Hoa Nghiem received his PhD in Economics from School of Economics, Finance and Marketing, College of Business, RMIT University, Melbourne, Australia. He is currently a Lecturer in Economics at the International School, Vietnam National University, Hanoi, Vietnam. His research interests include macroeconomics, monetary policy, international economics, environmental economics, and energy economics. He has published various Scopus-indexed journals such as *Technological Forecasting and Social Change*, *Renewable Energy*, *Finance Research Letters*, *Utilities Policy*, *Economic Analysis and Policy*, *Management of Environmental Quality*, *Journal of Environmental Management*, etc. He also acts as a reviewer for various journals such as *Finance Research Letters*, *Heliyon*, *Economic Change and Restructuring*, *Journal of Asia Pacific Economy*, *International Journal of Manpower*, and *Journal of International Economics and Management (JIEM)*. Before that, he used to work as a Lecturer at Foreign Trade University, Hanoi, Vietnam; in the banking sector, as a Researcher at Deposit Insurance of Vietnam (DIV) – a Bank Regulator and as a Risk Management Officer at PVcomBank – a commercial bank in Vietnam.

Phuong Mai Nguyen is an Associate Professor at the International School, Vietnam National University, Hanoi. Her main research interests include sustainable development, consumer behaviors, corporate governance, and entrepreneurship. She obtained her MBA from National Kaohsiung University of Applied Sciences in Taiwan. In addition, she gained her PhD in Business Administration from University of Economics and Business, Vietnam National University, Hanoi, and a postdoctoral fellowship at Chung-Ang University, Republic of Korea.

Thi Kim Duyen Nguyen is a Lecturer at the International School – Vietnam National University, Hanoi (VNU-IS). She holds a Bachelor's degree in international business and a Master of Business Administration (MBA), demonstrating her commitment to academic excellence and leadership in the business world. In her career as an educator, Ms Duyen has focused primarily on a diverse range of subjects, including international business, management, tourism, marketing, and e-commerce. Her research interests align closely with her teaching areas. In the field of tourism, Ms Duyen is particularly interested in how global events and market trends influence consumer preferences and the overall industry landscape.

Thu Giang Nguyen is currently a PhD candidate in Agricultural and Resources Economics at Kangwon National Univeristy (Korea). She has obtained a Master's Degree in Development Economics from Paris 1 Pantheon-Sorbonne University (France).

Jackie Lei Tin Ong is an experienced Tourism and Hospitality Consultant and an educator. She has lectured in her expertise for universities, government, and United Nations agencies with deep dedication in nurturing the next generation of leaders in tourism and hospitality industry in enhancing the quality and sustainability of this dynamic sector. She is currently a Senior Lecturer in the RMIT University based in Vietnam. Her professional expertise focuses on managing sustainability in the tourism and hospitality sector in the Asia Pacific region. Her skills extend across various domains including community-based tourism, coastal tourism development, urban tourism planning, corporate social responsibility, policy analysis, and cultural heritage tourism. Strong program and project management professional, Dr Ong graduated with a doctorate in Sustainable Development Planning from the University of Queensland, a Master's degree in Urban Planning from the University of Melbourne, and a Bachelor's degree from the National University of Singapore.

Ana Teresa Pedreiro holds a PhD in Public Health from the National School of Public Health of the Nova University Lisbon. She is the author and coauthor of several papers, and her research interests include health promotion, health management, health marketing, and public health. She is a researcher at the Health Sciences Research Unit: Nursing.

Huong Trang Pham is a Lecturer at the Business School, RMIT University Vietnam. She holds a doctoral degree in Tourism from Vietnam National University Hanoi, a Diploma in Tourism Management in Austria, and two MBA degrees from the University of Applied Sciences Bremen, Germany, with distinction. She got Austrian and German scholarships for her studies. Her primary areas of teaching and research encompass international strategic management, sustainable tourism, destination marketing, and tourism management. Dr Trang consistently engages with individuals possessing significant expertise and experience across diverse backgrounds.

Aishwarya Singh Raikwar is currently working as an Assistant Manager in the Vibrant Village Programme (VVP) Cell of the Ministry of Tourism, Government of India. She is a TEDx Speaker and a recipient of the 'Young Researcher Award' at the Asian Tourism Research Conference held at Christ University, Bangalore. She is also a UGC-NET qualified candidate and a certified Tourism Professional. Her research interests include blue economy and maritime tourism in the context of changing global order and diplomacy in the Indo-Pacific region. She is passionate about oceans and wants to work on some of the biggest challenges of our times: Climate Change, Maritime Security, Sustainable Development Goals, and Geopolitics in the Indo-Pacific region. A traveler with a strong academic background, a diplomatic, personable, highly organized, and self-motivated person with a maritime vision. She aspires to contribute to the field of research and wishes to positively impact people's lives with her work.

Sandhya H. is an Assistant Professor at Christ (Deemed University) Bangalore and is currently pursuing PhD from the school of Business and Management at Christ University, Bangalore. has successfully completed her MPhil in Tourism and Master's in Tourism Administration (MTA) from Christ University Bangalore, India, by securing first rank. She has industry experience of one year having worked with one of the leading multinational tour operations companies, Kuoni SOTC, Bangalore, and teaching experience of over 7 years in various colleges and institutions in Kerala. She has worked as a Research Associate of the Major Research Project on Strategic Intervention of Destination Management Organizations – State of Karnataka. Her research interests include Tourism Marketing, Destination Management, Smart Tourism, and Destination Branding.

Parag Shukla is an Assistant Professor in Commerce at Maharaja Sayajirao University of Baroda, India, specializing in Marketing Management. He earned his Bachelor's and Master's degrees from the same university, focusing on Marketing Management. Dr Shukla's research centers on retailing, and he has a background in content analysis within the television and media research industry. He teaches management courses at various levels and has published extensively in national and international journals and conferences. His current research project is titled 'An Empirical Investigation of Experiential Value vis-a-vis Usage Attitude of Selected Mobile Shoppers in Gujarat'. Dr Shukla is notable for receiving the Silver Medal at the 68th International All India Commerce Conference for his research, earning the Best Business Academic of the Year Award, a significant recognition in Indian Education and Retail Industry.

Bruno Barbosa Sousa, PhD, is a Professor in Polytechnic Institute of Cavado and Ave (IPCA, Portugal) Head of Master Program – Tourism Management and Head of Master Program Tourism Marketing (IPCA, Portugal). PhD Marketing and Strategy (University of Minho, Portugal). He is the author and coauthor of several papers, and his research interests include tourism management, marketing, and strategy. Editorial board member of several peer-reviewed scientific journals and ad hoc reviewer of several peer-reviewed scientific journals. Member of the scientific committee of several national and international congresses and conferences. He has published in the *Journal of Enterprising Communities*, *Tourism Management Perspectives*, *Current Issues in Tourism*, *Journal of Organizational Change Management*, *World Review of Entrepreneurship, Management and Sustainable Development*, *Wine Economics and Policy*, among others.

Fathimah Nachia Syed completed her Master of Science in International Business at the Ghazali Shafie Graduate School of Government (GSGSG), Universiti Utara Malaysia. The title of her project paper was 'The Effects of Muslim Tourist Perceived Values Towards Muslim Tourist Satisfaction in Langkawi'. Her research interests include cross-cultural management and selected issues in international business. Ms Fathimah has experience working in the corporate sector and has also published book chapters.

Pankaj Kumar Tripathi is working as an Assistant Professor in the Department of Accounting and Financial Management, Faculty of Commerce, the Maharaja Sayajirao University of Baroda, Vadodara. He has vast experience in academics, worked in different states of India adding varied multicultural experience to his career. He holds a PhD in Finance and Accounting. His areas of expertise include financial accounting, financial economics, business economics, and business administration. He has published various papers in national and international journals.

Bindi Varghese is currently affiliated with CHRIST (Deemed to be University), Bangalore, as an Associate Professor in the tourism department. With 20 years of experience instructing undergraduate and graduate students, Bindi Varghese holds a doctorate in Commerce specialized in Tourism Management. She is also the campus research coordinator and associate editor of ATNA, Journal of Tourism Studies which is a UGC peer-reviewed journal. Along with her academic expertise, she is closely associated with several NGOs working in the field of Tourism livelihood. Dr Bindi has completed a major research project on Strategic Intervention of Destination Management Organizations for the State of Karnataka. She has authored and edited books in the area of tourism and destination management, and her active researches undertaken include: Impact Assessment Studies, Medical Tourism, Destination Management Organization, and Ecological Studies. Currently, she is undergoing two funded projects under CHRIST university.

Amirah Sariyati Mohd Yahya is a PhD candidate in the Department of Geography Universiti Malaya. Her research focuses on e-waste, and she already published her findings in both local and international publications together with both her supervisors Dr Tengku Adeline Adura Tengku Hamzah and Associate Professor Dr Aziz Shafie. Her research focuses on the Malaysian public's knowledge, attitude, and practices regarding e-waste. She graduated with a Bachelor of Science (Geography) from Universiti Malaya in 2016 and a Master of Environment from Universiti Putra Malaysia in 2018.

Kashifa Yasmeen holds a PhD in Psychology and serves as an Assistant Professor at the University of Sahiwal. Her expertise lies in organizational psychology and personality development. An active researcher, she has published many academic papers. She has played the role of a reviewer for numerous publications and is a member of various committees within the University of Sahiwal. Additionally, she has a 5-year tenure as a practicing psychologist.

Kausar Yasmeen is an esteemed academic currently serving as an Assistant Professor at the University of Nizwa since 2017. Previously, she was the Head of the Department at Preston University, Kohat, Pakistan, and a Lecturer at University Utara Malaysia. She boasts an extensive publication record on topics like economic happiness, entrepreneurship, and microfinance. She has authored a book, a book chapter, completed two research projects, and received awards for her contributions.

Do Hai Yen got a PhD in Cultural Studies at Hanoi University of Culture, Viet Nam. She graduated with a BA and an MA degree in Tourism Studies at Hanoi National University, Vietnam. She also has more than 15 years of working experience in tourism businesses in different positions in travel agencies and 16 years of working experience in the field of teaching tourism in Vietnam, in the positions of tour guide tour program operator, and salesman. She is currently a Lecturer at the Faculty of Tourism, East Asia University of Technology, Hanoi. Fields and research interests are sustainable tourism development, tourism culture, culture tourism, and tourism studies.

Rosmiza Mohd Zainol is a Senior Lecturer at the Geography Programme, Faculty of Social Sciences and Humanities, Universiti Kebangsaan Malaysia (UKM). She graduated in Master in Geography from UKM. She completed her PhD in Agricultural Geography from the same university. Her research interest area includes the agricultural geography, small farmers and agripreneur development, crop residue management, community development, and sustainable agriculture development. At the national level, her research group has worked closely with government agencies such as the Ministry of Agriculture and Food Industries, Muda Agricultural Development Authorities, and other agencies to evaluate socioeconomic program outcomes and impact studies of agriculture development.

Preface

The *Emerald Handbook of Tourism Economics and Sustainable Development* is a meticulous effort of many minds. Before finalizing the topic of the book, the editors conducted a lot of research. We received constant support from Emerald Publishing's team during the whole process, and the contributors were very committed. The face of the tourism industry is vastly changed. Tourism economics is experiencing several unpredictable changes around the world, especially after COVID. Chapters have been thoroughly organized and revised, yet greatly edited to emphasize the key points. There are several specialization chapters in the handbook, including Bridging the Nexus: A Comprehensive Review of Tourism, Technology, and Climate Change – Uncharted Territories and Future Avenues, Green Horizon, Qualitative Insights into Harvesting Sustainability: The Role of Organic Agriculture in Advancing Sustainable Development Goals, Chapter on Agrotourism, Chapter on Transshipment Troubles, Challenges and Socio-economic Impacts of Proposed Road Development to Gunung Mulu National Park, Sarawak, Malaysia: A Case Study of Local Community Perspectives in Sustainable Tourism, The Role of ICT and Infrastructure in Tourism Development: New Evidence from Asian Economies, The Tourism-Climate Nexus, Tourists' Behavior Post-Covid 19: Do Education, Perceived Knowledge and Perceived Risk Matter? Branding India as a Sustainable Blue Economy Destination in the Indian Ocean Region, Fostering Resilience through Travel: Prioritizing Well-being, Accessibility, and Sustainability, The Importance of Relational Marketing in Health and Wellness Tourism: Study applied to the Euroregion Northern of Portugal and Galicia, The Economics of Market and Inclusive growth in tourism: A Narrative of Rural artisans' Socio-economic Well-being, Halal Tourism as a New Strategy for Sustainable Tourism Economic Contribution and Employment Opportunities of Tourism and Hospitality Sectors, Revenge or Renewal? Nexus of Revenge Tourism, Boomerang Effect, and Sustainable Development Goals, Revitalization and Adaptive Re-use of Heritage Buildings for Event Venue in Rajasthan: Opportunities and Challenges. Thus, this handbook serves as an indispensable guide for navigating the complex terrain of tourism economics and sustainability. Drawing upon the collective wisdom of leading experts in the field, this book offers a comprehensive framework to drive sustainable development, making it a valuable resource for scholars and practitioners. Within the pages of this handbook, you will discover an invaluable resource for those seeking a deeper understanding of tourism economics and its pivotal roles in forging a path toward sustainable development. We wish you all a happy reading.

Editors
Ahmed Imran Hunjra
Rabat Business School, International University of Rabat, Morocco
Anukrati Sharma
Department of Commerce & Management, University of Kota, India

Chapter 1

Bridging the Nexus: A Comprehensive Review of Tourism, Technology and Climate Change – Uncharted Territories and Future Avenues

Kausar Yasmeen[a], *Mustafa Malik*[a], *Kashifa Yasmeen*[b], *Muhammad Adnan*[c] *and Naema Mohammed Al Bimani*[a]

[a]University of Nizwa, Oman
[b]University of Sahiwal, Pakistan
[c]LINCOLN University College, Malaysia

Abstract

Tourism, Technology and Climate Change: The tourism industry is indispensable both for its socio-cultural offerings and its profound economic implications. The economic multiplier effects inherent in the drivers of tourism can stimulate the regional economy even before these areas emerge as tourism meccas. While vast amounts of research have detailed tourism's overarching significance, there is an evident void in understanding its multifaceted impacts, particularly where technological advances, environmental performance (EP) and economic benefits converge. A thorough examination of 907 research records led to this chapter, which identifies these gaps by referencing nine observational and 11 intervention studies. Achieving a Cohen's kappa value of 0.75, the authors note a strong consensus among reviewers, adhering to Cohen's (1940) standards. The findings from the first quarter highlight several areas within the tourism industry that have been under-researched. Particularly, the integration of technology, from ATM infrastructures enhancing tourist financial experiences to digital platforms elevating traveller education and awareness, and tech-driven solutions addressing demographic and ethical considerations in tourism, remains insufficiently explored. Additionally, the authors recognise an existing gap in knowledge regarding the nexus between tourism

The Emerald Handbook of Tourism Economics and Sustainable Development, 1–19

Published under exclusive licence by Emerald Publishing Limited
doi:10.1108/978-1-83753-708-220241001

development and its climatic repercussions, especially before tourism ventures are fully realized. This chapter aims to channel future research into these lesser-trodden areas, fostering a comprehensive grasp of tourism's evolution in the face of rapid technological advancements and its interplay with environmental shifts.

Keywords: Tourism multiplier effects; environmental performance; infrastructure development; economic implications; traveller awareness; ethical tourism; climate impact

1. Introduction

The interplay between tourism, technology and climate change has witnessed increased scrutiny in recent years. However, this scrutiny, more often than not, is skewed towards recognising the threats rather than exploring sustainable, technology-driven solutions. While there is an undeniable global consensus on the adverse effects of climate change – often amplified by rapid industrialisation and sectors like tourism – a nuanced understanding remains lacking. The world's leading economies, such as the USA, China, Russia and India, although major contributors to global pollution, also have the resources and technological prowess to innovate and spearhead global initiatives towards sustainable tourism. The Paris Agreement and United Nations' guidelines have attempted to set a global framework, but the actualization of these guidelines remains scattered, with sectors like tourism often eluding stringent oversight.

This chapter delves into a comprehensive literature review, aiming to bridge existing research gaps, particularly in the tourism industry's technological and environmental facets. Several recent studies have highlighted the escalating emissions from tourism, with the aviation sector being a prime contributor (Sid Perkins, 2018; Davide, 2020). The relationship between global tourism and climate change is multifaceted. It is not merely about the carbon footprint but also about issues like sewerage contamination, which has direct implications for public health (Seabloom et al., 1998). We delve deep into research trends, including studies on the negative repercussions of infrastructure development on biodiversity, the rampant carbon dioxide (CO_2) emissions and the often-overlooked aspects of tourism like the effects of financial management systems such as ATMs on the environment. Recent literature, for instance, points towards the increasing carbon footprint of global tourism and its projected growth (Avi & Hassan, 2022; Huseynova & Mazanova, 2022; Zulvianti et al., 2022).

The originality of this chapter lies in its amalgamation of disparate research areas, bringing together technological, environmental and touristic perspectives to provide a holistic view of the ongoing challenges and potential solutions. Instead of merely identifying problems, this review highlights areas ripe for innovative, tech-driven solutions, making it a beacon for future research endeavours. While various qualitative and quantitative approaches have dissected the tourism-climate conundrum, we identify the spatial panel model (SPM) as a promising tool, outclassing traditional panel procedures that could lead to skewed parameter approximations. In aligning with the reviewer's feedback, we have incorporated a wider range of studies, ensuring the literature review is robust, comprehensive and up-to-date. This chapter leans heavily on recent studies, ensuring our discourse remains relevant to the current global scenario and setting the tone for future explorations in the domain.

2. Research Methodology

For the systematic review, the selection criteria were executed step-by-step to identify gaps in prior research and provide guidance for upcoming research. The research methodology for conducting a systematic review encompasses a search strategy, selection criteria and ultimately synthesising the search results.

2.1 Search Strategy

In April 2022, this research meticulously scoured prototypical articles published in major databases, including PubMed/Medline, PsycINFO, SportDiscus and Embase. Various terms and their combinations were used to identify suitable articles, such as carbon footprints of tourism, environmental performance (EP) of tourism, CO_2 and tourism and tourism and climate change. The exact term combinations utilized were as follows: [(CO_2*) OR (carbon*) OR (pollution*) OR (climate change)]. Titles and abstracts that did not align with the selection criteria were excluded. Full versions of the chosen articles were procured and subsequently reviewed by two independent assessors for validity. Additional keywords and phrases employed included climate change effects on tourism; climate change effects on travel; global warming and travel; global climate change and tourism; climate change tourism in North America; climate change and outdoor recreation; climate change, nature and travel; ecology, tourism and economy; climate change, economy and tourism; climate change's impact on winter tourism, oceans, mountains and skiing.

Furthermore, references related to climate change were sourced from academic researchers worldwide. Cohen's kappa (Cohen, 1940) was used as a measure of agreement between the reviewers see Table 1.1. The reference lists of included articles were also scrutinized to ensure a comprehensive search. Cohen's kappa coefficient is a statistical measure employed to gauge inter-rater reliability for qualitative data. With a resultant value of 0.625, as determined by Cohen's Kappa, there is assurance that the literature was meticulously reviewed by the author, and identified gaps were duly reported. A P-value less than 0.005 indicates significant concordance between

Table 1.1. Cohen's Kappa.

	Value	Asymp. Std. Error	Approx. T	Approximately Significance
Measure of kappa agreement	0.624	0.147	3.957	0.000

Source: Author.

the two reviewers. The observed *P*-value stands at 0.000, signifying a significant level of agreement. However, it is noteworthy to mention the variation between the values obtained in Excel (0.75) and SPSS (0.625). This slight discrepancy arises from SPSS's kappa correction for probability, as indicated in Table 1.2. McHugh (2012) posits that an optimal approach for researchers might involve calculating both percent agreement and kappa concurrently. In cases where the probability of guesswork among raters is elevated, employing the kappa statistic would be advisable. Conversely, if raters have undergone rigorous training and minimal guesswork is anticipated, researchers can confidently utilise percent agreement to ascertain inter-rater reliability.

In Table 1.2, 'count' refers to observed occurrences, differentiating between actual count and expected outcomes based on chance. If an expected value surpasses 0.2, it is indicative of a result above mere chance. The data suggest that Reviewer 2's evaluations are closely aligned with another reviewer's sentiments.

2.2 Selection Criteria for the Systematic Review

The selection criteria for the systematic review were meticulously executed to identify gaps in earlier research and guide future inquiries. These criteria encompass

- Types of studies;
- Participants;
- Outcome measures;
- Data extraction;
- Search results and
- Characteristics of included studies.

Table 1.2. Cross-tabulation of Reviewer 1 and Reviewer 2.

			Reviewer Two			
			1	**2**	**3**	**Total**
Reviewer one	1	Count	5	1	0	6
		Expected	2.1	2.1	1.8	6.0
	2	Count	2	5	1	8
		Expected	2.8	2.8	2.4	8.0
	3	Count	0	1	5	6
		Expected	2.1	2.1	1.8	6.0
Total		Count	7	7	6	20
		Expected	7.0	7.0	6.0	20.0

Source: Author.

2.2.1 Types of Studies

The focus was on research studies featured in peer-reviewed English-language journals, though select books were also incorporated. The analysis included both observational (cross-sectional and longitudinal) and intervention studies (randomized and non-randomized). Notably, apart from two references (Sid Perkins, 2018; Davide, 2020), all cited materials were sourced from reputable books and scholarly publications. These references are widely recognized within the academic community and serve as foundations for theses and dissertations across both qualitative and quantitative research domains. Future researchers would benefit from adhering to guidelines provided in this chapter.

2.2.2 Participants

One primary objective of this review was to discern variables pertinent to carbon footprints and EP in tourism. As such, global case studies addressing these metrics were incorporated.

2.2.3 Outcome Measures

Included studies needed to provide comprehensive evaluations of the carbon footprints and EP within the tourism sector. This encapsulated both observational and intervention studies, particularly those highlighting relationships among various components of carbon footprints and EP in tourism.

2.2.4 Data Extraction

Data from prior research were systematically extracted using a structured template. Two independent reviewers handled this process, focussing on

- Author details;
- Year of publication;
- Nature and methodology of the study;
- Participant demographics and
- Geographical context relevant to assessing tourism's environmental impact.

Researchers operated autonomously and with discretion. No attempts were made to contact original authors for supplementary or missing details from the articles under review.

3. Results

3.1 Search Results

Approximately 907 records were initially identified through systematic research. After removing duplicates, 443 unique records remained for the filtering process.

The study inclusion procedure is depicted in Fig. 1.1. Upon reviewing titles and abstracts, 24 studies were shortlisted for potential inclusion. Full-text articles were then accessed and assessed for suitability. As a result, 20 studies were ultimately included. A Kappa value below 0.4 is considered subpar (with a Kappa of 0 indicating agreement no better than random chance). Kappa values ranging from 0.4 to 0.75 are deemed moderate to good, while a Kappa exceeding 0.75 indicates excellent agreement. The observed Cohen's kappa was 0.89, signalling a strong concurrence between reviewers, consistent with the benchmarks set by Cohen (1940). The methodology review is supported by Yasmeen et al. (2019), Yasmeen and Adnan (2023) and Yasmeen and Viswanathan (2017).

3.2 TheCharacteristics of Included Studies

Within the curated collection of 20 studies, nine were classified as observational, seven as cross-sectional, one as longitudinal and 11 as intervention studies. Notably, only two articles originated from the year 2000, with the remainder being disseminated post-2007. Contributions from Sid Perkins (2018) and David (2020) were also integrated for comprehensive analysis. These studies, with their

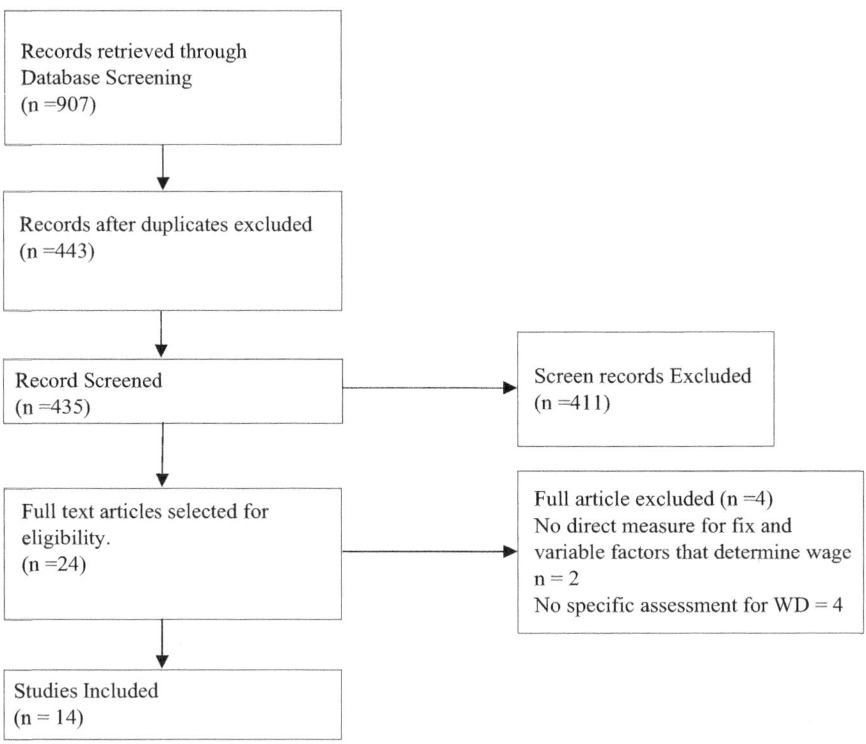

Fig. 1.1. The Flowchart of Study Selection. *Source:* Author.

global scope, sought to elucidate the existing lacunae in conventional research pertaining to the carbon footprints and environmental efficacy of the tourism sector. Fig. 1.1 shows the flowchart of study selection.

4. Results

In this chapter, two pivotal facets of previous research are examined. The first facet delineates the considerable carbon footprint preceding the inauguration of the tourism industry to visitors, while the second underscores the substantial carbon emissions post-visitor accessibility. A thorough analysis reveals notable gaps in the literature presented herein. For the benefit of forthcoming scholars, various methodologies related to environmental studies and climate change are elaborated upon in the appendices.

5. Prior Research and Remarks

Write in an academic language and incorporate technological actors that impact on the environment with coherence: Enormous volume of carbon is emitted before we embark onto the tourism industry, quote GhulamRabbany et al. (2013). They further add that tourism is among the largest and most rapid developing industries at the global economic arena and contains a substantial amount of environmental, cultural, social and economic impacts, both constructive and deleterious. Tourism potentially can play a significant role in the economic amelioration of a country. Contrary to that, the meagerness of appropriate planning can cause adverse influences on biodiversity and a pristine environment. In addition to that, the augmentation of tourism facilities such as accommodation, water supplies and recreational resorts can involve sand and soil erosion, extensive paving and land degradation which can evidently cause the distinction of wildlife and deterioration of scenery (Oad et al., 2022).

The developments of marinas and breakwater have been responsible for causing changes in currents and coastlines, deforestation, extraction of building material such as sand affects coral reefs leading to erosion and destruction of habitats. Hence, tourism undermines the ecosystem and depletes the potential to encounter natural calamities such as floods, drought and hurricanes and with human-created stresses, such as pollution can cause climate change GhulamRabbany et al. (2013). Evidence suggest that nature tourism is meticulously integrated with bio-diversity and the fascination developed by the rich and wide-ranging environment can also cause diminishing of biodiversity when land and resources are strained by exaggerated exploitation and effects on vegetation, wildlife, mountain, marine and coastal environment, and water resources surpass their carrying capacity.

The diminishing of biodiversity leads to extinction of tourism potential. Tourists and suppliers can carry/import exotic species such as insects, wild and cultivated plants and contagious diseases that the natives are not associated with can cause immense disruption and even annihilation of the ecosystem. The tourism industry/destinations contribute to global warming. Prior to inaugurating destinations for tourism, appropriate measures and policies documented may be implemented, failing to comply with the policies, these destinations can prove to be a continual threat to the world's biodiversity and ecosystem, which can lead to extinction (Davide, 2020; Hunjra et al., 2022; Mohanty et al., 2022). Hence, it is evident that in this chapter, several devoted researchers are focussing on the impacts of the tourism industry on environmental change and global warming before travelers' influx towards tourists' destinations. There is meagreness in the implementation of the Paris agreement and climate change guidelines while, they are adopted when the developmental work is under process.

Moreover, the carbon is emitted while clearing and levelling the natural areas before the construction work is proceeded (Gössling, 2000). Ecosystems such as forests perform as carbon sinks through the absorption and storing process while CO_2 is emitted. Once this carbon-rich vegetation is eliminated, CO_2 is discharged back into the atmosphere. The mangrove forests which grow in many tropical areas are an important source not only to absorb but also have the capacity to store huge amounts of carbon resulting in an eco-friendly environment. Studies suggest that mangrove forests have the capacity to store four times more than any other tropical forests around the globe.

In the contemporary global landscape, the tourism sector stands as a prominent contributor to both economic development and environmental impact. GhulamRabbany et al. (2013) emphasise that while the tourism industry holds potential for economic upliftment, it simultaneously emits a substantial volume of carbon even before it fully establishes itself. Notably, tourism, with its expansive reach, exerts multifaceted repercussions, encompassing environmental, cultural, social and economic domains, each carrying the potential for both beneficial and detrimental outcomes. The double-edged sword that is tourism can stimulate economic progress. However, the lack of strategic planning can inflict deleterious effects on biodiversity and environmental sanctity. Infrastructure developments catered to tourism, like accommodations, water facilities and leisure centres, often necessitate land modifications, leading to issues like soil erosion, extensive paving and land degradation. Such alterations can result in biodiversity diminution and landscape deterioration, as noted by Oad et al. (2022).

Technological interventions, such as the establishment of marinas and breakwaters, have been observed to perturb natural coastlines and water currents, leading to challenges like deforestation and coral reef destruction due to excessive material extraction. The interplay of these factors not only weakens the ecosystem's resilience against natural disasters like floods, droughts and hurricanes but also heightens human-induced stresses like pollution, further exacerbating climate change. There exists a nuanced relationship between nature tourism and biodiversity. While the allure of diverse ecosystems can promote tourism, undue exploitation can strain the very resources that attract visitors, affecting vegetation, wildlife and water resources beyond their thresholds. This strain can lead to biodiversity loss, nullifying the potential for sustainable tourism. Another critical aspect involves the inadvertent introduction of non-native species by tourists, which can destabilise local ecosystems. Such anthropogenic interferences, if not monitored, can pose substantial threats to global biodiversity and ecosystem stability, echoing sentiments by Davide (2020).

It is paramount to emphasise that many of the detrimental environmental effects arise from preparatory phases of establishing tourist destinations. Activities such as land clearing and levelling release significant amounts of carbon, as highlighted by Gössling (2000). Ecosystems, particularly forests, play a pivotal role as carbon sinks, absorbing and storing CO_2. When these carbon-dense vegetation's are disrupted, CO_2 is reintroduced into the atmosphere. Among these, mangrove forests in tropical regions emerge as vital carbon storehouses, possessing the capacity to store carbon at quadruple the rate of other tropical forests. Given the multifaceted implications highlighted in this chapter, it becomes evident that adherence to international guidelines, such as the Paris Agreement, is paramount during developmental phases. Ensuring compliance can pave the way for a balanced coexistence of tourism and environmental sustainability.

The repercussions of tourism infrastructure development on ecological systems, notably mangrove forests, have emerged as a subject of pressing concern. These crucial ecosystems are frequently felt to pave the way for facilities such as marinas, seaside resorts and recreational areas (Self et al., 2010). The establishment of resorts, airports and other tourism-centric infrastructures invariably leads to substantial carbon emissions even before they become operational for tourists. Further exacerbating the carbon footprint is the construction of skyscrapers to accommodate tourists. Such edifices contribute significantly to CO_2 emissions throughout their lifecycle, from material manufacturing to transportation and building processes (Goodwin, 2011, pp. 1–256). Literature suggests that stringent adherence to environmental policies during the development of these tourist amenities could prevent the indiscriminate clearing of natural landscapes. While some nations sacrifice vast swathes of greenery in the quest for infrastructural growth, others meticulously craft tourist attractions ensuring minimal environmental degradation. The latter approach ensures that the aesthetic appeal of greenery remains omnipresent, from resort locations to urban cores, with a reduced risk of adverse climate impacts.

Upon the inauguration of the tourism sector, a further surge in carbon emissions becomes evident. A saturation point is reached when tourist influx exceeds the environmental threshold, triggering several deleterious consequences. Regions subjected to mass tourism face heightened risks, from soil erosion and increased pollution to habitat destruction and mounting pressures on endangered species. Moreover, these areas often grapple with water scarcity, leading to competition among local communities for this essential resource (Shan, 2022). The implications of the tourism sector on climate change, thus, warrant comprehensive analysis.

Water contamination, as evidenced by Seabloom et al. (1998), is another side-effect of the burgeoning tourism sector. Construction activities, coupled with maritime recreation, exert undue stress on water ecosystems. Peak tourism seasons exacerbate pressures on sewage treatment systems, leading to overflows. One particularly ominous aspect associated with recreational boating is the direct discharge of sewage into water bodies with inadequate flushing capabilities. Such discharges in proximity to shellfish beds elevate risks of diseases like typhoid, dysentery and infectious hepatitis upon human exposure or consumption.

Moreover, Seabloom et al. (1998) emphasise that the inadequacies in sewage management contribute not only to water pollution but also to air quality degradation, engendering a slew of health complications. Surprisingly, despite the gravity of these issues, they remain relatively underexplored in academic research. This chapter underscores the exigency for future studies to delve deeper into the relationship between sewage systems, air quality and health, potentially examining sewage systems as an independent variable and air quality as a mediator or moderator, with health outcomes as the dependent variable.

Antonova et al. (2021) provide an illuminative discourse on water exploitation, categorising it into four distinct clusters: water usage, water consumption, water management and good practices. These categories are methodologically and thematically examined in isolation. The study also underscores existing research gaps and their origins. Its findings underscore the potential of academic research to bolster transnational methodologies, striving for sustainability. Notably, this study, as asserted by the authors, represents a pioneering systematic analysis centred on water exploitation within the hotel sector. Given its comprehensive nature, the author advocates for its replication across global tourism destinations to address the existing literature lacuna.

There is an emerging awareness about environmental ramifications among tourists, enhancing their travel experiences. Yet, scholarly endeavours addressing this dimension remain sparse. Over the past few decades, heightened apprehensions regarding global warming trends have been palpable. The United Nations Charter propounds the global imperative to mitigate CO_2 emissions (Seo, 2017). The foremost determinants of global warming include pollution and CO_2 emissions. The nexus between organizational actions and the environment can inadvertently exacerbate global warming. Aspects such as strategic planning, resource extraction and processing, infrastructure development, environmental repercussions, consumption of natural assets and waste generated by entities significantly influence EP. Khan et al. (2019) delved into the correlation between education and carbon production, discerning an inverse relationship. Concurrently, Mishra et al. (2020) scrutinized the effects of education on carbon and methane emissions across 181 nations using panel data methodology. Their findings indicated that while education negatively impacts methane emissions, its influence on carbon emissions is relatively neutral when considering other control variables. Environmental challenges, though extensively studied in sectors like extraction and manufacturing, have received scant attention in the tourism industry. Specifically, there's a paucity of knowledge about tourist cognisance regarding the ecological footprint of their activities and their conscientiousness about EP.

Primarily, in developing nations, the allure of natural landscapes and resources beckons tourism development, often encroaching upon fragile ecosystems. Tourism expansion leads to a slew of environmental ramifications, including degradation of ecosystems, parks and forests, excessive water extraction, solid waste disposal issues, reduced arable land and water sources, vegetation clearance, soil erosion and introduction of alien species, resulting in habitat and wildlife depletion (Sasidharan et al., 2002). Notably, in areas marked by significant tourist activities and natural attractions, waste management emerges as a critical challenge. For instance, cruise ships in the Caribbean produce a staggering 70,000 tons of waste annually. Such waste, when not managed aptly, mars natural environments, aquatic ecosystems, scenic locales and infrastructure. Improper waste disposal, especially by trekking tourists, despoils pristine mountainous terrains, as they frequently abandon leftover supplies, oxygen cylinders and camping gear. Such behaviours litter these regions with refuse reminiscent of developed nations, even when local waste management infrastructure is often rudimentary at best.

Transportation is another premier cause of global warming; tourism produces around 5% of global CO_2 emissions. Aeroplanes are among the leading sources of producing pollution, the other means of transportation such as tourist busses, trains and ferries impart a minimal impact on environment. Highest amount of tourism related air pollution is produced by automobiles (Al-Mulali et al., 2015; Jiaqi et al., 2022) According to Jiaqi et al. (2022) automobiles are the highest sources of emission of carbon monoxide among all transportation means. transportation is the leading contributor to CO_2 emissions in the tourism sector, accounting for 75% of emissions, with half of that attributed to air transport. Accommodation follows closely behind, representing 22% of emissions, while other activity-related tourism contributes 4%. Projections indicate a significant increase in emissions from tourism activities, expected to more than quadruple by 2035. In Nepal, tourism activities contribute around 10% of total CO_2 emissions, with transportation being the primary culprit, responsible for 89% of tourism-related emissions. Accommodation and food and beverage services each contribute about 5%. Notably, accommodation services experienced the most rapid increase in CO_2 emissions, with a growth rate of 48.4% between 2000 and 2008, followed by food, beverage, recreational services and cultural sports at 34% and transportation at 25%. Studies have found that education negatively affects methane emissions, while carbon emissions remain unaffected.

A pivotal study indicates that a singular Transatlantic return flight accounts for roughly 50% of the CO_2 emissions produced from all other sources by an average individual in a year (Guo et al., 2022). Rossi et al. (2020) identify transportation as the primary contributor to environmental pollution. When delving into the tourism sector, it becomes imperative to dissect various elements impacting the environment, such as accommodation facilities, tour and climbing operations, transportation modalities, lake and sea ports, road infrastructure, ATM facilities, recreational venues and sewage systems and discern their environmental compatibility. Sid Perkins (2018) underscores the CO_2 production and emission of other greenhouse gases (GHGs) attributed to the construction and upkeep of infrastructures like airports, hotels and tourist resorts, along with the procurement processes for items like food and souvenirs. Elkhwesky (2022) posits that there's a concerning dearth of initiatives towards proactive environmental strategies within the tourism sector.

Upon meticulous examination, several determinants with potential environmental repercussions have been spotlighted, such as lodging amenities, tour and climbing services, transportation systems, ports, tourist-centric road networks, ATMs, recreational hubs and their interplay with climate change. Yet, research gaps persist, especially regarding the environmental implications of sewage systems, the ecological impact of financial structures like ATMs and their collective contribution to climate change. It's anticipated that forthcoming research endeavours will bridge these gaps.

Gössling and Lund-Durlacher (2021) and Pavlatos (2021) assert that the energy consumption patterns of hotels and tourism-related infrastructures have significant implications for global climate change. Visual data detailing the energy utilization of operational resources in Barbados suggests that certain facilities, like heating, are predominant energy consumers, closely followed by kitchen appliances, food catering and electricity usage. However, the environmental impact varies considerably depending on the energy sources employed. Infrastructure that harnesses renewable energy, such as solar power through photovoltaic systems, tends to operate more efficiently and significantly curtails CO_2 emissions. Emerging tourism infrastructures, parking spaces and road networks consume land areas that could otherwise serve as green spaces, planting CO_2-absorbing trees. Therefore, it's crucial to prioritise eco-friendly architectural designs when developing tourism infrastructure, aiming to mitigate pollutants and various emissions.

Amelung et al. (2007) elucidate the looming repercussions of climate change worldwide, attributing them, in part, to cyclical tourism activities. The narrative posits that while certain tourist destinations grapple with dwindling visitor counts, potentially losing their allure, others are witnessing an ever-increasing influx. Gössling et al. (2017) delineate that the key environmental ramifications of tourism, contributing to GHG emissions, predominantly stem from energy consumption patterns, the rampant use of fossil fuels and transportation modes favoured by tourists. These converging factors underscore the paramount environmental challenges intertwined with tourism, propelling developed nations to strategise for CO_2 emission reductions – a primary culprit behind global warming. Yet, despite these pressing issues, a relative dearth in comprehensive research persists on the interplay between climate change and the tourism sector, notwithstanding the significant imprint of tourism activities on global environmental shifts (Scott et al., 2012).

Current empirical endeavours probing into the nuanced influence of tourism on CO_2 emissions yield findings that, at times, border on the ambiguous (Carr-Harris and Lang; Lee and Brahmasrene; Sigala and Leslie; Sun). Nonetheless, these statistics serve as a rudimentary yardstick, hinting at the potential carbon footprint attributable to the tourism sector and facilitating nuanced estimations of tourism-induced CO_2 emissions. In a contrasting approach, scholars such as Arai and Goto (2017), and Simkhada et al., adopt a qualitative lens to assess the environmental ramifications of tourism. A prevailing sentiment within the academic discourse is the elevated concern regarding CO_2 emissions stemming from tourism endeavours.

As nations worldwide hinge on tourism for economic buoyancy, there emerges a discernible narrative detailing environmental degradation. Activities buttressing economic growth often exacerbate CO_2 trajectories, culminating in heightened waste generation. This cascade effect underscores a nuanced relationship wherein nations grapple with an inverse correlation between burgeoning tourism, energy conservation, economic development, and escalating carbon emissions. It's imperative to recognise that the evolution within the tourism sector is intricately woven with a tapestry of dynamic inter-regional cultural and trade exchanges, transcending mere managerial paradigms.

6. Methodologies Used in Tourism Industries and Technological in Relation to Climate Change

In recent years, a plethora of empirical studies have investigated the econometric relationship between economic growth and carbon dioxide (CO_2) emissions. Comprehensive examinations, such as those by Iqbal et al. (2019) in Europe, Khan and Ozturk (2021) in Central America, Iwata and Okada (2011) spanning 28 countries (both OECD and non-OECD), and studies specific to regions like China by Jalil and Mahmud (2009) and Turkey by Ozturk and Acaravci (2010), have added depth to the discourse. A more nuanced exploration by Ozturk and Dincer (2021) employed decomposition techniques, delving into CO_2 emission intensity and its multifaceted components across Portugal's 36 economic sectors. Notwithstanding these advances, a conspicuous gap persists in utilising decomposition analysis methodologies to assess the intricate impacts of various determinants on the fluctuations in carbon dioxide emissions, especially those engendered by tourism. Although several studies have discerned a nexus between carbon dioxide emissions and the broader energy-economic interplay, few, like Adedoyin et al. (2021) in China, have incorporated decomposition analysis.

Expanding on this, Iqbal et al. (2019) postulated that the scalar impact becomes significant for countries such as Ireland, especially when juxtaposing rising emissions in the transport sector against the notable energy efficiency strides in the domestic and service sectors. Fu et al. (2021) contended that the pivotal levers to attenuate CO_2 emissions pivot around transport services and intensity factors. In a quest for methodological robustness, Baloch et al. (2020) juxtaposed several decomposition analytical frameworks and vouched for the Logarithmic Mean Divisia Index (LMDI) owing to its solid theoretical underpinning, user-friendliness and transparent interpretability. Globally, however, extant literature applying the LMDI at sectorial and sub-sectorial levels remains sparse. Zhang et al. (2021) cast a wider net across European nations, probing the nexus between economic advancement and environmental degradation. Further refining methodological instruments, Anser et al. (2020) employed the Geographic Panel Data Model (GPDM) to elucidate Environmental Kuznets Curves (EKC), deducing that the Spatial Panel Model (SPM) often outperforms traditional panel methodologies, given that the latter can occasionally lead to spurious parameter estimations. The application of spatial panel data models in elucidating regional variances in carbon intensity in China was adeptly tackled by Khokhar et al. (2020).

Turning to Latin America, Sovacool et al. (2021) harnessed data to dissect the ramifications of foreign direct investments and human capital on pollution generation within the Environmental Kuznets Curve (EKC) framework. The panel technique revealed a dichotomous outcome: a negative correlation between human capital and emissions for economically advanced nations juxtaposed against a positive correlation for their less affluent counterparts. The EKC paradigm, when applied by You and Lv (2018) using Australian data, unveiled a U-shaped correlation between education levels and emissions, as deciphered from the autoregressive distributed lag (ARDL) findings. Khan et al. (2019) further explored this dynamic, assessing the relationship between educational attainment and CO_2 emissions, leveraging ARDL econometric techniques and employing a gamut of proxies.

7. A Need of the Study and Directions for Future Research

Despite the rise of significant challenges, both the tourism sector and academia have undertaken limited research on the effects of climate change on tourism. Conversely, there's a focus on the impact tourism and associated activities have on global environmental shifts. Seabloom et al. (1998) underscore the pollution stemming from sewage systems, leading to air contamination. Such deteriorated air quality can lead to serious health problems. Notwithstanding the warning signs related to climate change, efforts to address this pressing issue remain scant. Upcoming researchers ought to prioritise studying the sewage system and health-related matters as independent variables, while treating air quality as a dependent one. Rosa et al. (2021) segmented water usage into four categories: water consumption, water management, water usage and health practices. The author claims this to be a pioneering study on water use within the hotel industry. Based on various data points, the author believes Antonova et al. (2021) possesses several shortcomings.

In conclusion, targeted research examining the tourism industry and its associated elements – like accommodation, tour operators, transportation, lakes and seaports, tourist routes, ATMs and recreational venues – all indicate a substantial contribution to climate change. Seo (2017), Khan et al., and Mishra et al. (2020) focused on education and the environment, determining that while there's extensive exploration in sectors like manufacturing and extraction, there's limited awareness about the environmental footprint of the tourism sector, especially regarding tourists' awareness of their actions' environmental implications. Due to cultural variations globally, findings from Mishra et al. (2020) should not be universally applied. Linking tourism to climate change necessitates considering education as a key factor in future studies. While greenery and nature have historically been vital for environmental health, they're increasingly overlooked in the pursuit of tourism development. Research quantifying the environmental damage caused by tourism – particularly deforestation and habitat destruction – is notably sparse. Methodologies, both qualitative and quantitative, have been employed to gauge tourism's environmental impact. However, decomposition analysis – a key tool to study CO_2 emission fluctuations resulting from tourism – has been inadequately emphasized in research.

8. Conclusion

The confluence of technology, tourism and the environment is at a pivotal juncture. As tourism, a vital global industry, thrives on technological advancements for efficiency and innovation, it simultaneously encounters challenges related to environmental sustainability. Today, we assemble to deliberate on the findings of our recent study, a critical exploration into the symbiotic relationship between these elements, identifying gaps in current research, and proposing avenues for future inquiry. Our meticulous review of 907 records selected nine observational and 11 intervention studies for detailed examination. This process, marked by a commendable Cohen's kappa of 0.75, suggests significant consensus among reviewers. Our study primarily sheds light on the expansive impact of tourism, intensified by technological drivers, and the myriad factors steering its trajectory.

Notwithstanding the vast corpus of literature addressing tourism's environmental footprint, there is a conspicuous lack of detailed analysis of water usage within the tourism industry. Specifically, the technology-driven management, consumption, and impacts of water use and best practices demand further attention. The current literature often omits a comprehensive understanding of how technology-fuelled systems like sewage treatment and financial infrastructures (e.g. ATM machines) intersect with tourism and its environmental ramifications. Alarmingly, the focus on the environmental repercussions of technological advancements in tourism, especially concerning the degradation of natural habitats and greenery, remains notably sparse.

Given the technological advancements in water management, it is imperative for future research to adopt a comprehensive lens, dissecting water usage within the tourism sector and its subsequent environmental implications. Subsequent studies should prioritise understanding how emerging technologies in the sectors of waste management, finance and transportation impact tourism's environmental footprint. Technology plays a pivotal role in monitoring and improving air quality. As such, upcoming studies could explore how technological solutions in the tourism industry can be harnessed to address sewage, health-related challenges and air quality concomitantly. The dynamic interplay between these elements can provide pivotal insights into environmental sustenance. As we aim to fathom the EP of the tourism industry, it is essential to recognise that technological advancements, alongside other variables like demographics and policy frameworks, can function as critical mediators or moderators.

The dire consequences of neglecting nature and green spaces in tourism endeavours are evident. Future research should emphasise the role of green technologies in promoting sustainable tourism practices, ensuring a harmonious balance between growth and environmental preservation. As we navigate the intricate tapestry of tourism, technology and the environment, it is clear that a holistic, interdisciplinary approach is crucial. This study, while unearthing significant gaps in current understanding, serves as a beacon for future research, urging scholars to delve deeper into the synergies and frictions at this triad's intersections. It is our collective aspiration that future endeavours in this realm are guided by sustainability, innovation and a deep respect for the environment.

Declaration of Conflicting Interests

The author(s) declared that there are no potential conflicts of interest with respect to the research, authorship, and/or publication of this article. The authors also declare that the authors have no human participants and have not used human data or human tissues.

Acknowledgements

The authors would like to express gratitude to dignified colleagues, including Dr. Zain Muhammad Abdu Karim for providing them with diligent information and sincere advice on the first draft of this chapter. All views expressed in this paper entirely belong to the authors.

References

Adedoyin, F. F., Ozturk, I., Agboola, M. O., Agboola, P. O., & Bekun, F. V. (2021). The implications of renewable and non-renewable energy generating in sub-Saharan Africa: The role of economic policy uncertainties. *Energy Policy*, *150*, 112115. https://doi.org/10.1016/j.enpol.2020.112115

Al-Mulali, U., Fereidouni, H. G., & Mohammed, A. H. (2015). The effect of tourism arrival on CO_2 emissions from the transportation sector. *Anatolia*, *26*(2), 230–243. https://doi.org/10.1080/13032917.2014.934701

Amelung, B., Nicholls, S., & Viner, D. (2007). Implications of global climate change for tourism flows and seasonality. *Journal of Travel Research*, *45*, 285–296. https://doi.org/10.1177/0047287506295937

Anser, M. K., Iqbal, W., Ahmad, U. S., Fatima, A., & Chaudhry, I. S. (2020). Environmental efficiency and the role of energy innovation in emissions reduction. *Environmental Science & Pollution Research*, *27*. https://doi.org/10.1007/s11356-020-09129-w

Antonova, N., Ruiz-Rosa, I., & Mendoza-Jiménez, J. (2021). Water resources in the hotel industry: A systematic literature review. *International Journal of Contemporary Hospitality Management*, *33*(2), 628–649. https://doi.org/10.1108/IJCHM-07-2020-0711

Arai, T., & Goto, M. (2017). A survey on present tourism in Nepal and its ripple effects on other industries. *Environment and Ecology Research*, *5*, 467–475. https://doi.org/10.13189/eer.2017.050701

Avi, M. A. R., & Hassan, A. (2022). Medical tourism in Bangladesh and innovative technology application. *Handbook of technology application in tourism in Asia*, 953. https://doi.org/10.1007/978-981-16-2210-6

Baloch, Z. A., Tan, Q., Iqbal, N., Mohsin, M., Abbas, Q., Iqbal, W., & Chaudhry, I. S. (2020). Trilemma assessment of energy intensity, efficiency, and environmental index: Evidence from BRICS countries. *Environmental Science & Pollution Research*, *27*, 34337–34347. https://doi.org/10.1007/s11356-020-09578-3

Cohen, M. R. (1940). Some difficulties in Dewey's anthropocentric naturalism. *Philosophical Review*, *49*(2), 196–228. https://doi.org/10.2307/2180800

Davide, F. (2020). *How tourism contributes to global warming*. ecobnb.

Elkhwesky, Z. (2022). A systematic and major review of proactive environmental strategies in hospitality and tourism: Looking back for moving forward. *Business Strategy and the Environment*. https://doi.org/10.1002/bse.3076

Fu, F. Y., Alharthi, M., Bhatti, Z., Sun, L., Rasul, F., Hanif, I., & Iqbal, W. (2021). The dynamic role of energy security, energy equity and environmental sustainability in the dilemma of emission reduction and economic growth. *Journal of Environmental Management*, *280*. https://doi.org/10.1016/j.jenvman.2020.111828

GhulamRabbany, M., Afrin, S., Rahman, A., Islam, F., & Hoque, F. (2013). Environmental effects of tourism. *American Journal of Environment, Energy and Power Research*, *1*(7), 117–130. www.ajeepr.com

Goodwin, H. (2011). *Taking responsibility for tourism*. Goodfellow Publishers Limited. www.goodfellowpublishers.com

Gössling, S. (2000). Sustainable tourism development in developing countries: Some aspects of energy use. *Journal of Sustainable Tourism*, *8*(5), 410–425. https://doi.org/10.1080/09669580008667376

Gössling, S., & Lund-Durlacher, D. (2021). Tourist accommodation, climate change and mitigation: An assessment for Austria. *Journal of Outdoor Recreation and Tourism*, *34*, 100367. https://doi.org/10.1016/j.jort.2021.100367

Gössling, S., Peeters, P., Ceron, J.-P., Dubois, G., Patterson, T., & Richardson, R. B. (2017). The eco-efficiency of tourism. *Disponıvelem*. http://www.datasheets.tips/culture-and-the-arts/the-eco-efficiency-of-tourism/Acessoem:junhode2017.http://www.datasheets.tips/culture-and-the-arts/the-eco-efficiency-of-tourism/

Guo, B., Wei, C., Yu, Y., Liu, Y., Li, J., Meng, C., & Cai, Y. (2022). The dominant influencing factors of desertification changes in the source region of Yellow River: Climate change or human activity?. *Science of the Total Environment*, *813*, 152512. https://doi.org/10.1016/j.scitotenv.2021.152512

Hunjra, A. I., Azam, M., Bruna, M. G., Verhoeven, P., & Al-Faryan, M. A. S. (2022). Sustainable development: The impact of political risk, macroeconomic policy uncertainty and ethnic conflict. *International Review of Financial Analysis*, *84*, 102370. https://doi.org/10.1016/j.irfa.2022.102370

Huseynova, A., & Mazanova, O. (2022). The leading role of digital technologies in the development of the smart city concept in Azerbaijan. https://doi.org/10.20944/preprints202207.0050.v2

Iqbal, W., Altalbe, A., Fatima, A., Ali, A., & Hou, Y. (2019). A DEA approach for assessing the energy, environmental and economic performance of top 20 industrial countries. *Processes*, *7*(12), 902. https://doi.org/10.3390/pr7120902

Iwata, H., & Okada, K. (2011). How does environmental performance affect financial performance? Evidence from Japanese manufacturing firms. *Ecological Economics*, *70*(9), 1691–1700. https://doi.org/10.1016/j.ecolecon.2011.05.010

Jalil, A., & Mahmud, S. F. (2009). Environment Kuznets curve for CO_2 emissions: A cointegration analysis for China. *Energy Policy*, *37*(12), 5167–5172. https://doi.org/10.1016/j.enpol.2009.07.044

Jiaqi, Y., Yang, S., Ziqi, Y., Tingting, L., & Teo, B. S. X. (2022). The spillover of tourism development on CO_2 emissions: A spatial econometric analysis. *Environmental Science & Pollution Research*, *29*, 26759–26774. https://doi.org/10.1007/s11356-021-17026-z

Khan, M., & Ozturk, I. (2021). Examining the direct and indirect effects of financial development on CO_2 emissions for 88 developing countries. *Journal of Environmental Management*, *293*, 112812. https://doi.org/10.1016/j.jenvman.2021.112812

Khan, M. K., Teng, J. Z., & Khan, M. I. (2019). Effect of energy consumption and economic growth on carbon dioxide emissions in Pakistan with dynamic ARDL simulations approach. *Environmental Science & Pollution Research*, *26*, 23480–23490. https://doi.org/10.1007/s11356-019-05640-x

Khokhar, M., Hou, Y., Rafique, M. A., & Iqbal, W. (2020). Evaluating the social sustainability criteria of supply chain management in manufacturing industries: A role of BWM in MCDM. *Problemy Ekorozwoju*, *15*, 185–194.

McHugh, M. L. (2012). Interrater reliability: The kappa statistic. *Biochemia Medica*, *22*(3), 276–282. https://hrcak.srce.hr/89395

Mishra, S., Sinha, A., Sharif, A., & Suki, N. M. (2020). Dynamic linkages between tourism, transportation, growth and carbon emission in the USA: Evidence from partial and multiple wavelet coherence. *Current Issues in Tourism*, *23*, 2733–2755. https://doi.org/10.1080/13683500.2019.1667965

Mohanty, P., Sharma, A., Kennell, J., & Hassan, A. (Eds.). (2022). *The Emerald handbook of destination recovery in tourism and hospitality*. Emerald Publishing Limited. https://doi.org/10.1108/978-1-80262-073-320221030

Oad, S., Jinliang, Q., Shah, S. B. H., & Memon, S. U. R. (2022). Tourism: Economic development without increasing CO_2 emissions in Pakistan. *Environment, Development and Sustainability*, *24*(3), 4000–4023. https://doi.org/10.1007/s10668-021-01601-y

Ozturk, I., & Acaravci, A. (2010). CO_2 emissions, energy consumption and economic growth in Turkey. *Renewable and Sustainable Energy Reviews*. https://doi.org/10.1016/j.rser.2010.07.005

Ozturk, M., & Dincer, I. (2021). A comprehensive review on power-to-gas with hydrogen options for cleaner applications. *International Journal of Hydrogen Energy*, *46*(62), 31511–31522.

Pavlatos, O. (2021). Drivers of management control systems in tourism start-ups firms. *International Journal of Hospitality Management*, *92*, 102746. https://doi.org/10.1016/j.ijhm.2020.102746

Perkins, S. (2018, May 7). *Tourism is four times worse for the planet than previously believed.* Science. https://www.science.org/content/article/tourism-four-times-worse-planet-previously-believed. Accessed on May 11, 2023.

Rosa, L., Sanchez, D. L., Realmonte, G., Baldocchi, D., & D'Odorico, P. (2021). The water footprint of carbon capture and storage technologies. *Renewable and Sustainable Energy Reviews*, *138*, 110511. https://doi.org/10.1016/j.rser.2020.110511

Rossi, R., Ceccato, R., & Gastaldi, M. (2020). Effect of road traffic on air pollution. Experimental evidence from COVID-19 lockdown. *Sustainability*, *12*(21), 8984. https://doi.org/10.3390/su12218984

Sasidharan, V., Sirakaya, E., & Kerstetter, D. (2002). Developing countries and tourism ecolabels. *Tourism Management*, *23*(2), 161–174.

Scott, D., Gossling, S., & Hall, M. (2012). International tourism and climate change. *Wiley Interdisciplinary Reviews: Climate Change*, *3*, 213–232.

Seabloom, E. W., Van Der Valk, A. G., & Moloney, K. A. (1998). The role of water depth and soil temperature in determining initial composition of prairie wetland coenoclines. *Plant Ecology*, *138*(2), 203–216. https://doi.org/10.1023/A:1009711919757

Self, R. M., Self, D. R., & Bell-Haynes, J. (2010). Marketing tourism in the Galapagos Islands: Ecotourism or greenwashing?. *International Business & Economics Research Journal*, *9*(6). https://doi.org/10.19030/iber.v9i6.590

Seo, S. N. (2017). Beyond the Paris Agreement: Climate change policy negotiations and future directions. *Regional Science Policy & Practice*, *9*(2), 121–140. https://doi.org/10.1111/rsp3.12090

Shan, X. (2022, May). Intelligentization of tourism Japanese environmental scenic spots based on intelligent translation hardware system, *2022 6th international conference on intelligent computing and control systems (ICICCS)* (pp. 548–551). IEEE. https://doi.org/10.1109/ICICCS53718.2022.9788174

Sovacool, B. K., Griffiths, S., Kim, J., & Bazilian, M. (2021). Climate change and industrial F-gases: A critical and systematic review of developments, sociotechnical systems and policy options for reducing synthetic greenhouse gas emissions. *Renewable and Sustainable Energy Reviews*. https://doi.org/10.1016/j.rser.2021.110759

Yasmeen, K., & Adnan, M. (2023). Zero-day and zero-click attacks on digital banking: A comprehensive review of double trouble. *Risk Management Journal*, *4*(25). https://link.springer.com/article/10.1057/s41283-023-00130-4Springer

Yasmeen, K., Alruways, N. M. J., & Mondal, S. (2019). Influence of microfinance and empowerment on satisfaction: Case study of Pakistan. *Studies in Business and Economics*, *22*(1), 65–77.

Yasmeen, K., & Viswanathan, K. (2017). Influence of country factors on entry mode through knowledge and transactional cost economics: Market entry: Evidence from construction firms. *International Journal of Economics and Financial Issues*, *7*(3), 581–585.

You, W., & Lv, Z. (2018). Spillover effects of economic globalization on CO_2 emissions: A spatial panel approach. *Energy Economics*, *73*, 248–257. https://doi.org/10.1016/j.eneco.2018.05.016

Zhang, L., Godil, D. I., Bibi, M., Khan, M. K., Sarwat, S., & Anser, M. K. (2021). Caring for the environment: How human capital, natural resources, and economic growth interact with environmental degradation in Pakistan? A dynamic ARDL approach. *Science of the Total Environment*, *774*, 145553. https://doi.org/10.1016/j.scitotenv.2021.145553

Zulvianti, N., Aimon, H., & Abror, A. (2022). The influence of environmental and non-environmental factors on tourist satisfaction in halal tourism destinations in West Sumatra, Indonesia. *Sustainability*, *14*(15), 9185. https://doi.org/10.3390/su14159185

Chapter 2

Green Horizon: Establishing a Biomass Recycling Facility in the Tirupur–Tuticorin Region for Sustainable Biofuel Supply to Maritime Shipping

M. Bina Celine Dorathy

Alliance University, India

Abstract

Ocean transportation is not only the cheapest and the best mode of bulk transport but also the most polluting form of transportation. The International Maritime Organization (IMO) has set strict targets to cut down carbon dioxide (CO_2) emissions, following which several initiatives have been taken by the shipping industry to embrace new technologies that can make the industry greener. Significant investments have been made into research and development (R&D) to develop alternative marine fuels. This chapter explores the feasibility of setting up a Biomass Recycling Facility (BRF) in the Tirupur–Tuticorin region in Tamil Nadu. The region was chosen because Tirupur being a textile valley generates tonnes of textile wastes every year. It can become good feedstock for biofuel generation, and it is also near Tuticorin Port, which is one of the major ports in Tamil Nadu. On an average, every year 1,000 vessels of medium and large size call at this port. There is a high probability that a BRF established in the vicinity can generate and supply bioethanol for the ships calling at Tuticorin Port. It is apparent from the findings of the study that the feedstock generated by textile industry alone may not be sufficient to meet the huge volumes of biofuel requirements of vessels, more over considerable investments into infrastructure and technology are required. But the study points out that still it could become a viable option because of the government support and favourable Foreign Direct Investment (FDI) policies. The growing demand for biofuel and the increasing price in the world market can become an added advantage.

The Emerald Handbook of Tourism Economics and Sustainable Development, 21–40

doi:10.1108/978-1-83753-708-220241002

Keywords: Biofuel; bioethanol; BRF; MARPOL; textile wastes

1. Introduction

Shipping plays an important role in international trade and it is one of the most energy efficient modes of transportation. However, it is also a significant and growing source of carbon emissions. The shipping industry emits around 940 million tonnes of carbon dioxide (CO_2) each year, accounting for at least 2.5% of total global CO_2 emissions (Maritime Supply Chains, n.d.). A container ship is reported to be as polluting as 50 million cars, while emissions from 15 largest ships would be equivalent to that of 760 million cars worldwide it is fore casted that this would increase by 4% annually under business-as-usual conditions. This equates to a 50%–250% increase by 2050 (Different Ways to Reduce CO_2 Emissions from Sea Freight, n.d.)

In addition to CO_2, cargo ships emit dangerous pollutants such as nitrous oxide (NOx) and sulphur oxide (SOx). These pollutants have an impact on air quality and contribute to global warming. Mitigation measures must be implemented as soon as possible for all of the reasons stated above. The IMO has set a goal of reducing emissions by 50% by 2050. The IMO has approved necessary steps to minimise greenhouse gas emissions from international shipping. The Energy Efficiency Design Index (EEDI) is required for new ships under the IMO's pollution prevention treaty (MARPOL), and every ship must have a Ship Energy Efficiency Management Plan (SEEMP).

Table 2.1 below depicts the timeline set by IMO for vessels to reduce their CO_2 emissions.

There are several strategies to reduce vessel CO_2 emissions:

- Route planning optimization;
- Switching over to renewable and greener fuel sources;
- Effective port operations;
- Steaming at a low temperature;
- Energy savings on board and
- The adoption of battery technology.

Table 2.1. IMO Recommendations for Reducing CO_2 Emissions 2015–2050.

2015 onwards	Energy Efficiency Design Index – by 'new ship'
2020 onwards	Improving the efficiency by 10% – by 'new ship'
2025 onwards	Improving the efficiency by 20% – by 'new ship'
2030 onwards	Improving the efficiency by 30% – by 'new ship'
2050 endpoint	CO_2 reduction by 50% (t/km)

Source: Adapted from Krčum et al. (2018).

Of all the above methods, transition to cleaner sources of fuel offers more scope for converting biomass to clean fuel. This would provide twin benefits: one, finding an efficient way to handle wastes which otherwise would end up in landfills causing huge environmental degradation and another, it offers a viable option to operate ships on green fuel, thereby making the shipping industry more sustainable. It is evident from many studies and researches that operating ships on biofuel is not a myth. In fact, ships are already being run on biofuel but in a small way. Prior to 2022, this was confined to a few demos, pilots and trials conducted onboard ships. However, this accelerated in 2022, with reports stating that around 930,000 tonnes of blended biofuel is being bunkered in Singapore and Rotterdam (Exploring the potential of biofuels in shipping - DNV, 2023). While this may appear to be huge Fig. 2.1, it represents only 0.1% of total maritime fuel use of 280 Million Tonnes of Oil Equivalent (Mtoe) every year. In 2023, the global production of advanced biofuels is expected to be 11 Mtoe. Between now and 2026, a substantial number of projects employing advanced biomass production are scheduled to come online, increasing total sustainable biofuel production levels up to 23 Mtoe per year (Exploring the potential of biofuels in shipping - DNV, 2023). While this implies significant development, it still falls short of the volume of biofuels required by shipping to have a significant influence on decarbonisation efforts. Because blended biofuels typically contain around 30% biofuel, we may determine that the statistics from Singapore and Rotterdam would represent approximately 280,000 tonnes of pure biofuels (Exploring the potential of biofuels in shipping- DNV, 2023).

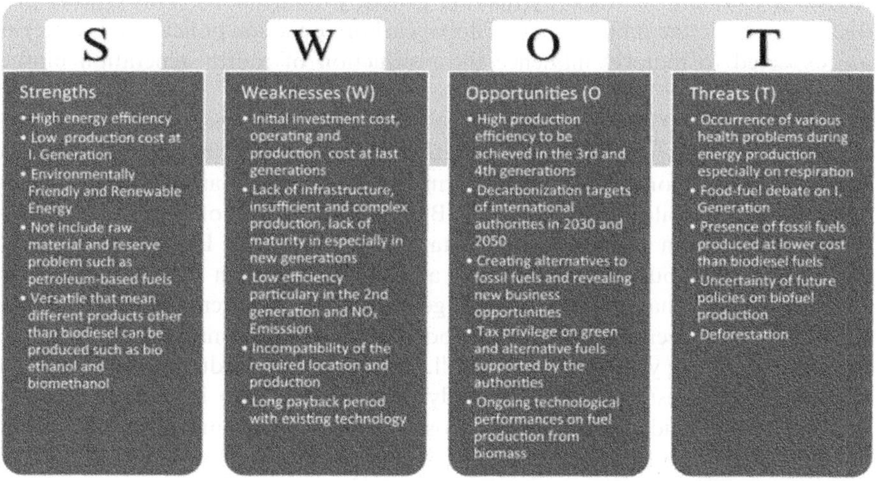

Fig. 2.1. SWOT Analysis for the Evaluation of Biodiesel as a Marine Fuel. *Source*: https://doi.org/10.1007/s10098-023-02501-7.

The IMO enacted new bunker fuel regulations in early 2020, lowering the maximum allowable sulphur content from 3.5% to 0.5%. The adjustment is expected to cut sulphur emissions by 77% or 8.5 million tonnes per year (Cutting sulphur oxide emissions). Following this, there has been an increase in demand for biofuels, which can significantly reduce CO_2 emissions from ships without requiring costly engine changes. Biofuels made from biomass are proving to be a viable choice for ships.

This study explores the possibility of establishing BRF at strategic locations to produce and supply biodiesel to ships. The study analyses the challenges in setting up such facilities and makes an attempt to suggest measures to handle them.

2. Objective of the Study

The following is the Strengths Weaknesses Opportunities and Threats (SWOT) analysis done on the possibility of generating and supplying biofuels to ships.

In the figure above, first generation biofuels refer to those generated from primarily human-consumed products such as corn, wheat, sugar cane, etc. Second generation biofuels refer to those generated from non-food products that have a minimum impact on arable land such as wood chips, straw, forestry wastes, urban solid waste, etc. The second generation biofuels are preferred to prevent fuel-food debate. The third generation biofuel production does not affect people's primary food needs. Waste oils and algae are considered the primary biomass, particularly waste cooking oil and palm oil. The fourth generation is intended to be a zero carbon economy which integrates the knowledge of other generations with engineering skills. In this process, micro-organisms and plants are utilized as feedstock to obtain biofuels by making genetic modifications. This type of production is still in the infancy stage. Global carbon reduction policies and strategies are expected to majorly influence the production of fourth generation biofuel (Bayraktar et al., 2023).

The above SWOT analysis, though presents some weaknesses and threats, does not discount the possibility of operating ships through biofuel. Taking the strengths and opportunities presented in the above figure as a base, this study explores the feasibility of setting up a BRF in the vicinity of Tirupur–Tuticorin region. The region is particularly suitable for locating a BRF because of its proximity to Tirupur's textile industry and to the Tuticorin Port as well. Many studies point out that textile waste is a good feedstock for generating bioethanol. Bioethanol has been accepted as a good alternative for marine diesel and its demand is growing year on year, but still, the large-scale production of bioethanol has not yet been established. This study explores if a BRF in this region could achieve it. The study evaluates the pros and cons of this initiative and recommends measures to handle the pitfalls.

3. Methodology

The study is qualitative in nature, and both primary and secondary sources of information have been used in the study. The research happening in the field of converting biomass into biofuel with particular reference to ships have been reviewed. The opinion of people associated with the shipping industry on the feasibility of the concept was collected through a questionnaire.

3.1 Research Questions

The study addresses the following research questions:

- Is it a viable option to generate bioethanol from textile wastes?
- Is it a viable option to establish a BRF in the Tirupur–Tuticorin region to supply bioethanol to ships calling at Tuticorin Port?
- What can be the challenges in implementing the project?

3.2 A Review of Literature

IMO agreement on technical regulations to reduce CO_2 emissions: MARPOL, Annex VI, chapter 4 adopted July 2011, entered into force in January 2015.
The IMO is critical in resolving environmental concerns relating to the shipping industry, particularly CO_2 emission reduction. This survey of the literature delve into the IMO's attempts to develop and apply technical regulations to reduce the environmental impact of maritime activities. The review examines the history of the IMO's engagement in environmental regulation. The evolution of IMO CO_2 emission reduction agreements, conventions and protocols were examined, highlighting major milestones, obstacles and triumphs.

Biofuels for the marine shipping sector an overview and analysis of sector infrastructure, fuel technologies and regulations (Hsieh & Claus, 2017)
The study looks into the role of biofuels in the maritime shipping sector, conducting a thorough examination of existing infrastructure, fuel technology and regulatory frameworks. The research is supported by the Innovation Fund Denmark and the International Energy Agency (IEA) Bioenergy Task 39, showcasing the joint efforts of academia industry and international organizations in creating sustainable maritime solutions. The participation of IEA Bioenergy Task 39 indicates a worldwide focus on bioenergy research. The article investigates the existing and growing regulatory frameworks controlling biofuels in the marine industry, including international regulations established by organizations such as the IMO and regional regulations influencing the adoption of biofuels.

The maritime shipping sector plays a crucial role in global trade and transportation, but its reliance on conventional fossil fuels raises environmental concerns. This detailed assessment examines the potential and obstacles of incorporating biofuels into the maritime shipping industry as part of ongoing efforts to reduce greenhouse gas emissions. Biofuels, derived from organic sources, offer a promising avenue to address environmental impact while maintaining the sector's critical role in the global economy.

Evaluating feedstock sources is part of determining the sustainability of biofuels. The review explores feedstocks such as algae, waste oils and agricultural leftovers, stressing their availability, environmental impact and potential to avoid rivalry with food production. The successful integration of biofuels into the maritime sector requires assessing their technical compatibility with existing ship engines.

This literature investigates any modifications that may be required to engines, fuel systems and other components in order to accommodate biofuels efficiently. Biofuel integration in the marine shipping sector is a possible path toward a more sustainable and ecologically friendly future. This in-depth examination highlights the technological viability, environmental benefits and obstacles related with biofuel adoption in maritime transportation. To fully achieve the potential of biofuels for ships, stakeholders such as governments, industry actors and research institutions must work together to address issues, develop supporting regulations and drive innovation in sustainable marine fuels. The switch to biofuels is a huge step toward a cleaner, more sustainable future for the worldwide shipping industry.

DNV's new white paper – Biofuels in shipping (Ovrom et al., 2023)
The white paper investigates the role of biofuels in the maritime industry, with an emphasis on sustainability, technical feasibility and regulatory issues. Authored by Eirik Ovrom and other experts, it provides insight into the current situation of biofuel usage in shipping as well as anticipated future developments. The paper explores the environmental benefits of biofuels in the maritime industry, including how these alternative fuels help to reduce greenhouse gas emissions and satisfy sustainability targets. The study goes into detail about the technical aspects of employing biofuels in maritime operations. There is also a discussion of market developments connected to biofuel use in the shipping industry.

Final report – Bio-Ethanol as an alternative fuel for vessels, Project number: MIIP 004 -2022, Maritime Knowledge Centre
The report investigates bio-ethanol as an alternative fuel for ships. It provides a detailed analysis of existing research on the use of bio-ethanol in the marine industry, including technological advances, environmental considerations and regulatory issues. The technical aspects of employing bio-ethanol as a naval fuel are examined. The report investigates engine compatibility and compares the environmental benefits and drawbacks of bio-ethanol to typical marine fuels. It investigates the economic feasibility of using bio-ethanol as a vessel fuel, including cost comparisons, prospective financial incentives and the long-term economic ramifications of switching to this alternative fuel.

A Study on the Production of biofuel from Textile Waste as a Feedstock (Shahi et al., 2022)
This review of the literature intends to provide insights into prior studies that studied the feasibility, problems and opportunities related with the manufacture of biofuel from textile waste. The manufacture of biofuel from textile waste is fraught with difficulties. The literature has addressed issues such as recalcitrance,

pollution and the requirement for pre-treatment procedures. To solve these obstacles, researchers have proposed novel solutions such as the use of specialized enzymes, genetically engineered microbes and enhanced pre-treatment techniques. It is critical to assess the environmental and economic viability of biofuel production from textile waste. The study evaluated the life cycle analysis, carbon footprint and economic feasibility of such technologies, providing vital insights into the overall impact of using textile waste as a biofuel feedstock.

A Methodology for the Optimization of Bioethanol Production via Biochemical Pathways (Raftery & Karim, 2014)
The increasing demand for renewable and sustainable energy sources have fuelled research into bioethanol production as a possible alternative to traditional fossil fuels. This paper presents a thorough technique for the optimization of bioethanol production, concentrating on the biochemical pathways involved in the conversion of biomass into ethanol. This methodology attempts to improve the efficiency, yield and economic viability of bioethanol production processes by methodically addressing critical elements and variables. This methodology offers a systematic and integrated approach to optimizing bioethanol production through biochemical pathways. By addressing key stages in the production process, from feedstock selection to fermentation and beyond, this methodology aims to contribute to the advancement of sustainable and economically viable bioethanol production.

This methodology aims to contribute to the advancement of sustainable and economically viable bioethanol production by addressing key stages in the production process, from feedstock selection to fermentation and beyond. Implementing this strategy could significantly boost bioethanol's competitiveness as a renewable energy source and promote its wider adoption in the global energy landscape.

Biofuel as an alternative shipping fuel: technological, environmental and economic assessment (Kesieme et al., 2019)
The maritime industry is under increasing pressure to switch to more ecologically friendly and sustainable fuel options. This evaluation delves into the many facets of biofuels as an alternative transport fuel, including technological improvements, environmental concerns and the economic sustainability of widespread use. This research looks on the technological landscape of biofuels in the maritime industry. It examines biofuel compatibility with existing ship engines and propulsion systems, as well as retrofitting possibilities and future upgrades needed for smooth integration. The analysis also addresses developing technologies that could improve biofuel consumption efficiency in maritime vessels.

The evaluation presents a complete analysis of the technological, environmental and economic implications of biofuels as a marine fuel alternative. Biofuels, on the one hand, have the potential to considerably lower the maritime industry's environmental footprint while also contributing to global efforts to address climate change. Addressing technological obstacles, guaranteeing economic sustainability and negotiating regulatory frameworks, on the other hand, are critical for a successful transition. Biofuels emerge as a possible solution as the industry explores sustainable alternatives. However, collaboration among stakeholders, politicians and industry leaders are required to promote innovation and support a seamless and sustainable transition to biofuels in the maritime sector.

A Case Study of the Textile Industry in Tirupur (Ramaswamy & Erkman, 1996) Tirupur, a bustling city in Tamil Nadu, India, has acquired the moniker 'Knitwear Capital' because of its dominance in the textile and garment industries. This case study goes into the evolution of Tirupur's textile sector, charting its growth, obstacles and innovative techniques used to attain sustainable practices. The case study begins with a historical overview of the development of the textile industry in Tirupur. This literature delves into the elements that contributed to Tirupur's development, from its humble beginning to its current status as a global hub for textile manufacture.

This case study investigates the critical influence of globalization in the growth trajectory of the textile sector in Tirupur. The study analyzes how Tirupur's businesses expanded their reach, established international collaborations and adapted to changing market dynamics to become a key player in the global textile market.

As the textile industry in Tirupur expanded, environmental concerns emerged. This study addresses the challenges related to water pollution, chemical usage and waste management. It explores how these challenges garnered attention from regulatory bodies, environmental activists and the industry itself. Faced with environmental scrutiny, the case study investigates the transformative initiatives taken by the textile industry in Tirupur. This includes the adoption of eco-friendly production processes, investments in water treatment technologies and the implementation of sustainable supply chain practices. The study also highlights the role of innovation in creating sustainable textiles.

The case study examines collaborative efforts within Tirupur's textile industry. It discusses how manufacturers, government bodies and non-governmental organizations worked together to address environmental challenges, improve working conditions and enhance the overall sustainability of the textile supply chain. Assessing the economic impact of sustainability initiatives, this section explores how embracing environmentally responsible practices affected the competitiveness and market positioning of Tirupur's textile industry. The study also considers the role of certifications and compliance in gaining consumer trust and access to international markets. The case study concludes by outlining the future outlook for the textile industry in Tirupur. It explores ongoing and potential future initiatives, challenges that may arise and the industry's role in shaping a sustainable and ethical future for textile manufacturing.

Tirupur's textile sector has grown from a local textile hub to a worldwide player, solving environmental concerns with sustainable techniques. This case study provides significant insights into the Tirupur textile industry's path, offering lessons and inspiration to other communities and sectors looking to balance expansion with environmental responsibility.

4. Biofuels in Shipping Industry

Three types of biofuels are relevant for maritime shipping:

(1) *FAME* (Fatty acid methyl ester) is produced from vegetable oils, animal fats or waste cooking oils via transesterification, a process that transforms various oils (triglycerides) to methyl esters. This is the most often used bio-diesel in the industry, and it is typically blended with regular marine diesel.
(2) *BTL* (Biomass to liquid) fuels are synthetic fuels derived from biomass via thermo-chemical conversion using the Fischer–Tropsch or methanol-to-gasoline processes. As a result of the technique, fuels that are chemically unique from traditional fuels such as gasoline or diesel but may also be used in diesel engines are possible.
(3) *HVO/HDRD* (Hydrogen vegetable oil/hydrogenation derived renewable diesel) is a byproduct of the fatty acid-to-hydrocarbon hydrotreatment process, which refines lipids or vegetable oils either alone or in combination with petroleum. To distinguish it from FAME biodiesel, diesel produced through this process is frequently referred to as renewable diesel. HVO/HDRD may be immediately integrated into distribution and refuelling facilities, as well as existing diesel engines, with no extra adjustments.

FAME is currently the most widely utilized biofuel in marine applications. It is either blended with standard petroleum fuels or used entirely as a biofuel (Use of biofuels in international shipping).

4.1 The Prospects of Bioethanol in Shipping

Bioethanol has been accepted as an alternate fuel for vessels, and there is growing demand for the production of bioethanol which can be understood from the following Fig. 2.2.

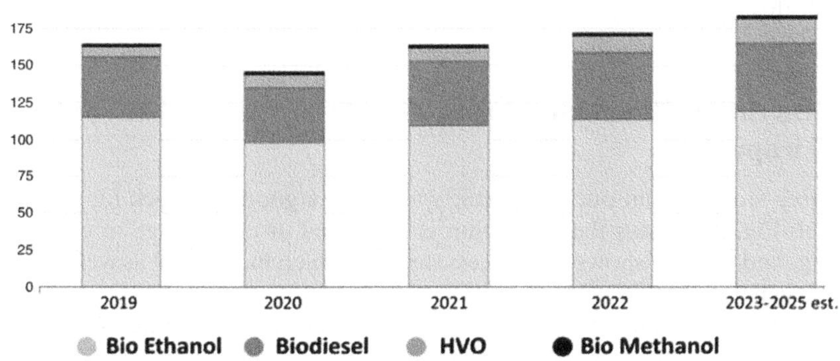

Fig. 2.2. Availability of Various Alternative Marine Fuels in Million Tons Per Year. *Source*: https://www.koersenvaart.nl/files/Report%20MIIP004-2022%20BioEOHpublic.pdf.

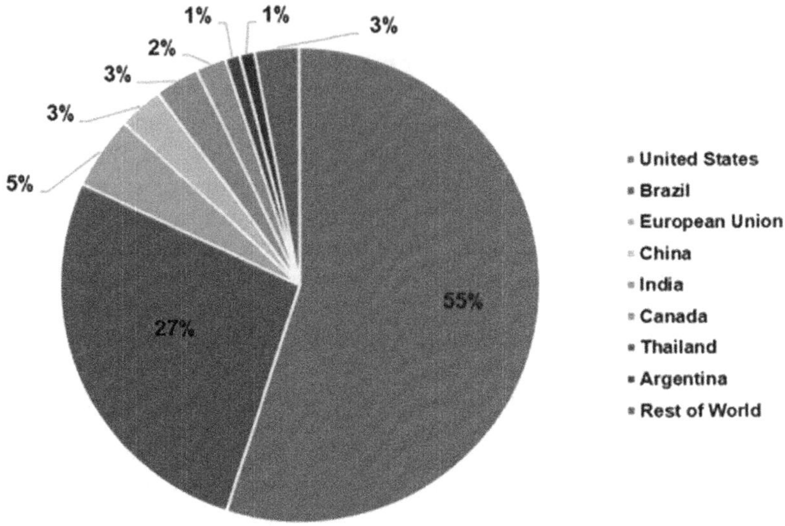

Fig. 2.3. Global Bioethanol Production Breakdown. *Source*: https://www.koersenvaart.nl/files/Report%20MIIP004-2022%20BioEOHpublic.pdf.

As illustrated in the above Fig. 2.3, global bioethanol production is over 100 billion litres, headed by the United States and Brazil, and accounted for USD 93 billion in 2020, with a CAGR of 5.2% predicted (Bio-Ethanol as an alternative fuel for vessels). In 2022, the price of ethanol was Rs 65.60 per litre, and the government of India may raise it by 3% in 2023–24 (Centre may hike cane-ethanol price by 3% for 2023–24 season, industry says).

From the above data, we can see that there is growing demand for bioethanol, and India's share is only 3% in the production of ethanol. Also, the pricing trend is favourable in India. Any BRF by meticulously carving out strategies can encash this opportunity.

5. Generating Bioethanol From Textile Wastes and Its Relevance for Tirupur

Research studies point out that textile wastes are a good feedstock for generating biofuel. Fig. 2.4 shows the waste that is generated at every stage of cotton processing, and Fig. 2.5 shows the process through which bioethanol as well as biogas can be generated from textile wastes.

Tirupur, predominantly being a cotton textile hub, generates more cotton wastes. This industry generates around 40 tonnes of combustible solid trash every

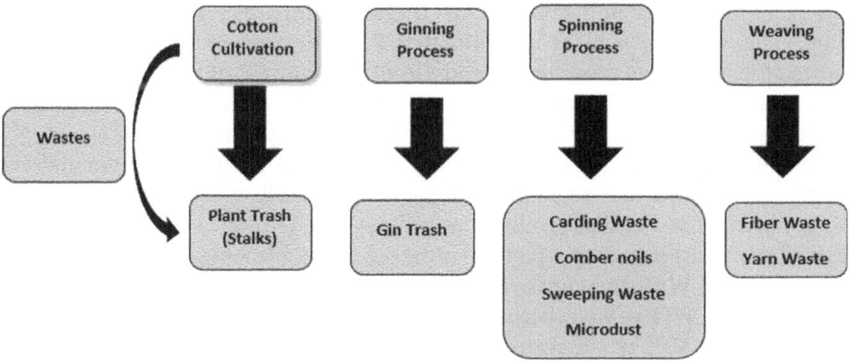

Fig. 2.4. Waste Generation During Cotton Processing. *Source*: https://www.researchgate.net/publication/361563703_A_Study_on_Production_of_biofuel_ From_Textile_Waste_As_a_Feedstock_-_A_Review.

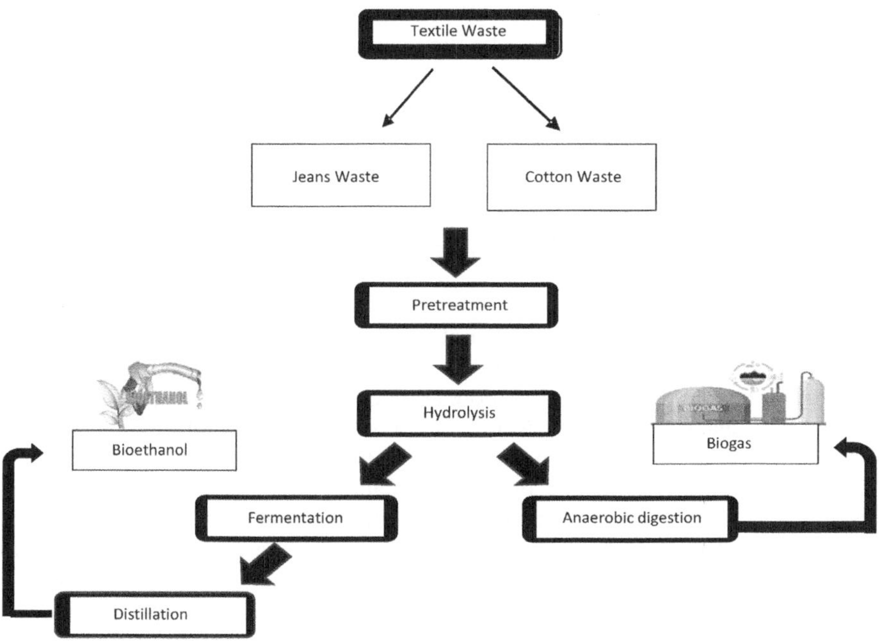

Fig. 2.5. Bioethanol Production From Textile Wastes. *Source*: https://www.researchgate.net/publication/361563703_A_Study_on_Production_of_biofuel_ From_Textile_Waste_As_a_Feedstock_-_A_Review.

day, including paper and textile scrap (rags and threads). This has a high fuel value. The sector generates an estimated 3,171 tonnes of paper waste, 9,430 tonnes of textile waste (rags and threads) and 59.25 tonnes of plastic waste (a Case Study of the Textile Industry in Tirupur) each year. Furthermore, houses generate an estimated 250 tonnes of municipal garbage per day (Preetha, 2022).

Tirupur being a textile manufacturing centre faces many environmental constraints. Textile waste is regularly dumped into landfills which is a serious cause for concern. The textile production process at Tirupur is highly pollutant. Particularly wet processing which involves scouring, bleaching and dyeing let out huge effluents. With the support of the government, Common Effluents Treatment Plants (CETPs) have been installed in Tirupur for treating effluents. The CEPTs recover 80% of the water required in processing, with the remaining 20% passing via reverse osmosis membranes (Preetha, 2022) and recovering waste mixed salt, bio sludge and chemical sludge. While chemical sludge is sent to cement plants for use as a co-processing ingredient, biosludge and waste salt are stored at the units. Nearly 60,000 tonnes of mixed waste salt and 50,000 tonnes of biosludge (Preetha, 2022) are stacked up at Tirupur's common and individual wastewater treatment plants.

From the above information, it is evident that Tirupur generates a good amount of feedstock for biofuel generation. Moreover the industry is into exports and Tuticorin happens to be the nearest port for Tirupur. The distance between Tirupur and Tuticorin by road is approximately 339 kms and takes about 5 ½ hours to reach. Tirupur is well connected by roadways to Tuticorin; therefore, locating a BRF in the Tirupur–Tuticorin vicinity may bring boundless benefits.

6. Setting Up Biomass Recycling Facilities (BRF) in the Tirupur–Tuticorin Region

With 13 major ports and almost 205 notified minor and intermediate ports, India can very well transform into a biofuel supplier for vessels. This can also be a strategy to make the Indian ports more competitive. Major ports in the southern region of India such as New Mangalore Port, Chennai and Cochin port are strongly involved in bunkering business due to their proximity to major international shipping routes. The bunkering facilities at these ports can be transformed into biofuel supply centres. India produces about 450–500 million tonnes of biomass per year. Biomass provides 32% (Biomass Energy in India - EAI, n.d.) of all the primary energy use in the country at present.

Now, to set up a BFR in the Tirupur–Tuticorin region, a road map needs to be developed. The following is an attempt to develop a road map for establishing BRF to serve Tuticorin Port (V.O. Chidambaranar Port). It is a transshipment gateway port that connects Colombo. Almost 70% of Tirupur's textile sector exports are brought to Tuticorin for direct export or transhipment (Tirupur exporters continue to prefer Tuticorin port over Chennai port - Apparel Resources).

The performance of Tuticorin Port for the financial year 2021–22 is depicted below in Table 2.2.

From the above table, it can be inferred that the performance of the port has been quite impressive during the year 2021–2022. It has achieved the highest container traffic, 7.81 lakh Twenty foot Equivalent Units (TEUs) during that period.

The above Fig. 2.6 presents data on three segments: the overall cargo traffic, container traffic and vessels handled at Tuticorin Port from 2017–2018 to 2021–2022. The overall cargo traffic has been calculated in million tonnes while cargo traffic has been measured in terms of Twenty - foot Equivalent Units (TEUs), a measurement used to indicate 20 feet long containers and the vessels handled by the port has been reported in terms of number. It is evident from Fig. 2.6 that there has been a decline in all the three segments between 2019–2020 and 2020–2021. The overall cargo traffic has decreased from 36.08 million tonnes in 2019–2020 to 31.79 million tonnes in 2020–2021, likewise the container traffic has come down from 8.04 TEUs in 2019–2020 to 7.62 TEUs in 2020–2021, and number of vessels handled has also dropped from 1588 in 2019–2020 to 1203 in 2020–2021. The decline may be attributed to Covid-19. But after this, the port shows improvement in all the three segments. The overall cargo traffic has increased from 31.79 million tonnes in 2020–2021 to 34.12 million tonnes in 2021–2022, the container traffic has gone up from 7.62 TEUs in 2020–2021 to 7.81 TEUs in 2021–2022 and number of vessels handled has also increased from 1203 in 2020–2021 to 1235 in 2021–2022. Furthermore, during the financial year 2022–2023, Tuticorin Port has handled Cargo Traffic of 38.04 million tonnes, against the previous year's achievement of 34.12 million tonnes, registering a growth of 11.5% (Annual Press Release, V.O. Chidambaranar Port Trust). All

Table 2.2. Tuticorin Port Performance (2021–22).

Traffic Handled	**34.12 Million Tonnes**
Cargo volume growth	7.33%
Achieved highest ever container volume	7.81 lakh TEUs
Container volume growth	2.49%
Last 5 years' CAGR containers	7.06%
Ships handled	1,235
Container moves/hour	29
Operating income	Rs. 636.29 crores
Net surplus after tax	Rs. 130.01 crores

Source: Adapted from Government of India. (2022, July 1). VOC Port Trust, Tuticorin - JAN. https://www.vocport.gov.in/port/UserInterface/PDF/JAN%20FINAL%2001.07.2022.pdf.

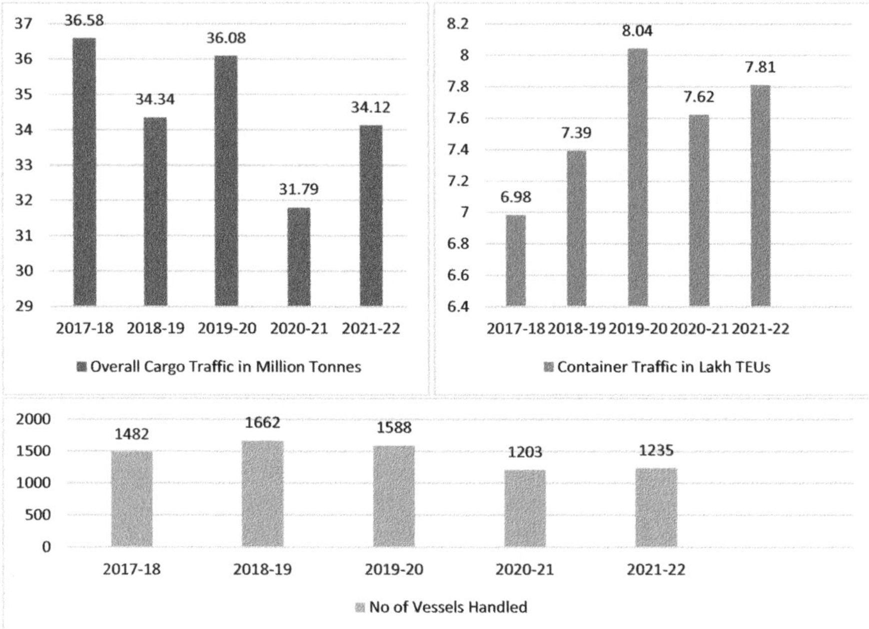

Fig. 2.6. Traffic Handled at Tuticorin Port Between 2017–2018 and 2021–2022 in Terms of Cargo, Containers and Vessels. *Source*: KORKAI, the e-Bulletin of V.O. Chidambaranar Port, Volume - 10 Jan–Mar 2022 https://www.vocport.gov.in/port/UserInterface/PDF/JAN%20FINAL%2001.07.2022.pdf.

this can be taken as evidence to the port's potential to attract traffic, which is an important factor to be considered for setting up a BFR in the region.

It is evident from the above data that a good number of vessels are calling at Tuticorin Port, and the port has 24/7 bunkering facility served by the Indian Oil Corporation (24*7 Bunkering, n.d.). Any investment in a BFR project in this region could turn out to be lucrative if the following aspects are scrutinized well before investing into the project.

- Location: The location of BRF must be in such a way that it is easily accessible to both Tirupur and Tuticorin. The textile wastes from Tirupur would be obtained and brought in to BRF to undergo the recycling process after which this may be supplied to the bunker facility at Tuticorin Port. BRFs proximity to both places are crucial. Depending upon the site identified, land acquisition and construction costs will have to be estimated. Minimum two hectares of land would be required for the project, and its cost depends on the location.

Construction cost may include plant base, boiler houses, warehouses, office building, etc.

- The Mode of Transportation: while textile wastes including the sludge from CETP may be transported through trucks, it is ideal to transport biofuel (bioethanol) from BRF to port through pipelines. This may require additional investment in pipelines.
- Technology: Cotton wastes as well as sludge require pretreatment to improve their digestibility to achieve high yield of biofuel. Pretreatment results in destroying the lignin, making it more digestible by exposing the amorphous region of the textile. This process is done through chemical, physical or biological method and this represents one of the main components of cost. After pretreatment, hydrolysis, fermentation and distillation are the other processes involved that convert the pretreated textiles to bioethanol. The cost of investment into technology has to be estimated.

  BRFs may have to invest heavily into plant and machinery required for recycling. Studies point out that in USA an optimal bioethanol plant can produces 42 million gallons of bioethanol per year from the processing of 770,000 tons of biomass per year using a dilute sulphuric acid pretreatment (Raftery & Karim, 2014). In India, the cost of setting up a bioethanol plant with a capacity of 10 kilo litres per day (KLPD) starts from 20 crores (approx.) (Aastha, 2023), and this might vary depending on the feedstock used.
- Blended fuel: Bioethanol is usually blended with marine diesel oil and then supplied to vessels. BRF may have to enter into agreements/contracts with the Indian Oil Corporation, which is currently supplying bunker to vessels to supply the blended mixture.

7. Discussion and Recommendations

It is evident from MARPOL regulations that sooner or later all the ships will have to be driven by cleaner fuel. The primary data collected point out the fact that ships have in fact started using Very Low Sulphur Fuel Oil (VLSFO), and the demand is increasing for it. 50% of the respondents felt that there is growing demand for VLSFO. Large shipping companies like MSC, AP Moller- Maersk and NYK are on the verge of floating their carbon neutral vessels.

While the feasibility is quite high, 87% of the respondents agreed that huge biofuel generating plants must be installed to commercially produce and supply biofuel to vessels. 50% of the respondents agreed that shipping companies can set up their recycling facilities at ports while another 50% said it is not a viable option. Out of the 50% who disagreed, 20% cited less no of vessels owned by shipping companies as a reason, remaining 30% cited regulatory compliances as a reason. 82% agreed setting up of independent BRF as a viable option to produce and supply biofuels to ships. 67% rated huge capital investment as the major constraint in setting up biofuel plants, followed by infrastructure bottlenecks (63%) and availability of feedstock (52%).

Huge recycling plants may have to be constructed as ships require huge quantities of biofuel. A single very large ship may consume the annual production from a single medium-sized biofuel facility which may be upto 100 million liters (Hsieh & Claus, 2017). Table 2.3 shows the classification of ships based on their Dead Weight Tonnage (DWT) (What are the 7 types of cargo ships?, 2022). The same has been compared with the ships calling at Tuticorin Port and was found that vessels with sizes ranging from Small Handy to Post -Panamax call at Tuticorin Port (Port of Tuticorin (India) - Arrivals, Departures, n.d.). The Port of Tuticorin's container terminal has a 370-meter long quay with alongside depth of 10.9 meters and can accommodate vessels to 50,000 DWT. There are also two shallow-water berths of a total length of 250 meters with alongside depth of 5.85 meters that can accommodate vessels upto 5000 DWT (Port of Tuticorin, n.d.).

It is apparent that a BRF established in Tirupur–Tuticorin region must have capacity to supply bioethanol to medium and large size vessels calling at Tuticorin Port. Which means, feed stock requirements will also be quite high. The facility must be equipped to handle large quantities of biomass. To cater to such huge requirements one BRF may not be sufficient therefore a network of BRFs may have to be located at strategic points. Likewise, textile wastes not only from Tirupur but also from its neighbouring regions such as Coimbatore and Karur may also be obtained. The textile industry at Tirupur has expanded its operations into surrounding regions too.

Biomass is highly diverse in nature and may be classified as follows:

- Field and plantation biomass;
- Industrial biomass;
- Forest biomass;
- Urban waste biomass and
- Aquatic biomass.

Besides textile wastes, urban waste biomass and field and plantation biomass from the region may also be obtained for recycling. BRF may also enter into contract with Port Reception Facility (PRF) at Tuticorin Port to collect ship residues and wastes which may be recycled. The cost of feedstock might vary depending on availability but in Tirupur, the industry associations and CEPTs are struggling to find an efficient way to dispose the wastes. BRFs could obtain such wastes at lower or even without any price. It can be understood from the above discussion that BRF is a large-scale business and economies can be achieved only through large-scale production. Therefore, it is important to diversify into other types of biofuels also. BRFs may establish buy back centres to collect different types of feedstock, which may be transported to the respective BRFs.

Given the situation, there is more room for many players to enter the biomass recycling sector, and the government also is encouraging green initiatives through policies and promotion measures. This would be a good business opportunity for companies engaged in or would like to engage in renewable energy. In India, BRFs may be developed through public–private partnership (PPP) mode also because of its huge investment requirements. Government support available in the form of capital subsidy and financial assistance may be utilized for that matter. The Ministry of New and Renewable Energy (MNRE) provides Central

Table 2.3. Classification of Vessels by Dead Weight Tonnage (DWT).

Size of the Vessel	Dead Weight Tonnage (DWT)
Small Handy	10,000–24,999 DWT
Handy size	28,000–40,000 DWT
Large Handy	35,000–39,999 DWT
Handymax	40,000–50,000 DWT
Traditional Supramax vessel	50,000–60,000 DWT
Panamax (it is the maximum size of the vessels that can sail through the Panama Canal)	60,000–80,000 DWT
Post-Panamax vessel	79,000–99,999 DWT
Aframax (the oil tankers of medium size, larger than Panamax ships but smaller than Suezmax vessels)	75,000–115,000 DWT
Suezmax (it is the maximum size of the vessels that can sail through the Suez Canal)	Approx. 150,000 DWT
Malaccamax (it is the maximum size of the vessels that can sail through the Malacca Straits)	280,000–300,000 DWT
Seawaymax (it is the maximum size of the vessels that can sail through the canal locks of the St. Lawrence Seaway)	10,000–60,000 DWT
Capesize (ships that cannot sail through the Panama Canal or Suez Canal but can pass through the Cape of Good Hope and Cape Horn)	100,000–129,999 DWT for Mini Capesize, 130,000–199,999 DWT for standard Capesize and 200,000+ DWT for large Capesize
VLCC (Very Large Crude Carrier)	Supertankers with 150,000–320,000 DWT capacity
ULCC (Ultra Large Crude Carrier)	Supertankers with 320,000–550,000 DWT capacity

Source: https://sinay.ai/en/what-are-7-types-of-cargo-ships/
Note: DWT - It is a measure of how much weight a ship can carry, so it includes the sum of the weights of cargo, fuel, fresh water, ballast water, provisions, passengers, and crew. This measurement does not include the empty weight of the ship and is measured in metric tonnes.

Financial Assistance (CFA) in the form of capital subsidy and financial incentives to the biomass energy projects in India (Biomass Programme | Ministry of New and Renewable Energy | India). The firms can utilise these schemes to reduce costs. Also, FDI up to 100% is permitted in the renewable energy sector under the automatic route, and no prior government approval is required (Joshi, 2022). This may also be utilized to attract foreign joint venture investments to set up BRF projects. FDIs from MNCs would help to tide over huge capital and technological requirements. Shipping companies, bunkering companies and ports may also expand their business into this promising segment.

Another significant cost is with respect to pipeline connectivity. These BRFs must be connected to Tuticorin Port through pipelines to transport bioethanol. Since ships would be using only the blended mixture, there should be pipeline connectivity to Indian Oil Corporation (IOC), which is at present supplying bunkers to vessels at Tuticorin Port. Bioethanol may be blended with marine diesel at the bunker facility operated by IOC and then supplied to ships. Pipeline projects are huge investments which BRFs may undertake jointly with IOC, or it may be developed by government through tenders.

It is evident that heavy costs may have to be incurred throughout the implementation of the project. Also, it may require contracts and agreements with other business parties such as IOC, textile units, logistics service providers, etc. Meticulous planning is required at every stage. Break even analysis may have to be done to determine the quantity of bioethanol to be produced and profit targets may have to be set accordingly.

8. Conclusion

Biofuel supply to ships is a booming sector where investments are awaited. Well-established companies with impressive financial statements can explore the field. What is needed is a clear road map. Also setting up BRF to supply biofuels to ships requires joint efforts from government, private businesses, infrastructure developers, ports and port reception facilities, bunker suppliers, shipping companies, logistics companies, regulatory bodies, etc. Given these findings, a concerted effort to reward research, invest in infrastructure and implement supportive legislation will be critical for realizing biofuels' full potential in reducing the environmental impact of shipping. The marine industry can play a critical role in promoting a more sustainable and resilient future for global transportation by embracing biofuels.

It is evident that setting up BRF is a challenging task. Despite the challenges. it appears to be a lucrative business going by the increasing demand for biofuels in transportation particularly in shipping industry. Besides, sustainable development is the order of the day, and any investment into greener technology is bound to pay off. The huge investments initially required can be recovered through increased price of ethanol. But it is important to be an early entrant into this sector to take advantage of price. As the global supply increases, the price is more likely to fall.

Setting up a BRF in the Tirupur–Tuticorin region can be a viable option provided a meticulous plan is chartered out with respect to location, investments, feedstock procurement, technology to be adopted to generate bioethanol and infrastructure to supply bioethanol to Tuticorin Port. Though it could be a project with challenges, it is worth investing not only because it is lucrative but it is also an investment for a better future.

However, further research and development are needed to increase the efficiency and cost-effectiveness of biofuel production. Furthermore, feedstock availability, land use and competition with food production must all be carefully examined to ensure biofuel supply networks' long-term viability.

References

24*7 Bunkering. (n.d.). https://www.vocport.gov.in/24_7_Bunkering.aspx. Accessed on February 12, 2024.

Aastha. (2023, May 15). *How to start ethanol production plant in India*. Corpseed ITES Private Limited. https://www.corpseed.com/knowledge-centre/how-to-start-ethanol-production-plant-in-india

Annual Press Release, V.O. Chidambaranar Port Trust. (n.d.). *V.O. Chidambaranar Port Trust*. https://www.vocport.gov.in/port/UserInterface/photos/Annual%20Press%20Release-2023742023142538.pdf

Bayraktar, M., Pamik, M., Sokukcu, M., & Yuksel, O. (2023). A SWOT-AHP analysis on biodiesel as an alternative future marine fuel. *Clean Technologies and Environmental Policy*. https://doi.org/10.1007/s10098-023-02501-7

Bio-ethanol as an alternative fuel for vessels. (n.d.). https://www.koersenvaart.nl/files/Report%20MIIP004-2022%20BioEOHpublic.pdf

Biomass Energy in India - EAI. (n.d.). https://www.eai.in/ref/ae/bio/bio.html#:~:text=India%20produces%20about%20450%2D500,in%20the%20country%20at%20present

Biomass Programme | Ministry of New and Renewable Energy | India. (n.d.). https://mnre.gov.in/bio-mass/

Centre may hike cane-ethanol price by 3% for 2023–24 season, industry says. (n.d.). https://www.business-standard.com/india-news/centre-may-hike-cane-ethanol-price-by-3-for-2023-24-season-industry-says-123101101000_1.html

Cutting sulphur oxide emissions. (n.d.). https://www.imo.org/en/MediaCentre/HotTopics/Pages/Sulphur-2020.aspx

Different ways to reduce CO_2 emissions from Sea Freight. (n.d.). https://sinay.ai/en/different-ways-to-reduce-co2-emissions-from-sea-freight/

Exploring the potential of biofuels in shipping - DNV. (2023). DNV GL. https://www.dnv.com/expert-story/maritime-impact/Exploring-the-potential-of-biofuels-in-shipping.html

Hsieh, C.-W. C., & Claus, F. (2017, October). *Biofuels for the marine shipping sector an overview and analysis of sector infrastructure, fuel technologies and regulations*. IEA Bioenergy. https://www.ieabioenergy.com/wp-content/uploads/2018/02/Marine-biofuel-report-final-Oct-2017.pdf

Joshi, A. (2022, June 6). *FDI in Indian renewable sector up by 100% to $1.6 billion in FY22*. Mercomindia.com. https://mercomindia.com/fdi-in-indian-renewable-sector-up. https://mercomindia.com/fdi-in-indian-renewable-sector-up

Kesieme, U., Chrysanthou, A., & Catulli, M. (2019). Assessment of supply interruption of rhenium, recycling, processing sources and technologies. *International Journal of Refractory Metals and Hard Materials*, *82*, 150–158.

Krčum, M., Gudelj, A., & Tomas, V. (2018). Optimal design of ship's hybrid power system for efficient energy. *Transactions on Maritime Science*, *7*(1), 23–32. https://doi.org/10.7225/toms.v07.n01.002

Maritime Supply Chains. (n.d.). *What is the impact on climate change*. Marine-Digital.com. https://marine-digital.com/article_maritime_supply_chains

Ovrom, E. A., Mostert, K. A., Khakhkhar, S., McKee, D. P., Yang, P., & Her, Y. F. (2023). A comprehensive review of the genetic and epigenetic contributions to the development of fibromyalgia. *Biomedicines, 11*(4), 1119.

Port of Tuticorin (India) - Arrivals, Departures. (n.d.). *Expected vessels - VesselFinder*. https://www.vesselfinder.com/ports/INTUT001

Port of Tuticorin. (n.d.). *World port source*. http://www.worldportsource.com/ports/commerce/IND_Port_of_Tuticorin_1426.php#:~:text=The%20Port%20of%20Tuticorin's%20Berth,vessels%20to%2050%20thousand%20DWT

Preetha, M. S. (2022, December 6). Tirupur textile processing sector struggles with mixed waste salt. *The Hindu*. https://www.thehindu.com/news/cities/Coimbatore/tiruppur-textile-processing-sector-struggles-with-mixed-waste-salt/article66230601.ece

Raftery, J. P., & Karim, M. N. (2014). A methodology for the optimization of bioethanol production via biochemical pathways. *Computer-Aided Chemical Engineering*, 597–602. https://doi.org/10.1016/b978-0-444-63433-7.50084-5

Ramaswamy, R., & Erkman, S. (1996). Case study of the textile industry in Tirupur. Socioeco.org. https://socioeco.org/bdf_fiche-document-478_en.html

Shahi, Z., Firoozi, Z., Mohammadisoleimani, E., Shahi, A., Naghizadeh, M. M., Mirzaei, E., Asad, A. G., & Mansoori, Y. (2022). Hsa_circ_0000479/Hsa-miR-149-5p/RIG-I, IL-6 axis: A potential novel pathway to regulate immune response against COVID-19. *The Canadian Journal of Infectious Diseases & Medical Microbiology*.

Tirupur exporters continue to prefer Tuticorin port over Chennai port - Apparel Resources. (n.d.). apparelresources.com. https://apparelresources.com/business-news/trade/tirupur-exporters-continue-prefer-tuticorin-port-chennai-port/

Tuticorin Port. (n.d.). www.thoothukudicorporation.com. https://www.thoothukudicorporation.com/tourism/tuticorin-port

Use of biofuels in international shipping. (n.d.). https://www.dnv.com/news/use-of-biofuels-in-international-shipping-240298

What are 7 types of cargo ships? (2022, April 14). https://sinay.ai/en/what-are-7-types-of-cargo-ships/

Chapter 3

Qualitative Insights Into Harvesting Sustainability: The Role of Organic Agriculture in Advancing Sustainable Development Goals

Manpreet Arora[a], *Sanjeev Gupta*[a] *and Amit Mittal*[b]

[a]Central University of Himachal Pradesh, India
[b]Chitkara University, India

Abstract

This chapter draws from bibliometric data and secondary sources to explore the interrelation between sustainable development and organic agricultural practices. The study contends that the adoption of organic farming methods holds the potential to address multiple Sustainable Development Goals (SDGs), thereby contributing to the well-being of all living organisms. The analysis of current trends in organic agriculture research has revealed prevalent themes pursued by academics. Notably, themes such as biotechnology, biodegradation, soil conversion, soil restoration, environmental sustainability, health, alternative agriculture and community development emerge as significant and burgeoning areas of study within the field. Additionally, other themes have come to the forefront in the findings, including agricultural emissions, cultivation practices, environmental regulations, erosion control, agricultural policy and sustainable agriculture. These themes present vital areas for further investigation and exploration, indicating areas where more research is warranted. This qualitative piece through content analysis provides a deep insight on the fact that organic farming in relation to sustainability is a very under researched area. Researchers and practitioners can explore this area as a tool to achieve the goals of sustainability in diverse dimensions. The study suggests some under-researched areas as future research agendas which include supply chain and distribution management, market development and consumer behaviour in relation to organic farming, investigating and creating novel commercial strategies for

The Emerald Handbook of Tourism Economics and Sustainable Development, 41–62

Published under exclusive licence by Emerald Publishing Limited
doi:10.1108/978-1-83753-708-220241003

organic farming, financial and investment mechanisms in the field of organic farming, technological developments in organic agriculture, analysing rules and regulations to see how they affect the development and sustainability of organic farming and social impact and stakeholder engagement for organic farming as tool to promote sustainability.

Keywords: Organic agriculture; organic farming; sustainability; Sustainable Development Goals; SDG; qualitative; economic growth; organic; organic produce; sustainable

1. Introduction

The goals for sustainable development were created as a road map for attaining a better and sustainable future for all living things on Earth. 193 nations worldwide came to an agreement on the 2030 United Nations (UN) Agenda for Sustainable Development in September 2015. These 17 goals address issues such as zero hunger, poverty reduction, climate change mitigation, education reform, health and well-being, access to affordable and clean energy, clean water and sanitation, decent work and economic growth, responsible consumption and production, life above and below the water and more. These objectives seek to improve the world by promoting ethical behaviour in all spheres and concentrating on protecting the environment, living things and the planet as a whole. These objectives, whether they relate to economic development or agricultural output, all have a green agenda that promotes protecting Mother Nature. The problems and worries surrounding the climate crisis have compelled us to consider our continuing actions that are negatively affecting the environment.

In order to meet the need for food throughout the world, modern technology has revolutionized agriculture in the last several decades. This has raised yield to unprecedented levels. However, the depletion of soil, air, water and fertilizer use have come at the expense of increased greenhouse gas emissions, soil deterioration and nutrient loss. The excessive use of chemicals and pesticides has harmed future land fertility and created ecological imbalances that affect all species on Earth. According to UN estimates from 2019, there will be 9.7 billion people on Earth by the year 2050, which means that there will undoubtedly be a significant increase in demand for agricultural products (https://news.un.org/en/story/2019/06/1040621). Food availability needs to keep up with the expanding population's demands for consumption in order to combat food scarcity. Maintaining and addressing the sustainability challenges for the future requires a strong focus on the quality of agricultural produce. In the next 15 years, the 2030 agenda for sustainable development – which is based on 17 goals – will tackle the world's problems, one of which is food security worldwide, which is closely tied to preventing climate change and preserving the planets ecological balances. To address the sustainability concerns of many industries, we need a revolutionary and holistic approach in every area. Thus, if applied intelligently, sustainable agriculture techniques can help achieve multiple Sustainable Development Goals (SDGs) at the same time, including poverty, unemployment, climate control and life under water. One such goal is food security.

The present piece argues that there is a relationship between sustainable development and sustainable agricultural produce, and it is based on bibliometric data. According to the author, organic farming methods have the potential to address a number of sustainable development objectives, including advancing the welfare of all living things. The foundation of all 17 SDGs is agriculture and food. The body of research shows that, in order to address the global challenge of feeding everyone, agricultural techniques must be changed. Specifically, organic farming must be adopted in order to maximize the benefits of food production for the environment. Organic farming can, in fact, offer solutions to a number of the issues listed in the SDGs. The Food and Agriculture Association states that the foundation of organic farming is the promotion of comprehensive production management systems that support and strengthen the agro-ecosystem. In addition, they address 'biodiversity', 'biological cycles' and 'social biological activities' so that they can be implemented to reduce adverse effects on society in addition to promoting the comprehensive development of sustainable practices linked to agricultural produce. 'Health technology, fairness and care' are the cornerstones of organic agriculture.

The world must transition to sustainable systems of consumption and production. Environmental deterioration has been significantly exacerbated by unsustainable production and consumption practices. Climate change, pollution, habitat damage, deforestation and biodiversity loss are all included in this. In order to mitigate these effects and protect ecosystems and natural resources for future generations, a shift toward sustainability is imperative. Because of the elevated rate of consumption, non-renewable resources, like fossil fuels, are running out quickly. In order to combat scarcity, sustainable practices encourage the use of renewable resources and energy-efficient technologies while minimizing waste and resource consumption. The adoption of unsustainable industrial practices results in significant greenhouse gas emissions which are contributing to climate change. Mitigating the effects of climate change and fostering a healthier ecosystem need a shift to sustainable practices, such as employing renewable energy sources and lowering carbon footprints. The sustainable patterns of consumption and production take society effects into account. Fair working standards, protecting workers' rights, and allowing everyone access to necessities are all part of this strategy. It also takes into account how production methods affect communities on a social and economic level. When unsustainable methods cause resource depletion, environmental harm and health costs, they frequently result in economic instability. Sustainable business practices encourage innovation, open up new employment prospects in the green sector and support long-term, steady economic growth. In general, the health of the earth and its people depend on the transition to sustainable patterns of production and consumption. In order to create a more resilient and balanced future, it entails the responsible use of resources, a reduction in environmental effect and respect for social and economic fairness.

Although the biggest challenge of organic farming is the high cost of production. Abebe et al. (2022) in their study found that in Lebanon, poorly designed organic value chain are the main causes of the rising expense of organic farming. Thus, as a local response to the sustainability of organic farming in developing nations with limited resources (land and labor) and distinguished by lengthy marketing channels, the authors propose a mixed approach of organic farming with other models, in particular agro-tourism. In such models, microfinancing or lending microcredit to promote sustainable ventures have proven to be important players (Arora & Singh, 2023a). Community development is highly related with rural development and which in turn promotes sustainability. Villages or rural areas should be valued when planning the development of rural areas, and their potential and regional features should be encouraged and utilized to support sustainable development (Cizler, 2013). The ideas of ecotourism can serve as a valuable framework for sustainable development for small-scale, organic farms that cater to tourists. This new kind of tourism can be used to support learning, well-being, social-cultural preservation, local development, sustainable agriculture and the environment (Choo & Jamal, 2009). On the same line, various business sectors can be motivated to promote organic products and do their bit via CSR activities. As CSR plays an important role in promoting sustainability (Arora & Walia, 2019), the entrepreneurial activity in organic production can pave towards economic development as models like providing microcredit to entrepreneurs are regarded to enablers of sustainability (Arora, 2023; Arora & Singh, 2023b). Consequently, while developing regional environmental policies, the varying effects of institutionally disaggregated components on sustainable development indicators across regions should be taken into account (Hunjra et al., 2023). Risk management factors also play a key role in achieving SDGs. The significance of risk management as a fundamental control strategy for creating sustainability plans that support reducing pollution emissions should not be undermined (Hunjra et al., 2022). Essentially each sector of the economy, digitalization and crisis management are primarily driven by sustainability and innovation (Arora et al., 2023). Sustainable actions promoted through sustainable reporting content can have a far-reaching impact on stakeholders, which can have ripple effect and further promotes sustainable actions (Arora & Rathore, 2022). To attain sustainable development, policymakers in developing nations should concentrate on the strength of their institutions that support the market Azam et al. (2022). The application and adoption of technology in the tourist and hospitality industries has been well covered in the literature, but the research structure in this sector has remained dispersed and fragmented (Hossain et al., 2022). In promoting sustainable development, various ontological perspectives can play an important role especially in multi-sectoral businesses (Arora & Sharma, 2023).

2. Bibliometric From Literature

There is a dearth of current research on organic farming and its relationship to the SDGs, and the literature presents a range of viewpoints. Future research has enormous potential, yet there is a significant research gap. On 26 October 2022, a search was conducted in the SCOPUS database using the terms 'Organic Farming' and 'Sustainable Development Goals' in order to examine the available literature. The top 15 papers on the topic were discovered. The widely cited paper came out in 2018. The top 15 articles from the dataset that was taken from the SCOPUS database using the search criteria are displayed in Table 3.1.

According to Keesstra et al. (2018), the primary methods for identifying soil and landscape solutions are those rooted in nature. Adopting nature-based solutions can contribute to the UN's SDGs being more fully realized. The potential for nature-based solutions as a very practical and long-term solution for hydrological risks and land degradation has been highlighted by the authors through an analysis of several case studies based on organic farming, rewinding land restoration, sediment trapping and wetland construction in various countries. According to Eyhorn et al. (2019), organic agriculture is regarded as a valuable component in the strategy for accomplishing the SDGs of the UN. A framework of policies has been suggested by the authors to initiate the transition. In order to accomplish the aims of sustainable development, Cerdà et al. (2020) made a strong case for the urgent necessity to employ environmentally friendly agriculture management practices. According to the authors, further research is necessary to fully understand organic farming management strategies in order to reap the benefits of sustainable practices. Based on their empirical research, they conclude that organic farming circumstances hold great promise for accomplishing sustainable development objectives.

The notion that biotechnology can be one of the intrusive methods for boosting sustainability is highlighted by Purnhagen et al. (2021). They believe that encouraging organic farming is another way to meet the objectives of sustainable development. According to the authors, policy rules and legislative reforms are necessary to achieve the targets set forth in the 2030 plan. One of the key instruments for achieving the SDGs is agriculture. One of the main factors in attaining environmental, economic and social sustainability is said to be organic farming. The objectives for organic agriculture are higher in countries like Germany, but the success rate is much lower. In order to achieve SDGs, organic farming can be encouraged by proper national awareness-raising campaigns and enhanced financial assistance (Siepmann & Nicholas, 2018). According to Haque et al. (2021), the paddy fields' highest levels of greenhouse gas emissions pose a significant risk to the sustainability of rice production and food security. In their research, the authors have put forth recommendations for beneficial soil additions that will increase production yield while preserving soil fertility over time.

According to Sarkar et al. (2021), organic farming is a concurrent method for enhancing the land and soil quality and restoring the environment. The authors contend that unsustainable agricultural practices are primarily to blame for the depletion of natural resources, such as soil, land, and water, and that organic agriculture can play a significant role in achieving significant global objectives like the SDGs set forth by the UN. Bieroza et al. (2021) discuss the real-world challenges associated with accomplishing the SDGs. The writers discuss the approaches that ought to be taken in order to address and comprehend the causes of the European Union's water quality issues. The authors believe that actively

Table 3.1. Top 15 Articles.

S. No	Title of the Paper	Authors	Year of Publication	Source of Publications	Citations as per SCOPUS as on 26 October 2022
1	The superior effect of nature-based solutions in land management for enhancing ecosystem services Open access	Keesstra, S., Nunes, J., Novara, A., (. . .), Kalantari, Z., Cerdà, A.	2018	*Science of the Total Environment* 610–611, pp. 997–1009	430
2	Sustainability in global agriculture driven by organic farming Open access	Eyhorn, F., Muller, A., Reganold, J.P., (. . .), Seufert, V., Smith, P.	2019	*Nature Sustainability* 2(4), pp. 253–255	98
3	Tillage versus no-tillage. Soil properties and hydrology in an organic persimmon farm in eastern Iberian Peninsula	Cerdà, A., Rodrigo-Comino, J., Yakupoğlu, T., (. . .), Hammad, H.M., Daliakopoulos, I.N.	2020	*Water* (Switzerland) 12(6), 1539	31
4	Europe's farm to fork strategy and its commitment to biotechnology and organic farming: Conflicting or complementary goals?	Purnhagen, K.P., Clemens, S., Eriksson, D., (. . .), Wesseler, J.H.H., Zilberman, D.	2021	*Trends in Plant Science* 26(6), pp. 600–606	26
5	German winegrowers' motives and barriers to convert to organic farming	Siepmann, L., Nicholas, K.A.	2018	*Sustainability* (Switzerland) 10(11), 4215	19

6	Organic amendments boost soil fertility and rice productivity and reduce methane emissions from paddy fields under sub-tropical conditions	Haque, M.M., Datta, J., Ahmed, T., (. . .), Ahmed, S., Sabagh, A.E.L.	2021	*Sustainability* (Switzerland) 13(6), 3103	9
7	Organic interventions conferring stress tolerance and crop quality in agroecosystems during the United Nations decade on ecosystem restoration	Sarkar, D., Dubey, P.K., Chaurasiya, R., (. . .), Parewa, H.P., Rakshit, A.	2021	*Land Degradation and Development* 32(17), pp. 4797–4816	4
8	What is the deal with the green deal: Will the new strategy help to improve European freshwater quality beyond the Water Framework Directive?	Bieroza, M.Z., Bol, R., Glendell, M.	2021	*Science of the Total Environment* 791, 148080	4
9	Drivers for the behavioural receptiveness and non-receptiveness of farmers towards organic cultivation system	Thakur, N., Nigam, M., Tewary, R., (. . .), Mahmoud, G.A.-E., Gupta, S.	2022	*Journal of King Saud University - Science* 34(5), 102107	3
10	Role of organic and conservation agriculture in ammonia emissions and crop productivity in China	Xu, P., Li, G., Houlton, B.Z., (. . .), Chen, A., Zheng, Y.	2022	*Environmental Science and Technology* 56(5), pp. 2977–2989	3

(Continued)

Table 3.1. *(Continued)*

S. No	Title of the Paper	Authors	Year of Publication	Source of Publications	Citations as per SCOPUS as on 26 October 2022
11	Organic farming: A prospect for food, environment and livelihood security in Indian agriculture	Manna, M.C., Rahman, M.M., Naidu, R., (…), Chaudhari, S.K., Subbarao, A.	2021	*Advances in Agronomy* 170, pp. 101–153	3
12	Towards sustainable organic farming systems	El Chami, D.	2020	*Sustainability* (Switzerland) 12(23), 9832, pp. 1–5	3
13	Soil security: A key role for sustainable food productivity	Thangavel, P., Sridevi, G.	2017	*Sustainable Agriculture Towards Food Security* pp. 309–325	3
14	Project education-based applied technologies for poverty alleviation, ending hunger, achieving food security and promoting sustainable Ag	MacLachlan, A., Pecen, R., Joye, N.	2017	IHTC 2017 - IEEE Canada International Humanitarian Technology Conference 2017 8058170, pp. 113–118	3
15	Application of SWOT-AHP to develop organic food marketing strategy	Najib, M., Sumarwan, U., Septiani, S., Fahma, F.	2021	*Academy of Strategic Management Journal* 20(1), pp. 1–8	2

Source: Compiled by authors from SCOPUS database accessed on 26 October 2022.

restoring agricultural ditches and streams is one of the issues that needs to be resolved in order to protect agricultural output.

According to Thakur et al. (2022), organic farming is urgently needed and involves a great deal of awareness, educational training and the application of scientific know-how in order to divert farmers who are currently using inorganic practices from these practices and help them understand the value of an ecologically sound environment. As a result, farmers begin to adopt organic farming as a substitute, which eventually aids in the achievement of the SDGs. In order to address the issue of ammonia emissions that harm the health of humans and other animals, Xu et al. (2022) advises using organic and conservation agriculture practices. The authors also argue that the implementation of such organic techniques can aid China in accomplishing the SDGs.

The authors would like to draw attention to the fact that, out of all the results of the search query conducted in the SCOPUS database, very few nations have made contributions to the literature about organic farming and sustainable development objectives. According to the dataset, Table 3.2 displays the total number of papers published by each nation. With six research publications published alongside the United States, India has taken the lead. Sweden and Germany have each released five. Italy, Egypt, the Netherlands, Australia, etc. come next. In this field, further research and scholarly involvement are required. The authors claim that despite the area's lack of investigation, it has a lot of potential.

Table 3.2. Countries Doing Research on Organic Farming and SDGs as Per Dataset Extracted From SCOPUS.

Country	No. of Documents Published	Country	No. of Documents Published	Country	No. of Documents Published
India	6	Spain	2	Portugal	1
US	6	United Kingdom	2	Saudi Arabia	1
Germany	5	Bangladesh	1	South Africa	1
Sweden	5	China	1	Switzerland	1
Italy	4	Czech Republic	1	Taiwan	1
Egypt	3	Greece	1	Thailand	1
Netherlands	3	Hong Kong	1	Turkey	1
Australia	2	Iceland	1		
Indonesia	2	Iran	1		
Pakistan	2	Poland	1		

Source: Compiled by authors from SCOPUS database accessed on 26 October 2022.

3. The Thematic Analysis of the Keywords From the Dataset

A thematic analysis of the extracted dataset's keywords was conducted to identify the main themes that the academic community has been addressing in relation to the theme of organic farming and sustainable development objectives. Organic farming and SDGs were clearly the two strings with the highest frequency of highlighted terms in the dataset. Furthermore, themes such as biotechnology, biodegradation, soil conversion, soil restoration, environment sustainability, health, alternative agriculture and community development have been included. Other themes that have surfaced include agricultural emissions, cultivation, environmental regulations, erosion control, agricultural policy and sustainable agriculture. A word cloud, displayed in Fig. 3.1 below, was created based on these themes that were taken from the dataset to highlight the authors' keyword analysis, which they conducted with the aid of word cloud.com.

Fig. 3.1. Word Cloud of Keywords. *Source*: Extracted from the dataset Compiled by authors from SCOPUS database accessed on 26 October 2022.

Around the world, organic produce has a far higher potential. Organic farming can help achieve a number of job-related objectives, such as food security, climate management, general well-being and fair employment for all. Both the potential for research and development in this field and the likelihood of achieving sustainable development objectives are very high. When these ideas are combined, there is potential to address a multitude of global issues and accelerate the 2030 agenda.

The potential is extremely significant if you look at the global organic agricultural land, which covers conversion areas and regions as well. Africa has 2.8% of the world's organic agricultural land, whereas Asia has 8.2%. Europe accounts for 22.8%, followed by 13.3% in Latin America, 5.0% in North America and the largest region, Oceania, at 47.9% (https://orgprints.org/id/eprint/37018/1/willer-lernoud-2019-world-of-organic-low.pdf).

The global area of organic agricultural land, measured in hectares, is displayed in Table 3.3, which is taken from the FiBL survey 2022. Additionally, it displays the percentage of organic agricultural land in the region worldwide. Oceania has the highest land available, second only to Europe. The top 10 nations with the greatest total amounts of organic agricultural land in 2020, as determined by this poll, were Australia, Argentina, Uruguay, India, France, Spain, China, United States of America, Italy and Germany. Table 3.4 below displays the total area of organic agricultural land in 2020, expressed in hectares for each country.

A crucial element in advancing organic farming is the global policy framework that has been embraced by numerous nations. It is imperative that specific laws and regulations be drafted with specific policy interventions to support organic

Table 3.3. World: Organic Agricultural Land (Including In-conversion Areas) and Regions' Shares of the Global Organic Agricultural Land 2020.

Region	**Organic Agricultural Land [Hectares]**	**Regions' Shares of the Global Organic Agricultural Land**
Africa	2′086′859	2.8%
Asia	6′146′235	8.2%
Europe	17′098′134	22.8%
Latin America	9′949′461	13.3%
Northern America	3′744′163	5.0%
Oceania	35′908′876	47.9%
World*	74′926′006	100%

Source: FiBL survey 2022. FiBL Statistics - Statistics.

Note: Agricultural land includes in-conversion areas and excludes wild collection, aquaculture, forest and non-agricultural grazing areas. *Includes correction value for French overseas departments.

Table 3.4. Top Ten Countries With the Largest Areas of Organic Agricultural Land 2020 as Per FiBL Survey 2022.

S. No	Country	The Area of Organic Agricultural Land 2020 in Hectares
1	Australia	35′687′799
2	Argentina	4′453′639
3	Uruguay	2′742′368
4	India	2′657′889
5	France	2′548′677
6	Spain	2′437′891
7	China	2′435′000
8	United States of America	2′326′551
9	Italy	2′095′380
10	Germany	1′702′240

Source: FiBL survey 2022.

farming. Otherwise, the excessive use of chemicals and pesticides is made in the rush to make money and meet production targets, which has a detrimental effect on the sustainability of the earth and all species combined.

We may see the repercussions in the form of different infections and diseases brought on by the food we eat, which is the outcome of harmful and inorganic activities. Based on the FIBL survey 2020–2022, the authors in Table 3.5 have

Table 3.5. Analysing the Regulatory Framework on Agroecological as Well as Organic, Worldwide.

S. No	Country	Policy/Regulations Fostering and Promoting Agroecology and Organic Agriculture and Achievement of SDGs
1	Madagascar	• First law of organic agriculture initiated in 2020 • In the process of adoption of national strategy for organic agriculture • Promoting decent work and focus on well being • Launching pilot projects • Initiating steps in the policy for climate control
2	Nicaragua	• Agroecological family farming' investment project is formulated for promoting law on agroecological and organic production

Table 3.5. *(Continued)*

S. No	Country	Policy/Regulations Fostering and Promoting Agroecology and Organic Agriculture and Achievement of SDGs
		• It also focuses on climate control, decent work overall wellbeing and removing hunger and poverty
3	Philippines	• Provision of safe, affordable and accessible food through Organic Agriculture Act • Promoting organic agriculture in country • Achieving SDGs through partnerships in various stakeholders
4	India	• Certain flagship programs have been initiated by government • In India, natural farming is promoted as Bharatiya Prakritik Krishi Paddhati Programme (BPKP) under a centrally sponsored scheme - Paramparagat Krishi Vikas Yojana (PKVY). • BPKP is aimed at promoting traditional indigenous practices which reduce externally purchased inputs. It is based mainly on on-farm biomass recycling with major stress on biomass mulching, using on-farm cow dung-urine formulations. • Periodic soil aeration and excluding all synthetic chemical inputs.
5	Uganda	• Uganda's Ministry of Agriculture, Animal Industry and Fisheries (MAAIF) has launched the National Organic Agriculture Policy (NOAP) in 2019. • Various steps have been taken by the government to induce agroecology and organic farming, focussing on supply chain and market development. • A focus on strengthening the policies and stakeholder involvement is there. • The design, implementation, management as well as financing of organic programmes is taken under strategic goals of the nation.
6	Sri Lanka	• President in 2019, intended to shift the farming practices of the country to be 100% organic. • In March 2021, the government released the National Agriculture Policy proposing a shift to

(Continued)

Table 3.5. *(Continued)*

S. No	Country	Policy/Regulations Fostering and Promoting Agroecology and Organic Agriculture and Achievement of SDGs
		organic fertilizers from 1 to 30% in three years though due to economic crises, the desired decisions could not be implemented for long
7	Togo	• In 2020, the Ministry of Agriculture, Animal Production and Fisheries (MAPAH) published a national strategy for developing organic and ecological agriculture for the period 2020–2030 • A strong political will can be found in policy documents to support organic agriculture
8	Bhutan	• In 2012, Bhutan was announced as the first country to fully convert to organic agriculture. • The National Organic Flagship Programme of the Ministry of Agriculture and Forestry (MOAF) focuses on accelerating the organic produce • Strong intention of the government can be seen in order to produce organic fertilizers, organic seeds and organic animal feed.

Source: FiBL & IFOAM-Organics International, the World of Organic Agriculture, Statistics and Emerging Trends 2022.

attempted to provide an overview of the regulatory framework for organic and agro-ecological practices that has been enacted in a few different countries.

The authors conducted a word cloud analysis of the entire text to gain an overview of the themes or keywords often used in the survey materials in order to determine the most frequently used terms in 'The World of Organic Agriculture, Statistics and Emerging Trends 2022'. Fig. 3.2 displays the word cloud created by the authors, emphasising the document's keywords such as organic food, farming, land and agriculture. In the shape of a tree, it provides us with a comprehensive image of the organic world and shows how organic farming is helping the world accomplish its SDGs.

4. Limitations of the Study

The major limitation of the study is that it relies on selective literature extracted from SCOPUS only and other secondary data. An opinion-based orientation can be subject to subjectivity also. The future research agenda thereby if explored can help the potential researchers to get deeper, objective and more rational insights into the field.

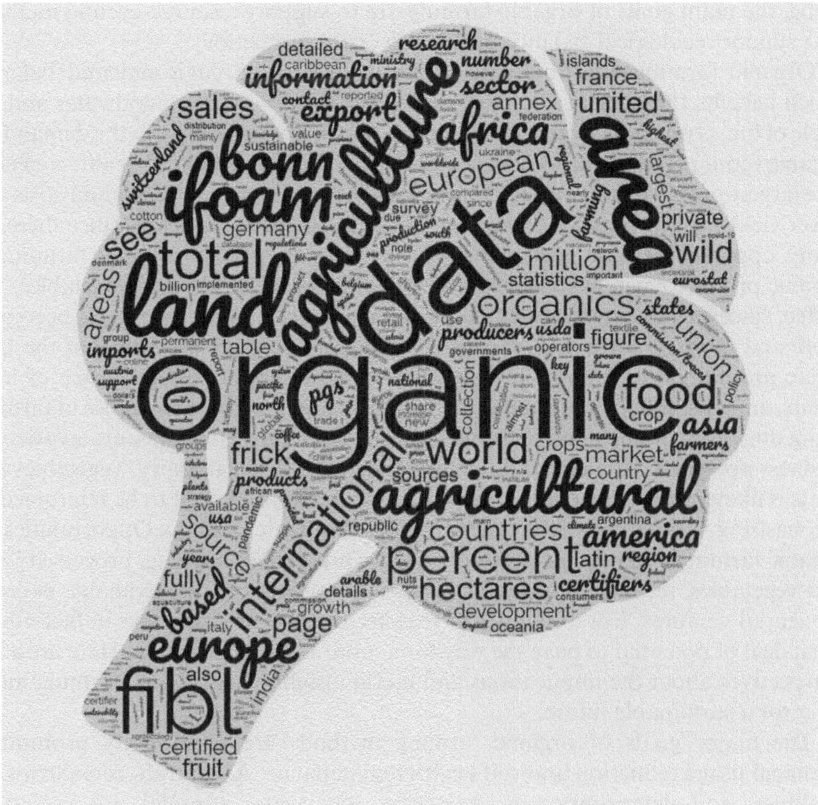

Fig. 3.2. Word Cloud 2. *Source*: Word Cloud of Keywords, created by author.

5. Discussion and Conclusion

Globally, the market for organic produce is growing every day. If produce is confirmed to have grown on land or soil that is not harvested using forbidden materials like synthetic fertilizers and pesticides, then it is considered organic. Due to the widespread belief that organic produce is superior for human consumption, it is always offered for a premium in the marketplace. Relying on conventional methods of soil production benefits health, nutrition and the general well-being of living things, but it also encourages sustainable activities that have historical roots. The most often purchased organic goods include meat, dairy products, fruits and vegetables. Numerous farming approaches worldwide advocate for using only natural substances to maintain the soil's nutrition and produce's quality, which not only promotes greater health but also a more sustainable planet. Although there are differences in organic farming regulations across the globe, the main goals of organic farming are to support resource-cycling methods that support ecological balance and biodiversity protection.

Organic farming contributes to both ecological and environmental balance, which benefits the planet's ecology, biodiversity, soil health, groundwater and the circle of life. Over the last 10 years, there has been a noticeable rise in the demand for organic food, perhaps due to people's growing environmental and health concerns. The largest problem with organic farming is that it has greater production costs and lower yields, which puts the farmer in a difficult position. However, since there are many opportunities for entrepreneurial ventures related to organic farming or organic produce, such as mushroom farming, processed fruits and vegetables, tea, coffee, spices, vineries and much more, the concept of organic farming is becoming more and more important as various governments support the SDGs of the UN. There are numerous options in agro-ecology that support the resilience of local plants, animals, soil, climate and humans. There are numerous examples of farming being done for climate control throughout the world, and this presents a variety of business prospects. The demand for processed organic meat, dairy, vegetables and fruits is likewise growing daily. Value chains in agriculture need to be reinforced by emphasising various industries that can support organic farming. Opportunities for organic farming and organic produce, such as mushroom farming, processed fruits and vegetables, tea, coffee, spices, vineries and much more, are available in entrepreneurial ventures. For a number of reasons, organic produce does in fact offer a great deal of potential to pave the way for a more sustainable future. Here are a few perspectives about the implications and useful insights that organic farming might offer for a sustainable future:

The major goals of organic farming methods are biodiversity promotion, chemical usage reduction and soil health maintenance. As a result, ecosystems are healthier, soil deterioration is decreased, and water supplies are preserved. Through the sequestration of carbon in the soil and the reduction of greenhouse gas emissions, these techniques help to mitigate climate change. Multiple possibilities exist for promoting environmental sustainability through organic farming. On the one hand, it maintains the excellent condition of the soil through preventing away with the usage of artificial chemicals and pesticides. The lack of these dangerous compounds from organic farming methods contributes to the preservation of soil microbial biodiversity, which is necessary for soil fertility and nutrient cycling. This approach lessens the ecological impact on the environment by giving priority to natural pest control through strategies including crop rotation, pairing crops together and the employment of natural predators. Organic farming reduces water pollution by not using chemical fertilizers, which eliminates dangerous materials from leaking into water bodies and guarantees cleaner water for human and animal use. Furthermore, greenhouse gas emissions are substantially reduced by organic farming. Conventional agriculture uses synthetic fertilizers, which emit enormous amounts of greenhouse gases, especially nitrogen oxide. Contrarily, organic farming maintains a strong emphasis on using natural fertilizers derived from compost, which emits fewer greenhouse emissions. This emission reduction is essential to the fight against climate change. Furthermore, improving biodiversity and soil health is a key component of organic farming, which helps sequester carbon. In order to mitigate the effects of climate change, healthy organic soils serve as carbon sinks, absorbing atmospheric carbon dioxide and reducing its release into the atmosphere. Preserving biodiversity is a further significant aspect. Natural habitat preservation and crop diversity are frequently encouraged by organic farming methods. Since there are no artificial chemicals present, an array of species, such as beneficial insects, birds and microbes, can live in harmony. In furtherance of enhancing the ecosystem's general health, biodiversity also helps with the natural management of ailments and pests. Organic farming promotes to the preservation of crop genetic diversity by avoiding genetically modified organisms (GMOs). This preserves heirloom varieties and ensures a larger gene pool, which could be important for eventual agricultural resilience in the midst of shifting environmental circumstances.

Finally, a focus on locally developed, small-scale organic farming practices strengthens the resilience of regional economies and communities. Smaller, more varied farms that practice organic farming place a high priority on sustainable land use. By promoting self-sufficiency and decreasing reliance on outside supplies, these farms improve the agricultural system. A more sustainable and environmentally friendly food system is also promoted by the emphasis on regional production and distribution, which lessens the negative environmental effects of transportation and storage. This support for local economies and communities encourages environmental sustainability by reducing the carbon footprint associated with large-scale industrial agriculture.

With its focus on environmentally friendly and sustainable practices, organic farming can greatly increase a nation's export potential. Due to their enhanced health and lower environmental effect, organic products are becoming more and more in demand worldwide. Countries that place a high priority on organic farming have the opportunity to access these global markets by exporting superior organic produce. Countries that prioritize organic farming have the chance to diversify their export industries and generate new sources of income by taking advantage of this demand. Organic certifications increase the marketability of these items globally by acting as a symbol of quality and sustainability. These exports not only create jobs but also improve the nation's reputation as a leader in sustainable agriculture, which may lead to new trade deals and collaborations. Furthermore, organic farming promotes health and well-being. Organic produce is grown without the use of artificial fertilizers or pesticides. Eating organic food improves both consumer and farmer health by lowering exposure to potentially dangerous toxins. In addition to this, organic farming frequently uses soil that is higher in nutrients, producing food that is higher in nutrients.

Another significant dimension of adopting using organic farming methods can significantly increase tourism. Within the larger tourism business, agro-tourism, sometimes known as agricultural tourism, has become more popular. There is a growing interest among tourists to visit organic farms, take part in farm-to-table events and discover sustainable farming methods. Nations that support organic farming have the opportunity to capitalize on this trend by creating agro-tourism experiences. Experiences that offer insight into organic farming practices, the creation of locally grown food and the chance to partake in eco-friendly activities are what attract tourists. In addition to providing revenue for nearby farmers and communities, these trips help tourists get a better understanding and appreciation of sustainable agriculture. Countries can establish themselves as pioneers in sustainable practices and attract travellers looking for genuine and eco-friendly experiences by endorsing organic agriculture and highlighting its relationship to the environment and community. Incorporating organic farming with sustainable tourism boosts local economies and presents the nation as an environmental steward, which may draw in a new generation of travellers looking for real, sustainable and instructive travel experiences.

Organic farming contributes to the SDGs and improves traveller well-being in a big way. To begin with, organic farming directly contributes to SDG 2 (Zero Hunger) by facilitating access to sufficient, healthy, and safe food. Tourists to communities that prioritize organic farming have the unique opportunity to sample fresh, chemical-free produce straight from the source. Because they are devoid of dangerous pesticides and artificial chemicals, these foods provide travelers with a healthier and more nutrient-dense option, improving their well-being by giving them access to high-quality, organic food. Further, because organic farming limits exposure to dangerous chemicals frequently used in conventional farming, it supports SDG 3 (Good Health and Well-Being). Travellers who participate in organic farm-to-table experiences or farm tours gain access to a cleaner atmosphere and less pesticide exposure, which enhances their general wellbeing. By using less harmful chemicals, organic agricultural practices help improve the health of farmworkers and the surrounding communities by fostering safer working conditions and healthier living situations.

Furthermore, organic farming contributes to SDG 3 (good health and well-being) since it reduces exposure to harmful chemicals that are often employed in conventional farming. Engaging in organic farm-to-table experiences or farm tours allows visitors to experience a cleaner environment and reduced pesticide exposure, both of which improve overall wellness. Organic farming practices promote better working conditions and healthier living environments for farmworkers and the surrounding community by utilising fewer toxic chemicals. Organic farming frequently promotes community development and supports local economies, which is in line with SDG 8 (Decent Work and Economic Growth). The economic development of these communities is facilitated by visitors who support and visit these organic farms. Through the creation of steady employment, the promotion of sustainable livelihoods, and the development of economic resilience, this support can improve the well-being of the local populace. In general, encouraging organic farming through tourism-related activities improves visitors' health by offering healthier food options, lowering their exposure to dangerous chemicals, and developing a greater awareness of environmentally friendly techniques. In addition to improving tourists' experiences, these initiatives support certain aspects of SDGs, fostering a healthier and more environmentally conscious future for everybody.

6. Future Research Directions

- ***Supply Chain and Distribution Management:*** Examining effective and long-lasting supply chain frameworks for organic agricultural products is an important area to be explored. This consists of ensuring fair trade practices are followed, cutting waste, and optimising distribution networks. In order to preserve the integrity and ability of organic products, research could concentrate on developing better logistics, storage, and transportation strategies.
- ***Market development and consumer behaviour:*** Understanding customer preferences, behaviour, and readiness to pay for organic products is essential for market development and consumer behaviour. Subsequent investigations may explore consumer attitudes, inclinations, and the influence of advertising tactics on consumer decisions. Analysing market trends across various geographies and demographic groups can help shape successful marketing plans for organic products.
- ***Investigating and creating novel commercial strategies for organic farming:*** This could entail investigating subscription-based farming, cooperative farming models, or online marketplaces that link customers with organic producers. It is essential to look into profitable business strategies that satisfy ethical and sustainable norms and guarantee farmers' profits.
- ***Financial and Investment Mechanisms:*** An analysis of organic farming's investment plans and financial support networks is a very underexplored area. The financial viability of organic farming could be the subject of research, with an emphasis on risk management, return on investment, and possible funding sources that could encourage further investment in environmentally friendly farming methods.
- ***Technological Developments in Organic agricultural:*** Studies could look into how technology is used into organic agricultural methods. This includes researching how to optimize organic agricultural procedures, increase productivity, and guarantee environmental sustainability through the use of artificial intelligence, the Internet of Things, and data analytics.
- ***Analysing rules and regulations to see how they affect the development and sustainability of organic farming.*** Making better policy decisions that promote the expansion of the organic sector can be aided by having a greater understanding of the impact that trade agreements, government regulations, and certification requirements have on the organic agricultural industry. This area has a bright chance of future research.
- ***Social Impact and Stakeholder Engagement:*** Investigating the effects of organic farming on society and interacting with different stakeholders can be one of the important areas of future research. Studying the advantages of organic farming for society and the environment, such as equitable employment practices, community empowerment, and rural development, may fall under this category, especially in relation to the achievement of SDGs.

Future studies in these areas will not only improve our knowledge of the viability of organic farming as an economy but also aid in the creation of innovative solutions, regulations, and strategies that promote the economy's growth and sustainability.

References

Abebe, G. K., Traboulsi, A., & Aoun, M. (2022). Performance of organic farming in developing countries: A case of organic tomato value chain in Lebanon. *Renewable Agriculture and Food Systems*, *37*(3), 217–226. https://doi.org/10.1017/S1742170521000478

Arora, M. (2023). Achieving sustainable developmental goals by empowering women entrepreneurs: Analyzing challenges and suggesting policy implications, *Women's empowerment within the tourism industry* (pp. 1–19). IGI Global. https://doi.org/10.4018/978-1-6684-8417-3.ch001

Arora, M., Kumar, J., & Valeri, M. (2023). Crises and resilience in the age of digitalization: Perspectivations of past, present and future for tourism industry, *Tourism innovation in the digital era* (pp. 57–74). Emerald Publishing Limited. https://doi.org/10.1108/978-1-83797-166-420231004

Arora, M., & Rathore, S. (2022). Analysing sustainability reporting content for creating value through engaging stakeholders: A qualitative approach. *Journal of Content, Community & Communication*, *16*. https://doi.org/10.31620/JCCC.12.22/04

Arora, M., & Sharma, R. L. (2023). Artificial intelligence and big data: Ontological and communicative perspectives in multi-sectoral scenarios of modern businesses. *Foresight*, *25*(1), 126–143. https://doi.org/10.1108/FS-10-2021-0216

Arora, M., & Singh, S. (2023a). Microfinance for achieving sustainable development goals: Pondering over Indian experiences for the preservation of magnificent African natural resources, *Research anthology on microfinance services and roles in social progress* (pp. 378–394). IGI Global. https://doi.org/10.4018/978-1-7998-7499-7.ch00

Arora, M., & Singh, S. (2023b). Microfinance and sustainability: Building perspective through literature and bibliometrics, *Transforming economies through microfinance in developing nations* (pp. 60–84). IGI Global. https://doi.org/10.4018/978-1-6684-5647-7.ch004

Arora, M., & Walia, S. K. (2019). From earning profits to sustainability: A critical evaluation of CSR initiatives in tourism sector, *Sustainable tourism: Breakthroughs in research and practice* (pp. 229–235). IGI Global. https://doi.org/10.4018/978-1-5225-7504-7.ch012

Azam, M., Ftiti, Z., Hunjra, A. I., Louhichi, W., & Verhoeven, P. (2022). Do market-supporting institutions promote sustainable development? Evidence from developing economies. *Economic Modelling*, *116*, 106023. https://doi.org/10.1016/j.econmod.2022.106023

Bieroza, M. Z., Bol, R., & Glendell, M. (2021). What is the deal with the Green Deal: Will the new strategy help to improve European freshwater quality beyond the Water Framework Directive?. *Science of the Total Environment*, *791*, 148080. https://doi.org/10.1016/j.scitotenv.2021.148080

Cerdà, A., Rodrigo-Comino, J., Yakupoğlu, T., Dindaroğlu, T., Terol, E., Mora-Navarro, G., .., & Daliakopoulos, I. N. (2020). Tillage versus no-tillage. Soil properties and hydrology in an organic persimmon farm in Eastern Iberian Peninsula. *Water*, *12*(6), 1539. https://doi.org/10.3390/w12061539

Choo, H., & Jamal, T. (2009). Tourism on organic farms in South Korea: A new form of ecotourism. *Journal of Sustainable Tourism*, *17*(4), 431–454. https://doi.org/10.1080/09669580802713440

Cizler, J. (2013). Opportunities for the sustainable development of rural areas in Serbia. *Problemy Ekorozwoju–Problems of Sustainable Development*, *8*(2), 85–91. https://ssrn.com/abstract=2296832

Eyhorn, F., Muller, A., Reganold, J. P., Frison, E., Herren, H. R., Luttikholt, L., .., & Smith, P. (2019). Sustainability in global agriculture driven by organic farming. *Nature Sustainability*, *2*(4), 253–255. https://doi.org/10.1038/s41893-019-0266-6

Haque, M. M., Datta, J., Ahmed, T., Ehsanullah, M., Karim, M. N., Akter, M. S., .., & El Sabagh, A. (2021). Organic amendments boost soil fertility and rice productivity and reduce methane emissions from paddy fields under sub-tropical conditions. *Sustainability*, *13*(6), 3103. https://doi.org/10.3390/su13063103

Hossain, M. R., Akhter, F., Sharma, A., & Hassan, A. (2022). Thirty years of research on application of technology in tourism and hospitality industry: A systematic literature review. *Technology Application in Tourism in Asia: Innovations, Theories and Practices*, 3–36. https://doi.org/10.1007/978-981-16-5461-9_1

Hunjra, A. I., Azam, M., Bruna, M. G., & Bouri, E. (2023). A cross-regional investigation of institutional quality and sustainable development. *Journal of International Financial Markets, Institutions and Money*, *84*, 101758. https://doi.org/10.1016/j.intfin.2023.101758

Hunjra, A. I., Azam, M., Bruna, M. G., Verhoeven, P., & Al-Faryan, M. A. S. (2022). Sustainable development: The impact of political risk, macroeconomic policy uncertainty and ethnic conflict. *International Review of Financial Analysis*, *84*, 102370. https://doi.org/10.1016/j.irfa.2022.102370

Keesstra, S., Nunes, J., Novara, A., Finger, D., Avelar, D., Kalantari, Z., & Cerdà, A. (2018). The superior effect of nature based solutions in land management for enhancing ecosystem services. *Science of the Total Environment*, *610*, 997–1009. https://doi.org/10.1016/j.scitotenv.2017.08.077

Purnhagen, K. P., Clemens, S., Eriksson, D., Fresco, L. O., Tosun, J., Qaim, M., .., & Zilberman, D. (2021). Europe's farm to fork strategy and its commitment to biotechnology and organic farming: Conflicting or complementary goals?. *Trends in Plant Science*, *26*(6), 600–606. https://doi.org/10.1016/j.tplants.2021.03.012

Sarkar, D., Dubey, P. K., Chaurasiya, R., Sankar, A., Chatterjee, N., Ganguly, S., .., & Rakshit, A. (2021). Organic interventions conferring stress tolerance and crop quality in agroecosystems during the United Nations Decade on Ecosystem Restoration. *Land Degradation & Development*, *32*(17), 4797–4816. https://doi.org/10.1002/ldr.4094

Siepmann, L., & Nicholas, K. A. (2018). German winegrowers' motives and barriers to convert to organic farming. *Sustainability*, *10*(11), 4215. https://doi.org/10.3390/su10114215

Thakur, N., Nigam, M., Tewary, R., Rajvanshi, K., Kumar, M., Shukla, S. K., .., & Gupta, S. (2022). Drivers for the behavioural receptiveness and non-receptiveness of farmers towards organic cultivation system. *Journal of King Saud University Science*, *102107*. https://doi.org/10.1016/j.jksus.2022.102107

Xu, P., Li, G., Houlton, B. Z., Ma, L., Ai, D., Zhu, L., .., & Zheng, Y. (2022). Role of organic and conservation agriculture in ammonia emissions and crop productivity in China. *Environmental Science & Technology*, *56*(5), 2977–2989. https://doi.org/10.1021/acs.est.1c07518

Chapter 4

A Review of Agrotourism on *Batagur* sp. Conservation Throughout Southeast Asia

Mohd Hairul Mohd Salleh[a,b], *Yuzine Esa*[a,c] *and Rozihan Mohamed*[a,*]

[a]Universiti Putra Malaysia, Malaysia
[b]Royal Malaysian Customs Department, Malaysia
[c]International Institute of Aquaculture and Aquatic Sciences, Malaysia

Abstract

Most Asians are directly or indirectly dependent on agriculture, and most live in rural areas. Since Asia is primarily an agricultural economy, rural tourism and agrotourism are critical in enabling growth. Agrotourism is a term that refers to the fusion of tourism and agriculture. The freshwater turtles (*Batagur* sp.) are a group of six turtles that once roamed southern Asia's great rivers and estuaries from the Indian Subcontinent to the Indonesian islands. Therefore, this review is to investigate the current state of agrotourism in *Batagur* sp. conservation and assess the associated challenges and opportunities. This review paper recognized the impacts and benefits for the rural site as an agrotourism industry. Moreover, three initiatives in *Batagur* sp. conservation programs were also discussed. Agrotourism, with a focus on the conservation of *Batagur* sp., is a vital driver of economic growth in rural Asian regions. It diversifies rural communities' income sources, enhancing economic stability and resilience.

Keywords: Tourism agriculture; freshwater turtle; rural site; breeding centre; economic stability; growth

*Corresponding author.

The Emerald Handbook of Tourism Economics and Sustainable Development, 63–80

Published under exclusive licence by Emerald Publishing Limited
doi:10.1108/978-1-83753-708-220241004

1. Introduction

1.1 Agrotourism

Agriculture is the lifeblood of most emerging economies, most notably Malaysia (Hassan et al., 2010). Green revolutions and lower plot ownership, as seen in Asian countries, have resulted in fast advancements in agriculture during the previous few years (Manjunatha et al., 2013). Asian countries face many socioeconomic, demographic and institutional issues that jeopardize their basic survival (Conway & Barbie, 1988). Most Asians directly or indirectly depend on agriculture, and most live in rural areas (Baloch & Haseeb, 1996, pp. 73–94). Tourism significantly gives a country's gross domestic product (GDP) and is a critical tool for job creation, poverty alleviation and sustainable rural development (Alola & Alola, 2018). Since Asia is primarily an agricultural economy, rural tourism and agrotourism play a critical role in enabling growth (Wimalaratana, 2014).

Tourism has become the world's largest industry (Hawkins & Hunt, 1988). The agricultural economy has been relatively unchanged over the last three decades (Beus, 2008). With agricultural yields declining due to human and external factors, agrotourism has been identified as a viable substitute for diversifying economic growth (Nematpour & Khodadadi, 2021). Farmers with small to medium-sized farms must diversify their income streams to remain in agriculture. One strategy that some farmers are considering is cultivating tourists and crops. Known as 'agriturismo' in Italy, 'sleeping in the straw' in Switzerland, 'farm stays' in New Zealand and 'farm holidays' in England, agrotourism is well-recognized all over Europe and some other countries (Banda et al., 2021; Chase et al., 2018; Rilla, 1999; Tiraieyari & Hamzah, 2012).

Today, new forms of tourism have emerged, including agrotourism, ecotourism, nature tourism, farm tourism and rural tourism (Bac, 2014). Agrotourism is a term that refers to the fusion of tourism and agriculture (Zhou et al., 2017). Furthermore, agrotourism is a subset of specialized tourism that utilises farms/conservation centres as educational and/or recreational locations (Fig. 4.1). In addition, agrotourism is described as attracting tourists or visitors to rural areas that are primarily agricultural (Tiraieyari & Hamzah, 2012). Agriculture and related activities are marketed as tourism products (Berweck et al., 2013). Integrating agriculture and tourism lets tourists benefit from agricultural and tourism operations. Although it has been performed for millennia, it was conceptualized theoretically for several decades (Park et al., 2011). This concept was born and refined in Europe and North America before spreading to several other countries (Tiraieyari & Hamzah, 2012).

1.2 Batagur sp.

Freshwater turtles (*Batagur* sp.) are a group of six species of enormous, magnificent and, in some cases, stunningly coloured turtles that once roamed Southeast Asia's great rivers and estuaries from the Indian Subcontinent to the Indonesian islands (Hudson, 2016). Five of six *Batagur* species are classified as

Fig. 4.1. One of the *Batagur affinis affinis* Conservation Centres in Malaysia is Located in Bota Kanan, Bota and Perak.

critically endangered (CR) on the International Union for Conservation of Nature (IUCN) Red List (Camara & Jamil, 2021; Salleh & Esa, 2022; Mohd Salleh et al., 2023). The six *Batagur* sp. have been a top conservation priority for Turtle Survival Alliance (TSA) since the organization's first field programmes began in Asia in 2005 with the Red-Crowned Roof Turtle, *B. kachuga*, and continued in 2006 with the Burmese Roofed Turtle, *B. trivittata*. TSA now operates recovery initiatives for five species throughout Bangladesh, India, Indonesia, Malaysia and Myanmar (Hudson, 2016).

Due to their consistent nesting habits and site fidelity, *Batagur* sp. has historically been one of the most exploited turtles in southern Asia (Stanford et al., 2020). Turtle hunters and egg collectors know that freshwater turtles return to the same beaches and sandbars yearly (Stanford et al., 2020). Due to the ease with which eggs and nesting females can be harvested, this extensive annual harvest of breeding females and eggs, exacerbated further by turtle capture in fishing nets, has resulted in chronic steep declines in most populations (Kasinathan et al., 2016; Salleh & Sah, 2019). In extreme cases, it may reduce natural recruitment to near zero (Hudson, 2016). Restoring wild *Batagur* sp. populations is one of the most challenging conservation tasks, and success will require much more financing.

1.3 Problem Statement and Knowledge Gap

Current research on agrotourism in the context of *Batagur* sp. conservation in Southeast Asia is limited and often lacks a holistic perspective. There is insufficient empirical data and analysis regarding the socio-economic and ecological impacts of agrotourism initiatives on these endangered turtle species (Mohd Salleh et al., 2022). Additionally, there is a dearth of studies that explore the best practices, challenges and community engagement models for sustainable agrotourism in *Batagur* sp. conservation (Mohd Salleh et al., 2023; Salleh et al., 2023). This lack of comprehensive research leaves a significant knowledge gap in understanding the full potential of agrotourism as a conservation tool and its broader implications for rural communities in Southeast Asia (Mukhlis et al., 2022). This lack of understanding hinders the development of evidence-based strategies for sustainable agrotourism practices that can simultaneously contribute to conservation efforts and benefit local communities. Closing this knowledge gap is essential to inform policy, conservation strategies and development efforts in the region (Hunjra et al., 2023). Therefore, there is a pressing need to investigate the current state of agrotourism in *Batagur* sp. conservation and assess the associated challenges and opportunities.

2. Economic Welfare and the Potential of Agrotourism

Agrotourism has a substantial economic and employment impact on the local community (Tiraieyari & Hamzah, 2012). It can contribute to rural development by creating new jobs and adding value. Indeed, it allows visitors to interact directly with the local community, experiencing the place and appreciating the local products, services and nature (Gao et al., 2009). Over the last century, agriculture has become more technologically advanced than ever (Drabenstott, 1994). The primary goals of agrotourism are to promote sustainable rural development and provide an additional revenue source for farmers and local communities (Karabati et al., 2009). Agrotourism has been identified as a possible source of many advantages. Agrotourism development can benefit the breeding centre, local communities and tourist administrators (Tiraieyari & Hamzah, 2012).

As a result, agrotourism has an economic, social and cultural influence and an environmental impact. Agrotourism initiatives that focus on *Batagur* sp. conservation in Southeast Asia have the potential to bring about several economic benefits. These include income diversification, job creation, increased revenue for conservation, community involvement, promotion of sustainable agriculture, entrepreneurship and enhanced local economies through tourist expenditure (Fig. 4.2). These economic welfare schemes underscore the significance of agrotourism as a tool for both conservation and socio-economic development in the region.

2.1 The Economic Impacts of Agrotourism on Local Community

The economic impact of studies on agrotourism in local communities may emphasise the positive features (Tiraieyari & Hamzah, 2012). But unfortunately, it has drawbacks, including lower-wage jobs and seasonality. Typically, agrotourism economic impact research focuses on area revenue and employment changes. However, as previously noted, agrotourism, a different type of industry,

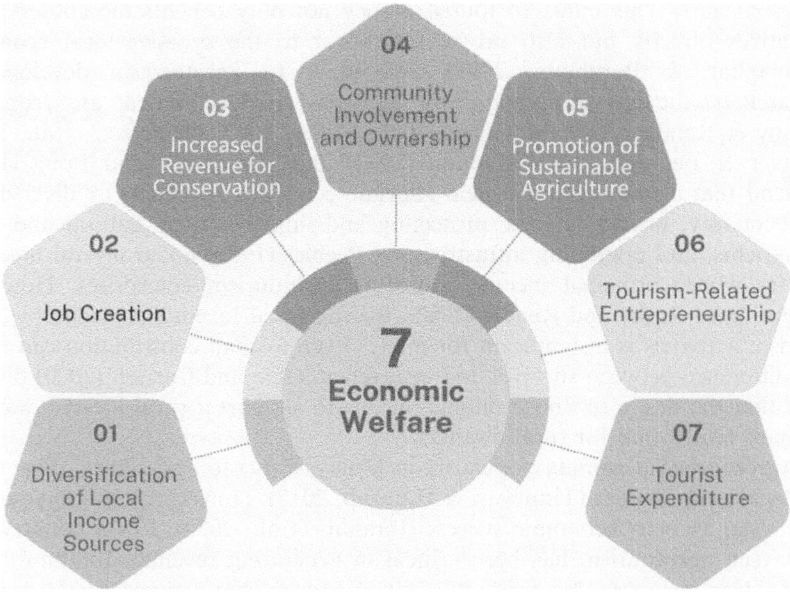

Fig. 4.2. Seven Factors of Economic Welfare: The Critical Impacts of Agrotourism on the Local Community.

has consequences for the local communities (Karabati et al., 2009). Agrotourism initiatives that focus on *Batagur* sp. conservation provide an opportunity for rural communities to diversify their income sources. This economic diversification is crucial in rural areas heavily dependent on agriculture, as it reduces their vulnerability to fluctuations in agricultural markets and climatic conditions. The income generated from agrotourism can serve as a supplementary source, thereby enhancing the economic stability of these communities (Delgado & Siamwalla, 2018; Evans & Ngau, 1991). Agrotourism projects aimed at *Batagur* sp. conservation often require the involvement of local residents as guides, caretakers, and staff at conservation centres. This job creation is not limited to the conservation centres themselves but also extends to related services, such as local accommodations, restaurants and transportation. These job opportunities contribute to reducing unemployment and poverty in rural areas (Cichowska & Klimek, 2011; Romanenko et al., 2020).

Agrotourism can stimulate entrepreneurship in rural areas. Locals may start their own small businesses, such as guesthouses, restaurants and craft shops, to cater to the needs of tourists. This, in turn, can stimulate economic growth in the region, create a competitive business environment and encourage innovation and diversification of services (Drăgoi et al., 2017; Stanovčić et al., 2018). Tourists who visit agrotourism sites for *Batagur* sp. conservation contribute to the local economy through their expenditure on accommodations, food, transportation and souvenirs. This influx of tourist money not only benefits the conservation initiatives directly but also provides a boost to the broader local economy (Chulaphan & Barahona, 2021). According to agrotourism development researchers, the key motivations for agrotourism development are economic (Busby & Rendle, 2000; McGehee, 2007; Miller, 1993; Nickerson et al., 2001; Weaver & Fennell, 1997). Gannon (1994) and Kieselbach and Long (1990) contend that agrotourism can help alleviate economic problems by diversifying the economy, providing jobs, protecting and improving the natural and built ecosystems, and upgrading infrastructure. Peebles (1995) noted several breeding centres whose principal income source is agrotourism enterprises. However, according to Busby and Rendle (2000), the increased revenue provided by agrotourism activities is insignificant for many. Even a small contribution can mean the difference between survival and insolvency. Getz and Carlsen (2000) discovered that the desire to find economic means to support a rural lifestyle was the primary motivation for rural dwellers.

Agrotourism is gaining popularity in Malaysia due to agricultural divergence for local communities (Tiraieyari & Hamzah, 2012). However, it is still young in Malaysia, as is its economic success (Ibrahim et al., 2023). For a country like Malaysia, agrotourism has been critical in generating revenue (Ibrahim et al., 2023). For instance, domestic and foreign tourists have increased (History of Agro-tourism, 2008). In addition, the increase in foreigners has increased foreign currency exchange in the country, generating revenue. In Malaysia, the majority of agrotourism destinations provide homestays. Homestays have grown in popularity in Malaysia due to the diversity of its culture, which stems from the country's various races. A homestay's objective is to give tourists a location to stay while participating in activities the owner organises (History of Agrotourism, 2008).

2.2 The Social and Cultural Impact of Agrotourism on Local Community

Agrotourism affects the indigenous people's behaviours, family relationships, lifestyles and traditions (Ismail et al., 2021). This activity connects visitors of many languages, cultures and lifestyles. As a result, agrotourism employs residents and can enhance social and cultural values (Huang & Stewart, 1996). Agrotourism has the potential to affect the socio-cultural fabric of the local community as well. Changing conditions have a socio-cultural impact on communities, not just rural residents but also urban residents (Topcu, 2007). Education activities in the *Batagur* sp. conservation centre were integrated into agrotourism development to provide new skills to the local communities. Agrotourism unites rural and urban communities; this way, rural communities can develop stronger ties to urban life, and urban residents can better understand the world's rural areas. This may result in more communication between urban and rural peoples (Kizos & Iosifides, 2007).

Agrotourism effectively reduces migration by providing job opportunities for indigenous people (Singh et al., 2016). Therefore, agrotourism initiatives assist communities in capitalising on their natural, historical and cultural resources, developing community pride and enhancing the local community's quality of life (Karabati et al., 2009). In contrast, tourists assisted with conservation centre activities on occasion. This assistance can be used as a justification for inviting guests. However, dealing with people from other cultures can occasionally result in complications, such as excessive drinking, family disagreements and unpleasant debates about politics, religion and culture (Karabati et al., 2009).

2.3 The Environmental Impact of Agrotourism on Local Community

Agrotourism can generate revenue that can be reinvested in *Batagur* sp. conservation efforts. Entrance fees, guided tours, souvenir sales and donations from tourists can provide a sustainable source of funding for research, protection and breeding programs. This financial support can significantly enhance the effectiveness of conservation initiatives (Ana, 2017). Agrotourism projects that engage local communities in *Batagur* sp. conservation efforts foster a sense of ownership and responsibility among the residents. This can lead to more active participation in protecting the turtles and their natural habitat, which, in turn, enhances the long-term sustainability of conservation programs. Communities that benefit economically from these initiatives are more likely to support and protect the species (Chemnasiri & Kaewmoung, 2008; Khawatkun & Thongma, 2015).

Agrotourism often highlights sustainable agricultural practices, promoting responsible farming and aquaculture. These practices not only contribute to *Batagur* sp. conservation by preserving their natural environment but also enhance the sustainability and productivity of local agricultural activities. This can lead to improved yields and higher-quality products, potentially increasing the income of local farmers (Daniela et al., 2008). Although agrotourism helps protect the environment and natural resources, it also brings a large influx of people, including tourists, employees and investors, into a geographically restricted area (Tiraieyari & Hamzah, 2012). As a result, the environment begins to deteriorate due to the growing population (Pimentel et al., 1995). Furthermore, agrotourism will also necessitate infrastructure, transportation and facilities, polluting the environment, impairing land solidity and depleting water resources, eventually degrading the tourist region (Tiraieyari & Hamzah, 2012).

3. The Benefits of Agrotourism

Three primary benefits are recognized for the rural site as an agrotourism industry capital stockpile. First, the quality of the rural location must influence the quality of the rural tourism experience. This leads to the second advantage of using the rural capital idea, which emphasises the importance of rural tourist enterprises developing additional durable links with the rural site they activate. Suppose rural capital is a critical input in rural tourism generating progression, a decline in quality is likely to be reflected in a similar reduction in the final product's quality. The decrease in the quality of rural capital assets could result from a lack of ongoing investment in their preservation. Rationally, if rural tourist enterprises protect and increase their rural capital stocks, their end products will likely be more valuable (Holloway & Humphreys, 2019; Ling et al., 2013). On the other hand, if rural tourist enterprises are hesitant to make these expenditures systematically, market failure is expected (Hall et al., 2003). In this situation, government involvement is required to make planned investments in sustaining or developing the available rural capital resource on behalf of society. This enables the rural resource as tourism capital to accomplish a third benefit: the profitable and sustainable development of economic activities, environmental programmes and community improvement (Ling et al., 2013).

The agrotourism industry spends significantly on rural capital for various reasons. For example, destination administrators may invest in bolstering or maintaining a modest edge over competitor locations offering comparable items. Meanwhile, the public sector may be more concerned with achieving sustainable rural development (Baloch & Haseeb, 1996; Marsden & Sonnino, 2008). For example, rural capital investment can improve local community living standards by providing recreational opportunities and critical utilities that would otherwise be unavailable. On the other hand, investment in rural capital may contribute to the protection or enhancement of the rural environment's value, improving the quality of life experienced by rural residents (Deininger & Byerlee, 2011; Garrod et al., 2006). Agrotourism focussing on *Batagur* sp. conservation offers a multi-faceted experience for travellers (Fig. 4.3). It goes beyond leisure and adventure,

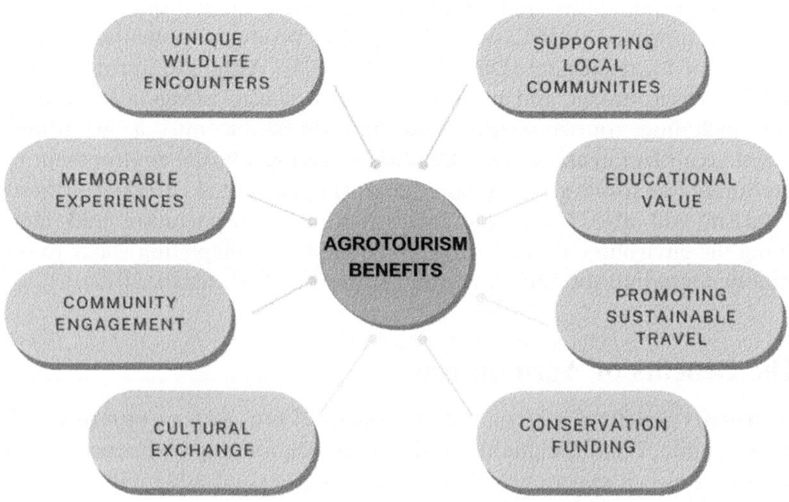

Fig. 4.3. These Agritourism Benefits for Travellers Were Reviewed in Several Research Articles.

providing a chance to impact the environment and local communities positively. By engaging with these insights and implications, travellers can appreciate the multifaceted benefits of agrotourism, making it a compelling and meaningful choice for their next adventure.

Agrotourism centred on *Batagur* sp. conservation offers an exceptional opportunity for travellers to engage with CR turtles in their natural habitat. This hands-on experience provides a profound connection with the natural world and a chance to witness the remarkable efforts being made to protect these species (Jeczmyk et al., 2021). When tourists participate in agrotourism activities, they directly contribute to the livelihoods of rural communities. Visitors play an active role in improving the economic well-being of these communities by staying in local guesthouses, dining at village restaurants and purchasing souvenirs crafted by locals. (Budiasa & Ambarawati, 2014; Chemnasiri & Kaewmoung, 2008). Moreover, agrotourism encourages local entrepreneurship. Visitors may encounter unique local products and services that are not readily available in mainstream tourist destinations. This promotes the growth of micro-businesses and artisanal crafts, contributing to the diversity of travel experiences (Hrymak et al., 2019). Agrotourism sites often serve as educational hubs where tourists can learn about the significance of *Batagur* sp. conservation. Visitors gain insights into the challenges these turtles face and the conservation efforts being undertaken. This educational experience promotes a deeper understanding of environmental issues and the importance of biodiversity conservation (Ionuţ et al., 2018; Simona et al., 2017).

Agrotourism aligns with the principles of sustainable and responsible travel. Tourists who choose such experiences are more likely to appreciate the need for eco-friendly practices and supporting initiatives that prioritize the preservation of natural environments. This aligns with the growing trend of eco-conscious travel (Ciani, 2017). Visitors to agrotourism sites may gain insights into sustainable agricultural practices. This can inspire travellers to adopt eco-friendly practices in their own lives and promote the use of sustainable farming methods, which have long-term benefits for both local agriculture and the environment (Phumrungruang & Baconguis, 2020). Agrotourism encourages local entrepreneurship. Tourist revenue generated through entrance fees and guided tours directly supports *Batagur* sp. conservation. By participating in agrotourism, travellers effectively become conservation allies, as their contributions help fund research, protection and breeding programs for these CR turtles (Mustika et al., 2012).

Agrotourism often provides opportunities for cultural exchange between tourists and local communities. Travellers can immerse themselves in the traditions and customs of the region, fostering a deeper appreciation for the diversity of human cultures (Chemnasiri & Kaewmoung, 2008; Karabati et al., 2009). Many agrotourism initiatives involve the local community in conservation efforts. This fosters a sense of shared responsibility and ownership of natural resources and wildlife. Such community engagement benefits conservation and enhances the well-being of the local population (Jean et al., 2023). Agrotourism experiences provide travellers with unique and memorable moments that go beyond typical tourist destinations. The opportunity to release turtle hatchlings into the wild or witness the nesting and hatching process can create lasting memories and a sense of accomplishment (Christou, 2010; Rong-Da Liang, 2017; Sharma & Arora, 2022).

4. The Conservation Centres of *Batagur* sp.

Table 4.1 shows six conservation centres of *Batagur* sp. distributed in Peninsular Malaysia. A place is managed by non-government organisations (NGOs). The Pasir Gajah, Kemaman, Terengganu conservation centres are organized by the Turtle Conservation Society of Malaysia (TCS). The rest of the conservation centre managed by the government body includes the Department of Fisheries of Malaysia (DOF), the Department of Wildlife and National Parks, Peninsular Malaysia (DWNP) and the Terengganu State Parks Management Council. Additionally, it may stimulate inbound investment for the rural community's economic and social benefit. Such impulses are not just the duty of public-sector organizations; NGOs and the corporate sector may have similar objectives. It will find that investing jointly in rural capital, either directly or indirectly, is an effective way to achieve them (Uysal et al., 2012).

5. Malaysia's Strategic Plan

The plan takes into account the following psychographic features of prospective visitors: Domestic and daily visitors include (a) Visitors interested in the biodiversity of flora and fauna; (b) Visitors interested in seafood and community traditions; (c) Educators, researchers and students interested in fisheries and agriculture and (d) Visitors interested in heritage and communities – visitors with a strong interest in history and community (Ling et al., 2013; Navarro-Martínez et al., 2020). Three conservation activities are significant to apply at conservation centres (Fig. 4.4). First, affection provides a haven for endangered species such as Painted Terrapins (*Batagur borneoensis*) and sea turtles. Protection: mangrove conservation in locations where they act as a natural flood control agent, slowing erosion and safeguarding the shoreline from storms and tsunamis. Attractions: rural tourism resources: People living at the site for a long time have relied on the site's natural resources to live. The fishing industry and small businesses make traditional seafood dishes like fish crackers, shrimp paste and fermented fish sauce, which bring in a lot of money (Ling et al., 2013; Williams et al., 2001).

However, due to agrotourism operations and the potential of the wetlands or rivers, the local population might make additional revenue by performing the following roles: Responsibilities by role river excursions to observe fireflies, sea cruises and recreational fishing are all available – eco-tour guides for forest trekking, turtle watching and wild animals and vegetation observation. Caterers and food stalls create a variety of employment opportunities for locals, contributing to the culinary industry and supporting the preparation and sale of regional delicacies, including both food and fruit. These roles often extend to employees

Table 4.1. Conservation Centre Records for *Batagur* sp. in Southeast Asia's Agrotourism Destinations.

Scientific Name	Common Name	Conservation Centre	Country
Batagur affinis	Southern river terrapin	Sre Ambel River, Koh Kong Reptile Conservation Centre	Cambodia
Batagur affinis	Southern river terrapin	Angkor Center for Conservation of Biodiversity	Cambodia
Batagur affinis	Southern river terrapin	Klong La-ngu River in Satun Province	Thailand
Batagur affinis	Southern river terrapin	Kedah River, Kepala Batas, Kedah	Malaysia
Batagur affinis	Southern river terrapin	Perak River, Bota Kanan, Perak	Malaysia
Batagur affinis	Southern river terrapin	Terengganu River, Kuala Berang, Terengganu	Malaysia
Batagur affinis	Southern river terrapin	Kemaman River, Kemaman, Terengganu	Malaysia
Batagur affinis	Southern river terrapin	Setiu Wetlands, Terengganu	Malaysia
Batagur baska	Northern river terrapin	Vawal National Park	Bangladesh
Batagur baska	Northern river terrapin	Sajnekhali Forest Station	India
Batagur borneoensis	Painted Terrapin	Langkat, North Sumatera	Indonesia
Batagur borneoensis	Painted Terrapin	Ujung Tamiang, Aceh	Indonesia
Batagur borneoensis	Painted Terrapin	Setiu Wetlands State Park, Terengganu	Malaysia
Batagur borneoensis	Painted Terrapin	Pengkalan Balak, Melaka	Malaysia
Batagur trivittata	Burmese roofed turtle	Yadanabon Zoological Gardens, Mandaley	Myanmar
Batagur trivittata	Burmese roofed turtle	Lawkananda Park, Bagan	Myanmar
Batagur dhongoka	Three-striped roofed turtle	National Chambal Sanctuary	India

(Continued)

Table 4.1. *(Continued)*

Scientific Name	Common Name	Conservation Centre	Country
Batagur kachuga	Red-crowned roofed turtle	National Chambal Sanctuary	India
Batagur kachuga	Red-crowned roofed turtle	Madras Crocodile Bank Trust and Centre for Herpetology	India

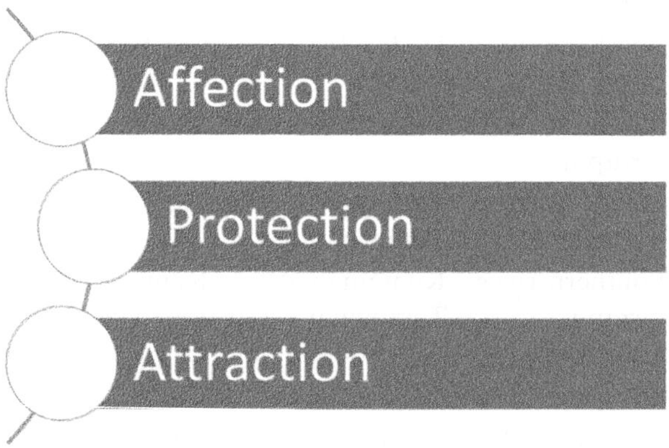

Fig. 4.4. Three Initiatives in Conservation.

and operators within resorts, transportation services, and general operations.Village Cultural Groups of cultural dancers and musicians such as 'wayang kulit' (shadow play) Village of Artisans promotion of indigenous handicrafts and souvenirs through the use of the indigenous 'mengkuang' or Pandanus atrocarpus (Ling et al., 2013).

6. Conclusion

In conclusion, agrotourism, with a focus on the conservation of *Batagur* species, is a vital driver of economic growth in rural Asian regions. It diversifies rural communities' income sources, enhancing economic stability and resilience. Additionally, the establishment of *Batagur* species conservation centres augments the economic impact, generating revenue and employment opportunities while fostering a sense of environmental responsibility within the local community. As such, agrotourism represents a win-win scenario, where economic development goes hand in hand with conserving unique wildlife species and promoting sustainable agricultural practices.

Acknowledgements

The authors want to acknowledge the Universiti Putra Malaysia for funding the study through a Graduate Research Fellowship. In addition, the authors would also like to express their gratitude to the Genetic and Breeding Laboratory, Department of Aquaculture, Faculty of Agriculture, Universiti Putra Malaysia, for the facilities and support of this research project. Furthermore, the authors thank the TCS and the Department of Wildlife and National Parks, Peninsular Malaysia, for their collaboration.

References

Alola, A. A., & Alola, U. V. (2018). Agricultural land usage and tourism impact on renewable energy consumption among Coastline Mediterranean Countries. *Energy & Environment*, *29*(8), 1438–1454. https://www.jstor.org/stable/26554641

Ana, M. I. (2017). Ecotourism, agro-tourism and rural tourism in the European Union. *Cactus Tourism Journal*, *15*(2), 6–14. https://www.cactus-journal-of-tourism.ase.ro/Pdf/vol16/1.pdf

Bac, D. P. (2014). The emergence of sustainable tourism - A literature review. *Quaestus Multidisciplinary Research Journal*, *4*, 131.

Baloch, U. K., & Haseeb, M. (1996). *Xenobiotics in the third world agricultural environment*. Taylor & Francis.

Banda, S., Mpolomoka, D. L., Pansho, M., & Musonda, L. (2021). Linking higher education to perceived benefits of agro-tourism for sustainable development in rural Zambia. *Mulungushi University Multidisciplinary Journal*, *2*(2), 61–69.

Berweck, S., Koohafkan, P., Cruz, M. J. R. D., Qingwen, M., Wenjun, J., Yehong, S., & Moucheng, L. (2013). Conceptual framework for economic evaluation of Globally Important Agricultural Heritage Systems (GIAHS): Case of rice-fish co-culture in China. *Journal of Resources and Ecology*, *4*(3), 202–211. https://doi.org/10.5814/j.issn.1674-764x.2013.03.003

Beus, C. E. (2008). *Agritourism: Cultivating tourists on the farm*. http://cru.cahe.wsu.edu/CEPublications/eb2020/eb2020.pdf

Budiasa, I. W., & Ambarawati, I. G. A. A. (2014). Community based agro-tourism as an innovative integrated farming system development model towards sustainable agriculture and tourism in Bali. *Journal of the International Society for Southeast Asian Agricultural Sciences*, *20*(1), 29–40. http://issaasphil.org/wp-content/uploads/2020/02/J-Issaas-v20n1-June-2014-Full-Journal.pdf#page=32

Busby, G., & Rendle, S. (2000). The transition from tourism on farms to farm tourism. *Tourism Management*, *21*(6), 635–642.

Camara, M., & Jamil, N. R. (2021). Evaluation of community-based species awareness programme: A case study of critically endangered freshwater river turtle Batagur affinis in Kg Pasir Gajah, Terengganu, Malaysia. In *Proceedings of the zoological society* (Vol. 74, No. 1, pp. 52–61). Springer. https://doi.org/10.1007/s12595-020-00359-x

Chase, L. C., Stewart, M., Schilling, B., Smith, B., & Walk, M. (2018). Agritourism: Toward a conceptual framework for industry analysis. *Journal of Agriculture, Food Systems, and Community Development*, *8*(1), 13–19. https://doi.org/10.5304/jafscd.2018.081.016

Chemnasiri, N., & Kaewmoung, D. (2008). Farm development to become agro-tourism area by community involvement in Saraburi, Thailand. *Amfiteatru Economic*, *10*, 184–194. https://www.amfiteatrueconomic.ro/temp/Article_63.pdf
Christou, P. (2010). *Achieving guest satisfaction through the agritourism experience: The case of Cyprus*. Nottingham Trent University.
Chulaphan, W., & Barahona, J. F. (2021). The determinants of tourist expenditure per capita in Thailand: Potential implications for sustainable tourism. *Sustainability*, *13*(12), 6550. https://doi.org/10.3390/su13126550
Ciani, A. (2017). Strengthening the sustainability of rural tourism and agrotourism in the twenty first century. *Metropolitan Commuter Belt Tourism*, 68–83.
Cichowska, J., & Klimek, A. (2011). The role of agrotourism in the development and conversion of rural areas. *Infrastruktura i Ekologia Terenów Wiejskich*, 11.
Conway, G. R., & Barbie, E. B. (1988). After the green revolution: Sustainable and equitable agricultural development. *Futures*, *20*(6), 651–670. https://doi.org/10.1016/0016-3287(88)90006-7
Daniela, C., Cecilia, N., & Dumitra, C. (2008). Sustainable development of the rural communities by promoting the agrotourist activities. *Scientific Papers*, *8*, 393.
Deininger, K., & Byerlee, D. (2011). *Rising global interest in farmland: Can it yield sustainable and equitable benefits?*. World Bank Publications.
Delgado, C. L., & Siamwalla, A. (2018). Rural economy and farm income diversification in developing countries. In *Food security, diversification and resource management: Refocusing the role of agriculture?* (pp. 126–143). Routledge.
Drabenstott, M. (1994). Industrialization: Steady current or tidal wave? *Choices*, *9*(4), 4–8. http://www.jstor.org/stable/43644635
Drăgoi, M. C., Iamandi, I. E., Munteanu, S. M., Ciobanu, R., Țarțavulea, R. I., & Lădaru, R. G. (2017). Incentives for developing resilient agritourism entrepreneurship in rural communities in Romania in a European context. *Sustainability*, *9*(12), 2205. https://doi.org/10.3390/su9122205
Evans, H. E., & Ngau, P. (1991). Rural-urban relations, household income diversification and agricultural productivity. *Development and Change*, *22*(3), 519–545. https://doi.org/10.1111/j.1467-7660.1991.tb00424.x
Gannon, A. (1994). Rural tourism as a factor in rural community economic development for economies in transition. *Journal of Sustainable Tourism*, *2*(2), 51–60. https://doi.org/10.1080/09669589409510683
Gao, S., Huang, S., & Huang, Y. (2009). Rural tourism development in China. *International Journal of Tourism Research*, *11*(5), 439–450. https://doi.org/10.1002/jtr.712
Garrod, B., Wornell, R., & Youell, R. (2006). Re-conceptualising rural resources as countryside capital: The case of rural tourism. *Journal of Rural Studies*, *22*(1), 117–128. https://doi.org/10.1016/j.jrurstud.2005.08.001
Getz, D., & Carlsen, J. (2000). Characteristics and goals of family and owner-operated businesses in the rural tourism and hospitality sectors. *Tourism Management*, *21*(6), 547–560. https://doi.org/10.1016/S0261-5177(00)00004-2
Hall, D. R., Roberts, L., & Mitchell, M. (2003). *New directions in rural tourism*. Ashgate.
Hassan, M. S., Shaffril, H. A. M., Samah, B. A., Ali, M. S. S., & Ramli, N. S. (2010). Agriculture communication in Malaysia: The current situation. *American Journal of Agricultural and Biological Sciences*, *5*(3), 389–396.

Hawkins, D. E., & Hunt, J. D. (1988). Travel and tourism professional education. *Hospitality & Tourism Educator*, *1*(1), 8–14. https://doi.org/10.1080/23298758.1988.10685361

History of Agro-tourism. (2008). http://www.scribd.com/doc/50731198/HISTORY-OF-AGROTOURISM-Prt-2008. Accessed February 03, 2022.

Holloway, J. C., & Humphreys, C. (2019). *The business of tourism*. Sage.

Hrymak, O. Y., Vovk, M. V., & Kindrat, O. V. (2019). Agrotourism as one of the ways to develop entrepreneurship in rural areas. *Scientific Messenger of LNU of Veterinary Medicine and Biotechnologies. Series Economical Sciences*, *21*(93), 31–34.

Huang, Y. H., & Stewart, W. P. (1996). Rural tourism development: Shifting basis of community solidarity. *Journal of Travel Research*, *34*(4), 26–31. https://doi.org/10.1177/004728759603400404

Hudson, R. (2016). River terrapins hatch throughout Asia. *Turtle Survival Alliance*. https://turtlesurvival.org/river-terrapins-hatch-throughout-asia/

Hunjra, A. I., Hassan, M. K., Zaied, Y. B., & Managi, S. (2023). Nexus between green finance, environmental degradation, and sustainable development: Evidence from developing countries. *Resources Policy*, *81*, 103371. https://doi.org/10.1016/j.resourpol.2023.103371

Ibrahim, M. S. N., Johari, S., Adam Assim, M. I. S., Wan Mohammad, S. K., Mohammad Affandi, S. H., Khan, W. R., & Hassan, S. (2023). Community well-being dimensions in Gunung Mulu National Park, Sarawak, Malaysian Borneo. *Humanities & Social Sciences Communications*, *10*(1), 1–15. https://doi.org/10.1057/s41599-023-01737-4

Ionuţ, R., Gina, T. A., Florin, T. I., Diana, M., Cornelia, P., & Ioan, P. (2018). Agrotourism education instrument for faculty students of different profiles. *Agricultural Management/Lucrari Stiintifice Seria I, Management Agricol*, *20*(3).

Ismail, S. S., Abd Hamid, Z., Fakri, S. N. S. M., & Ab Aziz, L. H. (2021). Rural tourism development in Cameron Highlands: Indigenous community's perception towards socio cultural and economic impacts. *International Journal of Academic Research in Business and Social Sciences* https://doi.org/10.6007/ijarbss%2Fv11-i13%2F8516

Jean, B. P., Stael, K. P. R., Strasser-King, D. E. U., Cephas, T. K. S. L., Brown, F., & Enyan, M. (2023). The socio-economic impact of rural and agrotourism on the economy of West African Rural Folks. *EPRA International Journal of Socio-Economic and Environmental Outlook (SEEO)*, *10*(4), 32–42. https://eprajournals.net/index.php/SEEO/article/view/1898

Jeczmyk, A., Uglis, J., & Steppa, R. (2021). Can animals be the key to the development of tourism: A case study of livestock in agritourism. *Animals*, *11*, 23–57.

Karabati, S., Dogan, E., Pinar, M., & Celik, L. M. (2009). Socio-economic effects of agri-tourism on local communities in Turkey: The case of Aglasun. *International Journal of Hospitality & Tourism Administration*, *10*(2), 129–142. https://doi.org/10.1080/15256480902851032

Kasinathan, Y., Rahman, S., & Hambali, K. (2016). A study on the factors affecting the hatching success of river terrapin (Batagur affinis) egg in artificial pond at wildlife conservation centre, Bota Kanan, Perak. In *Ecology Seminar, Malaysia*.

Khawatkun, K., & Thongma, W. (2015). Community participation in agro-tourism management in Koh Thepo, Mueang District, Uthai Thani Province, Thailand. *Agricultural Travel and Tourism*, *6*(1), 59.

Kieselbach, S., & Long, P. (1990). Tourism and the rural revitalisation movement. *Parks & Recreation*, *25*(3), 62–66. https://www.cabdirect.org/cabdirect/abstract/19901879451

Kizos, T., & Iosifides, T. (2007). The contradictions of agrotourism development in Greece: Evidence from three case studies. *South European Society & Politics*, *12*(1), 59–77. https://doi.org/10.1080/13608740601155443

Ling, S. M., Ramchandran, S., Shuib, A., Nair, M., Syamsul Herman, M. A., & Prabhakaran, S. (2013). Rural resources as tourism capital: The case of Setiu Wetlands, Terengganu, Malaysia. *Malaysian Forester*, *76*(1), 27–39.

Manjunatha, A. V., Anik, A. R., Speelman, S., & Nuppenau, E. A. (2013). Impact of land fragmentation, farm size, land ownership and crop diversity on profit and efficiency of irrigated farms in India. *Land Use Policy*, *31*, 397–405. https://doi.org/10.1016/j.landusepol.2012.08.005

Marsden, T., & Sonnino, R. (2008). Rural development and the regional state: Denying multifunctional agriculture in the UK. *Journal of Rural Studies*, *24*(4), 422–431. https://doi.org/10.1016/j.jrurstud.2008.04.001

McGehee, N. G. (2007). An agritourism systems model: A Weberian perspective. *Journal of Sustainable Tourism*, *15*(2), 111–124. https://doi.org/10.2167/jost634.0

Miller, B. (1993). Farm tourism-sustaining employment in a rural area. *Agricultural Manpower*, *1*(24), 8–11.

Mohd Salleh, M. H., Esa, Y., & Mohamed, R. (2023). Global terrapin character-based DNA barcodes: Assessment of the mitochondrial COI gene and conservation status revealed a putative cryptic species. *Animals*, *13*(11), 1720. https://doi.org/10.3390/ani13111720

Mohd Salleh, M. H., Esa, Y., Salleh, S. M., & Mohd Sah, S. A. (2022). Turtles in Malaysia: A review of conservation status and a call for research. *Animals*, *12*(17), 2184. https://doi.org/10.3390/ani12172184

Mukhlis, I., Rizaludin, M. S., & Hidayah, I. (2022). Understanding socio-economic and environmental impacts of agroforestry on rural communities. *Forests*, *13*(4), 556. https://doi.org/10.3390/f13040556

Mustika, P. L. K., Birtles, A., Welters, R., & Marsh, H. (2012). The economic influence of community-based dolphin watching on a local economy in a developing country: Implications for conservation. *Ecological Economics*, *79*, 11–20. https://doi.org/10.1016/j.ecolecon.2012.04.018

Navarro-Martínez, Z. M., Crespo, C. M., Hernández-Fernández, L., Ferro-Azcona, H., González-Díaz, S. P., & McLaughlin, R. J. (2020). Using SWOT analysis to support biodiversity and sustainable tourism in Caguanes National Park, Cuba. *Ocean & Coastal Management*, *193*, 105188. https://doi.org/10.1016/j.ocecoaman.2020.105188

Nematpour, M., & Khodadadi, M. (2021). Farm tourism as a driving force for socioeconomic development: A benefits viewpoint from Iran. *Current Issues in Tourism*, *24*(2), 247–263. https://doi.org/10.1080/13683500.2020.1711712

Nickerson, N. P., Black, R. J., & McCool, S. F. (2001). Agritourism: Motivations behind farm/ranch business diversification. *Journal of Travel Research*, *40*(1), 19–26.

Park, K., Phillips, W. J., Canter, D. D., & Abbott, J. (2011). Hospitality and tourism research rankings by author, university, and country using six major journals: The first decade of the new millennium. *Journal of Hospitality & Tourism Research*, *35*(3), 381–416. https://doi.org/10.1177/1096348011400743

Peebles, M. S. (1995). Cultivating the tourist/farm tourism. *Journal of the Tourism Society*, *86*(5).

Phumrungruang, B., & Baconguis, R. (2020). Agro-tourism as an extension strategy to promote agriculture in the northern region of Thailand. *Journal of Public Affairs and Development*, *7*, 75–101. https://orcid.org/0000-0002-3300-1208

Pimentel, D., Harvey, C., Resosudarmo, P., Sinclair, K., Kurz, D., McNair, M., Crist, S., Shpritz, L., Fitton, L., Saffouri, R., & Blair, R. (1995). Environmental and economic costs of soil erosion and conservation benefits. *Science*, *267*(5201), 1117–1123. https://doi.org/10.1126/science.267.5201.1117

Rilla, E. (1999). *Unique niches: Agritourism in Britain and New England*. University of California Small Farm Center, Cooperative Extension. http://www.sfc.ucdavis.edu/agritourism/insidecov.html

Romanenko, Y. O., Boiko, V. O., Shevchuk, S. M., Barabanova, V. V., & Karpinska, N. V. (2020). Rural development by stimulating agro-tourism activities. *International Journal of Management*, *11*(4), 605–613. http://www.iaeme.com/ijm/issues.asp?JType=IJM&VType=11&IType=4

Rong-Da Liang, A. (2017). Considering the role of agritourism co-creation from a service-dominant logic perspective. *Tourism Management*, *61*, 354–367. https://doi.org/10.1016/j.tourman.2017.02.002

Salleh, M. H. M., Esa, Y., & Pau, S. S. N. (2023). Conservation genetics of the critically endangered southern river terrapin (Batagur affinis) in Malaysia: Genetic diversity and novel subspecies distribution ranges. *Biology*, *12*(4), 520. https://doi.org/10.3390/biology12040520

Salleh, M. H. M., & Esa, Y. (2022). Conservation status of globally testudines terrapins based on COI mitochondrial markers. https://doi.org/10.3390/mol2net-08-12456

Salleh, M. H. B. M., & Sah, S. A. B. M. (2019). Twenty-first years breeding performance of southern river terrapin (Batagur affinis) at Perak River, Peninsular Malaysia. *International Journal of Nature and Life Sciences*, *3*(2), 52–58. https://dergipark.org.tr/en/pub/ijnls/issue/45384/579088

Sharma, A., & Arora, S. (2022). Virtual reality and blogging in tourism: Approached towards creating better tourist experience. *Journal of Tourism Insights*, *12*(1), 6. https://doi.org/10.9707/2328-0824.1252

Simona, A., Amelia, M., Sorin, M., Loredana, V., & Cornelia, P. (2017). Educational activities in animal farms which practice agrotourism. *Agricultural Management/Lucrari Stiintifice Seria I. Management Agricol*, *19*(3).

Singh, K., Gantait, A., Puri, G., & Swamy, A. (2016). Rural tourism: Need, scope and challenges in Indian context, *Hospitality and tourism: Challenges, innovation, practices and product*. Adhyayan Publishers and Distributors.

Stanford, C. B., Iverson, J. B., Rhodin, A. G., van Dijk, P. P., Mittermeier, R. A., Kuchling, G., Berry, K. H., Bertolero, A., Bjorndal, K. A., Blanck, T. E., & Buhlmann, K. A. (2020). Turtles and tortoises are in trouble. *Current Biology*, *30*(12), R721–R735. https://doi.org/10.1016/j.cub.2020.04.088

Stanovčić, T., Peković, S., Vukčević, J., & Perović, Đ. (2018). Going entrepreneurial: Agro-tourism and rural development in northern Montenegro. *Business Systems Research: International Journal of the Society for Advancing Innovation and Research in Economy*, *9*(1), 107–117. https://hrcak.srce.hr/ojs/index.php/bsr/article/view/12618

Tiraieyari, N., & Hamzah, A. (2012). Agri-tourism: Potential opportunities for farmers and local communities in Malaysia. *African Journal of Agricultural Research*, *7*(31), 4357–4361. https://doi.org/10.5897/AJARx11.035

Topcu, E. D. (2007). *Agri-tourism: As a new element of country planning*. A thesis submitted to the Graduate School of Natural and Applied Sciences of Middle East Technical University.

Uysal, M., Perdue, R. R., & Sirgy, M. J. (eds). (2012). *Handbook of tourism and quality-of-life research: Enhancing the lives of tourists and residents of host communities* (Vol. 1). Springer. https://doi.org/10.1007/978-94-007-2288-0

Weaver, D. B., & Fennell, D. A. (1997). The vacation farm sector in Saskatchewan: A profile of operations. *Tourism Management*, *18*(6), 357–365. https://doi.org/10.1016/S0261-5177(97)00039-3

Williams, P., Paridaen, M., Dossa, K., & Dumais, M. (2001). *Agritourism market and product development, status report*. Ministry of Agriculture, Food & Fisheries.

Wimalaratana, R. P. W. (2014). Promotion of agro tourism and rural development in South Asia: A coying strategy to global economic crisis. *Issues in India Sri Lanka Economies*, 105–124.

Zhou, J., Zhang, F. T., Gai, Y. J., Deng, B. K., Shao, J. X., & An, Y. Z. (2017). The connotation interpretation of high-efficiency agriculture under the perspective of a combined agriculture with tourism. In *IOP Conference Series: Earth and Environmental Science* (Vol. 81, No. 1, pp. 012065). IOP Publishing. https://doi.org/10.1088/1755-1315/81/1/012065

Chapter 5

Transshipment Troubles: India's Quest for Efficient Port Hubs

M. Bina Celine Dorathy

Alliance University, India

Abstract

India should hold a privileged position in maritime transportation due to its extensive coastline and advantageous location. However, the country heavily relies on other maritime nations like Singapore and Colombo for transshipment due to insufficient infrastructure and policy framework, and this has created disadvantage to Indian goods in the international market. The government has launched a significant drive to promote three transshipment ports in southern India with top-notch amenities in response to this worrying circumstance. It is anticipated that these ports would compete with the existing transshipment ports, particularly with Colombo, and divert the transshipped goods back to India. The move is expected to make Indian exports more competitive besides making India less prone to geopolitical and economic disturbances in the region. However, these initiatives have been met with many challenges. In fact, the first attempt of the Indian government to set up an International Container Transshipment Terminal (ICTT) at Vallarpadam in Cochin Port has met with failure despite its best infrastructure and connectivity. High-cost structure seems to deter the competitiveness of this port. Furthermore, the development of Vizhinjam into a transshipment hub in the same region has put additional pressure on Vallarpadam. This chapter draws attention to certain factors that might be considered to enhance the competitiveness of Cochin Port. Also, the study highlights the snags that could have well been avoided while implementing the project and might as well be avoided while implementing other projects in line.

Keywords: Vallarpadam ICTT; DP world; Colombo port; hub and spoke; relay network; ANP model

The Emerald Handbook of Tourism Economics and Sustainable Development, 81–97

Published under exclusive licence by Emerald Publishing Limited
doi:10.1108/978-1-83753-708-220241005

1. Introduction

Indian Ocean is one of the major shipping routes of the world. Ships plying along this route are heavily dependent on hub ports in Colombo, Singapore and Dubai. At the Maritime India Summit in 2021, the Prime Minister of India launched the Maritime Vision 2030, which aims at making Indian maritime sector to be on par with international standards. Among the various reforms and initiatives introduced, transforming India into transshipment hub is a crucial one. The recent Sri Lankan crisis has given another push to India's growing need for a transshipment hub. India's ports are located either on the east or west coastline. Since majority of ports on the west coast are gateway ports, cargo is delivered right away. On the eastern side, the country relies heavily on neighbouring maritime nations for transshipment because the ports on the east coast lack the necessary infrastructure. Containers are transported to an intermediate location for loading onto other ships sailing to distant ports as part of transshipment. The container transshipment terminal functions as a hub where goods from smaller ships from all over is brought in and either loaded onto larger ships for final destinations or the other way around (Manoj, 2016). Due of their weight, huge ships cannot access smaller ports, necessitating the use of specialized transshipment deep ports, which India unfortunately lacks. As a result, more than 80% of the cargo containers from India are shipped for transshipment to Colombo, Singapore, Port Klang (Malaysia) and Dubai; among them, Colombo alone accounts for over 45% (Container News, 2023).

The nation spends Rs. 1,500 crores annually on transshipment (Dasgupta & Thakurta, 2018). This has led to increased processing fees, which raise the price of exports and imports and reduce the nation's ability to compete internationally. To combat this, the government has embarked on an ambitious programme to promote three new projects namely Vallarpadam, Vizhinjam, and the International Container Transshipment Port (ICTP) at Galathea Bay in the Great Nicobar Island to promote transshipment facilities in the southern region of the nation. Subsequently, the government eased the Cabotage policy to allow foreign ships to navigate Indian waters (Dasgupta & Thakurta, 2018) and to increase ship calls. If all these efforts succeed, India is anticipated to become a container transshipment hub for the Asia-Africa, Asia-United States (US)/Europe container traffic besides bringing the transshipment commerce back to the country.

Out of the three projects, the Vizhinjam Port is being built on a Public – Private Partnership (PPP) mode with funding from the Kerala Government, Union Government and the Adani Ports with Adani Ports Private Limited being the developer. The project was started in 2015 and is expected to become fully operational in 2024. It is in fact the first greenfield port project to receive viability gap funding from the government of India. The Great Nicobar Port, which would come up at Galathea Bay in Great Nicobar Island of the Union Territory of Andaman and Nicobar Islands, is currently under pre-tender stage. The cost of the project is estimated to be around Rs. 48,000 crores (Mishra, 2023). The Kochi International Container Transshipment Terminal (ICTT), popularly known as the Vallarpadam Terminal, is the first to be commissioned among the three transshipment projects. The project was launched in 2005, and the first phase was commissioned in 2011 (Pti, 2011). It is also the first in the country to operate in an SEZ (A geographical area with more lenient economic regulations than the domestic laws of a nation is known as a Special Economic Zone [SEZ]) and also the country's first container transshipment terminal. The terminal is a part of the Cochin Port under the Union Government. The Dubai government-owned DP World Ltd. and the Cochin Port Trust have established it under the PPP mode. The estimated cost was around Rs. 3,200 crores, of which Rs. 1,700 crores came from the Indian Government and DP World invested around Rs. 1,600 crores (Pti, 2011).

Initially, the terminal's performance was impressive. It handled over 40,000 TEUs in April 2016, a 29% increase over the same month the previous year (2015). Compared to other ports in the area, the terminal could turn around vessels more quickly, thanks to its 31 moves per hour per crane (Pti, 2016). The number of vessel calls increased year on year, indicating consistent growth. The first ship of the weekly Far East container ship service (part of Galex service), connecting Kochi with ports in China and Korea, called at the ICTT in 2015. Galex (GLX) was a consortium of four vessel operators – Emirates Shipping Line, KMTC, RCL and Hanjin. The service handled 872 containers at ICTT in its first call itself (Press Trust of India & Business Standard, 2015). ICTT had best connectivity for all mainline carriers on the east-west shipping routes and excellent South Indian hinterland connectivity. The terminal was hopeful of attracting new services in the east west corridor through the attractive schemes of Cochin Port.

But the performance of this very promising venture deteriorated over the years. The failure of such an ambitious plan calls for an in-depth analysis as huge amounts have gone into this project and two other similar projects are also underway. This chapter makes an attempt to understand the causes and provide insights into the areas to be revamped and strategies to be drawn, which may help revive the existing project and avoid future failures as well.

2. Methodology

Both primary and secondary sources of information was used in the study. An extensive literature review was done to understand transshipment ports, liner shipping network, port performance, competitiveness of Colombo, Singapore and other transshipment ports in the region. Port websites and published reports on ICTT and Cochin Port were used to collect information on the performance and challenges of the terminal and the port. Primary information was collected by visiting the Cochin Port and ICTT by interviewing port officials.

2.1 A Review of Literature

The literature on 'The transshipment hub port competitiveness of the Port of Colombo against major Southeast Asian hub ports' offers a comprehensive understanding of the factors, challenges and strategies that shape its position in the region. While the Port of Colombo enjoys strategic advantages, it faces formidable competition from neighbouring hub ports. Further research is necessary to continually evaluate its competitiveness in this dynamic and highly competitive industry (Kavirathna et al., 2018).

Numerous investigations have examined the elements impacting port competitiveness. The study 'Exploring key variables of port competitiveness: evidence from Indonesian ports' discovered that ineffective government bureaucracy, customs clearance and strategic decision-making in Indonesia were the main causes of the discrepancy between policy expectations and realization of port facilitation. To improve operational performance, road connectivity, multimodal transportation and energy infrastructure must be reinforced. The lack of feasibility analysis, disregard for regional economic developments and sluggish adoption of standard technology were the root causes of these issues (Wahyuni et al., 2020).

The port environment of Southeast Asia is well-documented. Numerous studies have examined the tactics and development of the main hub ports in the area. These studies have shed light on the competition Southeast Asian ports pose to the Port of Columbo. According to the study 'The geography of maritime networks: A critical review', maritime network analysis did not pick up steam until the 2000s, when newly available shipping data, more computing power and updated conceptual frameworks for studying networks in general, supported the field. This contrasts with earlier cartographical and graph-theoretical analyses of maritime flows conducted in the 1940s and 1960s (Ducruet, 2020).

Scholars have examined the Port of Colombo's advantages and disadvantages to determine how competitive it is. Using the factor analysis model, the article 'A study of developing Colombo Port as a major multi-country consolidation (MCC) hub in South East Asia region with the help of improving the effectiveness of the MCC activities' concludes that it will be economical to develop Colombo Port as a major MCC hub and that its geographic location will provide it with a competitive advantage (Gajanayaka et al., 2015).

It was anticipated that the region's share of Southeast Asia's transshipment trade would rise in the near future. Major container terminal operators in Tanjung Pelepas, Singapore and Port Klang, as well as major container shipping lines operating along important east-west sailing routes, engaged in fierce competition for business, drawn by the potential of this sizeable and growing market. Data indicated that Singapore's lead in the transshipment industry appeared to be gradually eroding, despite the country's continued dominance in terms of market share as measured by annualized slot capacity and transshipment throughput. In general, Tanjung Pelepas is anticipated to present the biggest obstacle to Singapore's goals of becoming a transshipment hub (Lam & Yap, 2008).

When making decisions, shipping lines take into account various factors such as port competitiveness, efficiency, connectivity, and geographic advantages. The complexity of this choice is emphasized in the literature on 'Transshipment hub port selection criteria by shipping lines – the case of hub ports around the Bay of Bengal'. With its developing ports, the Bay of Bengal region presents opportunities as well as challenges, which makes it a focal point for this kind of research (Kavirathna et al., 2018). Further the changing standards and decision-making procedures used by shipping lines in this dynamic area should be the subject of ongoing research studies.

According to the literature on 'Influential Article Review – Understanding Transshipment Hub Port Selection Criteria', hub port competition has intensified across various regions and the transshipment hub port selection process has grown more complex due to the liner shipping industry's competitive nature. There are many factors to consider when making this decision. The selection of a hub port has become increasingly complex due to the rise of various liner service networks, such as 'hub and spoke' and 'relay'. Based on the finances, time, port traffic, location, operation and liner-related categories, the hub port selection criteria are assessed (Fox et al., 2019).

The body of research on 'Transshipment port's competitiveness forecasting using analytic network process modelling' highlights the importance of this tool for thorough and flexible decision-making. An Analytic Network Process (ANP) is especially well-suited for evaluating the intricate and multi-dimensional aspects of port competitiveness since it makes it easier to integrate various criteria while considering their interdependencies. The practical applications and case studies illustrate the actual use of ANP in the shipping and logistics sector. Transshipment ports are essential to international trade, and the effectiveness of supply chains can be strongly impacted by their competitiveness. ANP modelling is a potent tool for decision-making for those involved in the shipping and logistics sectors since it forecasts competitiveness by considering a number of factors (Munim et al., 2022). Using ANP to assess port competitiveness has been the subject of several studies. The competitiveness of container ports has been assessed using ANP, demonstrating the value of incorporating a variety of criteria into the evaluation procedure. For shipping companies and logistics suppliers, selecting the transshipment ports is an important choice. The ANP helps in transshipment port selection, highlighting the significance of factors such as infrastructure, accessibility and operational performance.

3. The Special Features of ICTT

Cochin Port, DP World and the Central Government have invested heavily into this terminal. The infrastructure at the terminal is one of the best. The following are the special features of ICTT.

- Faster Vessel Turnaround: The four Super-post Panamax Quay cranes and two mobile harbour Cranes, each of which are capable of dual lifts, allow for faster vessel loading and discharge.
- Faster Truck Turnaround: With 15 RTGCs, three reach stackers and automated yard management, container delivery is expedited.
- Simple documentation: The ICTT documentation department has been appropriately manned to provide faster clearance and a minimum truck wait time. Documentation procedures have been streamlined. To decrease effort and time at the documentation centre, eForm13 has been implemented.

- Terminal Operating System: ICTT employs an in-house built Terminal Operating System, Zodiac, which is managed and developed by DP World and is utilized across all DP World Terminals.
  - Zodiac container management software is used to collect real-time data on containers being discharged, loaded, gated in and gated out and moved to different positions in the yards.
  - The Zodiac planning tool also improves productivity by monitoring and maintaining terminal business transactions and data processing, as well as accurately recording and invoicing all transactions and services.
  - The Zodiac application automates billing, data sharing and reporting operations while keeping a complete record of all transactions and labour.
- Dynamic reporting tools provide terminal managers with important information, allowing them to track performance, improve customer service and maximize profitability.
- Terminal operations are available 24 hours a day and 365 days a year. The terminal is not closed on holidays.
- Adjustable cut-off time: Export containers can be gated four hours before the vessels berthing. Relaxation on a case-by-case basis is permitted for Special Service Requests (SSRs). Last-minute goods can now connect to the desired vessel.
- No theft/pilferage: Around-the-clock inside-terminal security assures no theft or pilferage at crucial places.
- Cutting-edge workshop: The cutting-edge engineering workshop ensures that the equipment operates at peak efficiency.
- SMS tracking: SMS can be used to track vessel data and containers.

3.1 Connectivity

- When compared to other terminals in India, ICTT has remarkable multi-modal connectivity.
- A four-lane expressway connects NH-17 through Mangalore, Goa to Mumbai, NH-47 through Coimbatore, Salem to Bangalore and NH-49 through Madurai.
- The longest rail bridge in India, stretching 4.62 kilometers, connects ICTT to markets such as Coimbatore, Bangalore, Chennai, Hyderabad, Nagpur and Delhi. The terminal inaugurated a regular weekly train service to Bengaluru in January 2020, resulting in a fivefold increase in rail throughput.
- It has good connections to other Indian ports on both coasts, including Mundra, Pipavav, Nhava Sheva, Goa, Mangalore, Tuticorin, Chennai, Katupalli, Krishnapatnam and Visakhapatnam

4. Why Did the ICTT Struggle?

ICTT had a good start and with its good infrastructure and connectivity, the port was expecting more vessel calls. But as years passed, ICTT struggled to attract enough business.

The following have been identified as the major bottlenecks.

One of the major reasons for its failure to compete with Colombo Port is its high cost, and most of it was due to high dredging expenditure. The terminal currently operates with a draught of 14.5 metres. With this shallow draught, it is difficult to berth vessels with a capacity of 10,000 TEUs or more. To become viable, the terminal will have to increase its draught further, Vessels with more than 18,000 TEU capacity require 18–20 metres draught. Heavy expenditure is being incurred towards dredging and its maintenance. While the cost of dredging is borne by the central government, the cost of maintaining the draught is borne by the Cochin Port. Based on the financial statements covering the last 10 years (2010–2011–2019–2020), the port has incurred Rs.938.72 crores for dredging and maintenance alone (Joseph, 2021), which has further strained the port's finances. Because of the port's vulnerability to siltation, yearly dredging expenses are inevitable.

High dredging costs are linked to high port tariffs, which contribute to the terminal's subpar performance. While Colombo charges only 3.3 lakhs, the vessel charges at Cochin are approximately 26 lakhs for a 1200 TEU ship (Joseph, 2021).

In addition, the fees for handling containers are among the highest in India. Tuticorin charges almost half as much as DP World at ICTT for handling a 40-foot container (FEU) while DP World charges about ₹ 13,000. In addition, DP World raised its charges by 7.27% (Joseph, 2021), citing a Tariff Authority for Major Ports (TAMP) order (Kumar, 2023), which certainly would have a negative impact on the terminal's ability to compete.

In order to compete with nearby container hubs, the port offers tariff concessions of up to 86% to mother vessels. The port has awarded concessions totaling 461.97 crores in connection to vessel-related costs during the last 10 years (Joseph, 2021). Only mainline vessels that meet rigorous criteria, such as making a certain number of calls within a certain amount of time and carrying a certain number of containers on each journey, are eligible for tariff concessions from the port trust. The concessions do not apply to ships that make ad hoc calls as well as regular service providers who do not meet the standards. Cochin Port charges over 55 lakhs for a visit on a 6,500-TEU ship for the India East Coast Express (IEX) service, which connects Vallarpadam to ports in the Mediterranean and Europe. If the service meets the conditions of the port, after three months, the shipowner of the consortium will receive a refund of about 40 lakhs. But even the 15 lakhs charged is far more than what Jebel Ali and Colombo are charging (Joseph, 2021).

Although Vallarpadam has a capacity of 12 lakh TEUs, only around half of them are now in use. The fundamental reason for this is that the majority of containers from Tuticorin and Chennai are still transshipped to Colombo due to their shorter transit time. While shipping from Tuticorin to Cochin takes 18 hours by sea, containers reach Colombo in 12 hours (Joseph, 2021). Furthermore, Vallarpadam is 76 nautical miles (nm) from the Suez route, which means that sailing to Cochin will take 8 hours longer than sailing to Colombo, which is only 19 nm away from Suez route and can be reached in an hour (Manoj, 2016). This appears to be yet another factor discouraging large maritime firms from transferring their transshipment to Vallarpadam.

The year 2021 saw extreme congestion at Colombo port because of supply chain disruptions following the Covid-19 pandemic. During that time, bigger ships from Mediterranean Shipping Company (MSC) and Maersk Line made a few ad hoc calls to ICTT. ICTT handled 71,543 TEUs in January 2021 – a record high since the terminal's founding. However, because of its high costs, ICTT was unable to sustain it and was unable to take advantage of the opportunity even when Colombo Port was in a terrible state.

Currently, only two mother vessels are regularly making calls at ICTT - Wanhia's China India Service (CS2) and IEX Service. Small ships that connect to the hub ports in the region of Jebel Ali and Colombo provide other services. In contrast to the ambitious plans to attract ships that carry more than 15,000 TEUs, there are currently about seven or eight regular services offered with smaller ships that have a capacity of 1,000–2,500 TEUs (Joseph, 2021).

5. Discussion and Recommendations

The terminal's hinterland connectivity is good, and there is no dearth of cargo. Also, the terminal's connectivity to mainstream shipping appears to be very good. But still the port is experiencing difficulty in attracting mother vessels owing to its high vessel related charges and poor draught. The port is located at a distance from Suez route which requires 8 hours more sailing time compared to Colombo Port. Under the circumstances, it is obvious that any major shipping line would prefer Colombo Port. The fact that Cochin Port was not able to attract business even when Colombo Port was in distress is a clear sign of the port's incompetence.

The terminal was constructed with the objective to reduce transshipment from India and also to divert transshipment from Colombo to Cochin. When such is the case a detailed understanding of the competitive strengths of Cochin Port is of paramount importance. It would help Cochin Port and ICTT to devise appropriate strategies to compete.

The key findings of some of the research papers on transshipment ports point out that Colombo Port's major strength is its low cost. One of the research studies has used the ANP model to assess the intra-port and inter-port competitiveness of four major transshipment ports (i.e.) Colombo port Singapore (SIG), Klang (PKG) and Tanjung Pelepas (TPP). The research has thrown light on the competitive strengths of the four ports which is used in this study for further assessment and to draw recommendations.

The ANP is a decision-making methodology and is designed to address complex decision-making problems involving multiple criteria and interdependencies among them. ANP starts with the creation of a decision hierarchy, which is a structured representation of the decision problem. The hierarchy is organized into a network of elements, which may include criteria, sub-criteria, alternatives and objectives. These elements are arranged in a way that reflects their relationships and dependencies. The core of ANP involves making pairwise comparisons between the elements at each level of the hierarchy. Decision-makers assign numerical values, known as 'pairwise comparison ratios', to express the relative importance or preference of one element over another.

The ANP model has identified seven criteria to evaluate intra port and inter port competitiveness, which is presented below (see Fig. 5.1).

The priority scores of each transshipment port's competitiveness factors are shown in Table 5.1. The Port of Colombo ranks first in terms of cost competitiveness (0.197), followed by information systems (0.155). The most competitive aspect for Port Klang is similarly cost (0.239), followed by port amenities (0.158). The most competitive element for the Port of Singapore is information systems (0.190), followed by port efficiency (0.181). Again, the Port of Tanjung Pelepas'

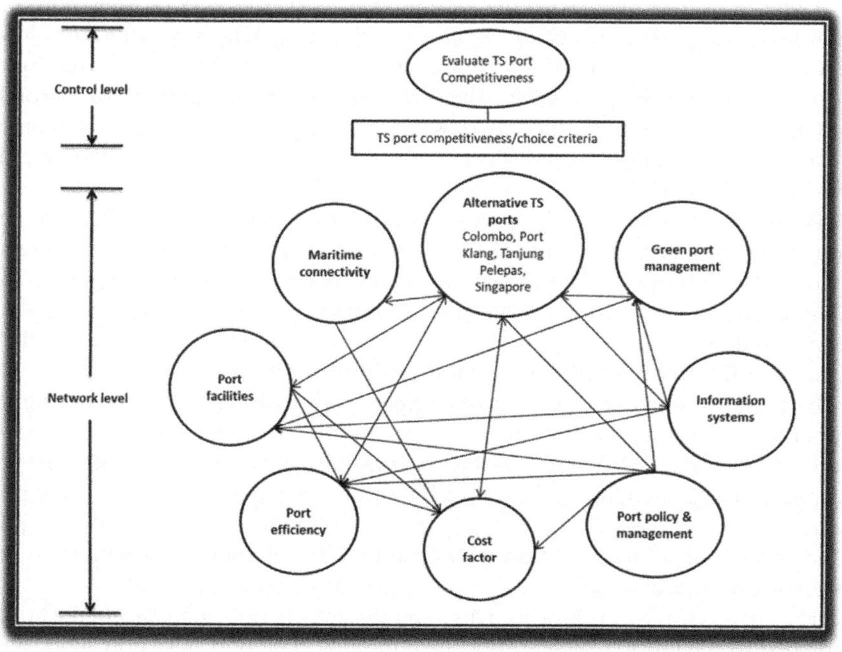

Fig. 5.1. The Transshipment Port Competitiveness Evaluation Using ANP Model. *Source:* Ziaul Haque Munim, Transport Policy, https://doi.org/10.1016/j.tranpol.2021.07.015.

Table 5.1. Intra-port Competitiveness of Colombo, Singapore (SIG), Klang (PKG) and Tanjung Pelepas (TPP) Ports.

Intra-port Competitiveness Evaluation of Each Transhipment Ports.				
Criteria/Port	**Colombo**	**Klang**	**Singapore**	**Tanjung Pelepas**
Maritime connectivity	0.145	0.141	0.101	**0.116**
Port facilities	0.141	0.158	0.178	**0.168**
Port efficiency	0.146	0.137	0.181	**1.175**
Cost factor	**0.197**	**0.239**	0.123	**0.213**
Policy and management	0.144	0.122	0.144	**0.110**
Information systems	0.155	0.124	**0.190**	**0.139**
Green management	0.072	0.078	0.082	**0.079**

Source: Ziaul Haque Munim, Transport Policy, https://doi.org/10.1016/j.tranpol.2021.07.015.

most competitive factor is cost (0.213), followed by port efficiency (0.175) (Munim et al., 2022).

Table 5.2 shows that, with the exception of cost, the Port of Singapore outperforms rival ports in six of the seven variables. Port Klang benefits from lower service costs and is the most competitive in terms of cost (0.168), followed by the Port of Colombo. In most cases, the differences between the Ports of Colombo and Klang are minimal, albeit Port Klang is marginally better than Colombo. Meanwhile, Tanjung Pelepas Port ranks last in all competitiveness metrics (Munim et al., 2022).

Based on the above analysis, it can be inferred that Colombo Port majorly derives its competitive advantage from cost factor. It is the criteria with highest score in the intra-port competitiveness segment. In the inter-port segment also, Colombo Port has got the second most competitive cost structure next only to Klang and is in the second position in terms of connectivity and in all the remaining five criteria, it is in the third position.

It can now be understood that to compete with Colombo Port, it is of utmost importance for Cochin Port to come out with a competitive tariff structure particularly for vessel related charges. It must also be noted that unless the port spends more on dredging, it is impossible to make bigger vessels call at the port. Attracting more vessels, particularly from Colombo is crucial for the port. Therefore, a tariff structure that would support the dredging expenditure as well as attract mainstream shipping companies needs to be worked out.

The port is currently following a discount system where around 80% of the tariff is refunded to the mother vessels after three months. The port may now work out a refund pattern where the refund percentage is tied up to the waiting period. The longer the days, the higher can be the refund percentage. A liberal eligibility criterion may be worked out and may also be extended to include feeder vessels. Adopting target pricing instead of cost-plus pricing may also help the port in developing a

Table 5.2. Inter-port Competitiveness of Colombo, Singapore (SIG), Klang(PKG) and Tanjung.

Pelepas (TPP) Ports

Port/Criteria	**Connectivity**	**Facilities**	**Efficiency**	**Cost**	**Policy**	**Information systems**	**Green Manager**
Colombo	0.214	0.086	0.087	0.145	0.172	0.155	0.078
Klang	0.211	0.095	0.095	**0.168**	0.188	0.184	0.098
Singapore	**0.454**	**0.258**	**0260**	0.098	**0.518**	**0.545**	**0.254**
Tanjung Pelepas	0.121	0.061	0.058	0.089	0.121	0.116	0.070

Source: Ziaul Haque Munim, Transport Policy, https://doi.org/10.1016/j.tranpol.2021.07.015

suitable tariff structure. The vessel related charges have to be determined taking into account the charges at Colombo Port. Fixing a competitive price first and then working backwards to determine the cost would be more appropriate. The number of port calls required to break even needs to be worked out. A tariff structure that balances high dredging cost and refund pattern combined with a suitable investment policy may help the port tide over its adversities.

Strategies may also have to be devised based on the factors the shipping lines use to select a transshipment port. The hub port selection is done based on the network model of shipping lines, which may be either the hub and spoke or relay. The Hub and Spoke network and Relay network are presented below (see Fig. 5.2).

Larger ships travel great distances and make scheduled calls at prearranged transshipment ports in a relay network. With the fewest possible ports of call, it is a combination of several mainline services to improve service coverage. When one mainline vessel receives cargo that is intended for a port that the vessel is not directly serving, transshipment procedures are carried out in relay networks between the two vessels with different ports of call patterns (Kavirathna et al., 2018). Conversely, in a hub-and-spoke network, relatively smaller ships, sometimes referred to as feeder vessels, link transshipment ports over shorter distances with fewer gateway ports. Additionally, there are specialized feeder carriers that primarily run vessels in the hub and spoke network (Kavirathna et al., 2018).

Studies point out that Colombo Port depends more on hub and spoke networks. Cargo from nearby gateway ports are brought through feeder vessels to Colombo. While this brings economies of scale to shipping lines, there is a disadvantage of increased cost through third party feeder operations. It might also result in port congestion. The Colombo Port has also experienced such unprecedented congestion in the recent past, which resulted in increased idle time and freight rates. The Colombo Port is highly dependent on the transshipment market in the Indian sub-continent. Understanding the risk of being highly dependent on the hub and spoke model, the port management is working on enhancing its relay transshipment operations by creating better infrastructure. It is also making

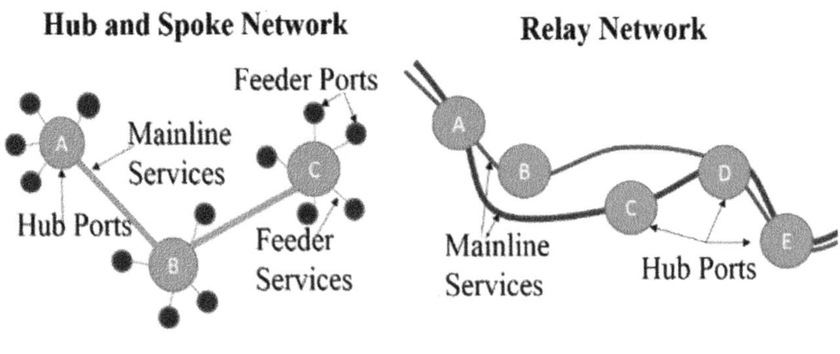

Fig. 5.2. Hub and Spoke and Relay Network Models. *Source:* https://www.sciencedirect.com/science/article/pii/S2092521218300221.

efforts to reduce overall cost, especially bunkering cost, which is currently comparatively higher in Colombo than in the South East Asian competitive hub ports. The port management is also planning to advance its information technology (IT) systems to streamline container operations.

Taking a cue from the above, Cochin Port can step up its capabilities to facilitate relay network liners. The port has good hinterland connectivity, cargo availability is not a problem, the infrastructure at ICTT is good enough to handle 12 lakh TEUs, IT infrastructure is also commendable and the vessel turnaround time has been reduced to less than 24 hours. All this means mainstream shipping firms with relay networks could prefer Cochin to Colombo if competitive rates are offered and enough draught is provided. It is important for the port to equip itself to handle Post Panamax vessels.

Also, it must be noted that though Colombo Port has scored high in terms of cost and connectivity, it is not very competitive in areas such as facilities, efficiency, policy, information system and green management. Perhaps the Cochin Port can work on these areas to compete with Colombo. With policies attuned towards green management, the port can become a sustainable transshipment hub. Steps such as installing recycling facilities for ship garbage and supply of biofuel can be considered.

The ICTT was established utilizing the PPP model, with DP World Ltd being awarded a 30-year contract to develop and run the terminal. Cochin Port is a landlord port. With the highest bid of 33.30% for the revenue sharing, DP World was awarded the contract. When the contract was awarded DP World was operating from the Rajiv Gandhi Container Terminal (RGCT) in Cochin Port. According to the terms of the contract, DP World was only to begin building the ICTT when traffic (volumes) at RGCT reached 400,000 TEUs annually. The bidder was not contractually required to build the ICTT, and after 8.5 years, RGCT was to be returned to Cochin Port Trust if this maximum traffic was not reached within 6 years of the project's award (Manoj, 2016).

Cochin Port Trust then discovered that the bidder could run the terminal for 8.5 years without going over the 400,000 TEU cap, avoiding the contractual requirement to build the ICTT (Manoj, 2015). As a result, Cochin Port Trust asked DP World to migrate from RGCT to ICTT early without tying it to the achievement of traffic at RGCT. DP World requested a departure from the tender criteria, and Cochin Port Trust granted it in exchange for concessions from the rules, which came with a ₹ 40.23 crore financial cost (Manoj, 2015). Among the concessions were

- An eight-year deferment of the 25 percent royalty payable to the port;
- A reduction in the upfront payment to offset the short-term use of existing equipment at RGCT;
- A relaxation of the height restriction at RGCT for the operation of cranes and
- A relaxation of the licence fee for the Q-7 berth.

It is confirmed that there was initially an unequal distribution of risk and incentives between the port and the PPP partner when Cochin Port Trust was forced to grant post-tender concessions to the bidder in order to facilitate an early migration to ICTT. The terminal operated at 35% capacity, so even after migration to ICTT, the additional benefits anticipated from the post-bid concessions could not be realized (Manoj, 2016).

One of the main causes of ICTT's poor performance is the flouting of terms and conditions. Such situations might lead to a lack of dedication or even a lapse in the concessionaire's ability to carry out their end of the bargain. Large sums of money are needed for the development of ports and terminals, and in a nation like India, PPP is used to carry out these projects. In such a situation, major deviations from contractual terms could hinder the development of projects in its true sense.

ICTT is currently only operating at 50% of its full capacity, despite having the ability to load one million TEUs annually (Gateway, 2022). In an effort to help ICTT compete with Colombo, the port has lowered its vessel charges. Additionally, the ICTT operates in a Special Economic Zone (SEZ) that offers financial breaks and administrative convenience on par with other international container terminals. However, none of this mattered, and the revenue shared is not even close to covering the annual maintenance costs of the draught that the Port Trust incurs.

Port choice also requires special attention. Initially, Enayam Port was identified as one of the three sites to be developed into a transshipment hub. But owing to land cost and public displacement issues, the proposed project was cancelled. A new site was identified at the Galathea Bay in the Great Nicobar Island. Likewise before developing the Vallarpadam Terminal into a transshipment hub, the cost of dredging should have been taken into account. The fact that the port is prone to siltation is a major disadvantage and this should have been considered while developing the proposal. Today, the port's inability to operate as a full-fledged transshipment port is majorly attributed to its high cost which is the result of dredging and maintenance. Transshipment port location must be chosen strategically after thorough evaluation of the pros and cons.

6. Recent Developments

The ICTP in Vizhinjam is expected to begin operations in 2024. It is expected to present ICTT intense competition because it is located just a few nautical miles from the port of Cochin. Under the PPP model, the Adani group would construct and manage Kerala's state-owned port of Vizhinjam for the ensuing 40 years. The group will not have to share its profits with the Kerala Government for the first 15 years. Additionally, the project is entitled for Rs. 1,635 crores viability gap funding (26). The Central and Kerala governments will contribute equally towards the fund. All of this could give the port the ability to outperform Colombo port and charge fees for vessels in a competitive manner. Unlike Cochin Port, Vizhinjam has a natural depth which can berth vessels carrying up to 24,000

TEUs because of its natural depth. The port is just 10 nautical miles from the international shipping route and is also at the same distance from Colombo which might fetch more cargo from Tuticorin and Visakhapatnam ports (Gateway, 2022).

Therefore, now, in addition to Colombo, the port of Vallarpadam has to get ready to compete with Vizhinjam also. In response, DP World is working hard to increase the scope and capability of its operations. Six new cranes, including two Shore to Shore mega max cranes, would be installed by the end of 2023. Furthermore, the DP World-owned Cochin Integrated Business Park, which will be the country's first free trade warehousing zone and the first in Kerala to be co-located alongside a terminal, is scheduled to open for business in the first quarter of 2024 (28). It is expected that the two would complement one another to offer outstanding chances for both local and international business.

A group of trade union leaders has requested the shipping minister to grant Cochin Port a viability gap funding of Rs. 450 crores (Kochi port union seeks Rs. 450-cr viability gap funding, 2018) in light of the port's unstable financial situation in the last few months. If granted, it would ease the financial constraints of the port.

7. Conclusion

India is developing a network of ports in strategic locations in an effort to position itself as a transshipment hub. It is anticipated that the three ports of Vallarpadam, Vizhinjam and Galathea Bay will be crucial in turning India into a maritime hub. The Vallarpadam experience shows that infrastructure facilities and good connectivity alone are not enough to compete. Appropriate strategies have to be devised taking into account the competitive environment. For example from the Vallarpadam case, it is evident that though the primary aim of ICTT and Cochin Port was to compete with Colombo and divert transshipment from there, appropriate strategies were not devised based on any SWOT analysis of Colombo port. This study has presented some recommendations taking a cue from the research findings on Colombo port. The study suggests ways to reduce cost as Colombo's competitive strength relies more on its cost effectiveness. The study also highlights that Colombo Port is more likely to experience problems in future because of its over dependence on hub and spoke network and congestion issues. Cochin Port needs to target such grey areas and gear up to compete. The development of Vizhinjam port in close proximity is expected to have major impact on Cochin Port. Unlike Cochin Port, Vizhinjam has got some unique advantages like natural draught and more closer proximity to both international shipping routes and a few other domestic ports in the region. The contractual agreement between Adani group and Kerala Government has conducive terms and conditions, besides the availability of a viability funding from the central and the state governments. All these are bound to create additional pressure on Cochin port. DP World has responded to this by increasing its infrastructure and investing into Free Trade Warehouse Zone adjacent to the terminal, which is expected to attract business. Cochin Port may have to develop a two-pronged approach – one to compete with Vizhinjam Port and the other to compete with Colombo Port. It is apparent that the future is going to be very challenging for Cochin Port.

References

Container News. (2023, January 27). India gears up to build US$5 billion container transhipment project off Bay of Bengal Island. https://container-news.com/india-gears-up-to-build-us5-billion-container-transhipment-project-off-bay-of-bengal-island/

Dasgupta, A., & Thakurta, P. G. (2018, May 29). How a change in shipping policy will cripple India's largest state-owned port. *The Wire*. https://thewire.in/economy/how-a-change-in-shipping-policy-will-cripple-indias-largest-government-owned-port

Ducruet, C. (2020). The geography of maritime networks: A critical review. *Journal of Transport Geography*, *88*, 102824.

Fox, L., Hogan, R., & Campbell, N. (2019). Influential article review-understanding transshipment hub port selection criteria. *American Journal of Management*, *19*(6).

Gajanayaka, H. G. A. R., Jayaweera, N. D., & Subasinghe, L. U. (2015). *Design and development of a novel re-configurable fixturing system (No. 2015-01-2495)*. SAE Technical Paper.

Gateway, M. (2022, April 4). Record transhipment traffic at ICTT Vallarpadam—Maritime Gateway. Maritime Gateway - South Asia's Premier Shipping and Logistics News Portal. https://www.maritimegateway.com/record-transhipment-traffic-at-ictt-vallarpadam.

Joseph, A. T. (2021, March 12). *How India's first container hub terminal failed*. https://www.fortuneindia.com/enterprise/how-indias-first-container-hub-terminal-failed/105292

Kavirathna, C. A., Kawasaki, T., Hanaoka, S., & Matsuda, T. (2018). Transshipment hub port selection criteria by shipping lines: The case of hub ports around the Bay of Bengal. *Journal of Shipping and Trade*, *3*(1). https://doi.org/10.1186/s41072-018-0030-5

Kochi port union seeks Rs. 450-cr viability gap funding. (2018, March 12). *Business Line*. https://www.thehindubusinessline.com/economy/logistics/kochi-port-union-seeks-rs-450-cr-viability-gap-funding/article64138024.ece

Kumar, B. V. S. (2023, April 11). Terminal handling fee hike in Kochi worries EXIM trade. *Business Line*. https://www.thehindubusinessline.com/economy/logistics/terminal-handling-fee-hike-in-kochi-worries-exim-trade/article66723822.ece

Lam, J. S. L., & Yap, W. Y. (2008). Competition for transhipment containers by major ports in Southeast Asia: Slot capacity analysis. *Maritime Policy & Management*, *35*(1), 89–101.

Manoj, P. (2015, December 20). Cochin Port flouted own tender terms, gave sops to thrust ICTT on DP World: CAG. *Mint*. https://www.livemint.com/Politics/Ia1otVmeCKBrEu7zfkGANL/Desperate-Cochin-Port-flouted-own-tender-terms-and-gave-so.html

Manoj, P. (2016, July 21). Can India's container transhipment hubs take on Colombo? *Mint* https://www.livemint.com/Politics/zUkTONWkFy8IczoI0rcBgK/Can-Indias-container-transhipment-hubs-take-on-Colombo.html

Mishra, A. (2023, October 20). *3 transhipment ports that are quickly taking shape in India: Analysing their impact, opportunities and challenges*. India Infra Hub. https://indiainfrahub.com/ports-aviation/3-transhipment-ports-that-are-quickly-taking-shape-in-india-analysing-their-impact-opportunities-and-challenges/

Munim, Z. H., Duru, O., & Ng, A. K. (2022). Transhipment port's competitiveness forecasting using analytic network process modelling. *Transport Policy*, *124*, 70–82. https://doi.org/10.1016/j.tranpol.2021.07.015

Press Trust of India and Business Standard. (2015, November 21). First ship of container ship service calls at ICTT. https://www.business-standard.com/article/pti-stories/first-ship-of-container-ship-service-calls-at-ictt-115112100613_1.html

Pti. (2011, February 11). PM inaugurates Kochi container terminal. *Business Today*. https://www.businesstoday.in/industry/infra/story/cochin-port-vallarpadam-island-19679-2011-02-11

Pti. (2016, May 2). Impressive performance by Vallarpadam ICTT. *The Times of India*. http://timesofindia.indiatimes.com/articleshow/52080991.cms?utm_source=contentofinterest&utm_medium=text&utm_campaign=cppst

Wahyuni, N. M., Surahman, I. G. N., & Yasa, P. N. S. (2020). The effect of service quality on customer loyalty mediated by customer satisfaction in tourism villages in Badung Regency. *Jurnal Ekonomi & Bisnis Jagaditha*, *7*(1), 46–52.

Chapter 6

The Challenges and Socioeconomic Impacts of Proposed Road Development to Gunung Mulu National Park, Sarawak, Malaysia: A Case Study of Local Community Perspectives on Sustainable Tourism

Mohamad Syahrul Nizam Ibrahim[a], *Shazali Johari*[b], *Amirah Sariyati Mohd Yahya*[c], *Rosmiza Mohd Zainol*[d] *and Suziana Hassan*[d]

[a]Universiti Malaysia Sabah, Malaysia
[b]Universiti Putra Malaysia, Malaysia
[c]Universiti Malaya, Malaysia
[d]Universiti Kebangsaan Malaysia, Malaysia

Abstract

Gunung Mulu National Park (GMNP) is one of the two protected areas with UNESCO World Heritage Site (WHS) status in Malaysia. Until now, this area can only be accessed by tourists via air and water transportation only. Recently, the government has proposed the construction of a road connecting GMNP to several areas, including Miri City through a high-impact infrastructure project. However, this project might instigate the potential benefits and challenges in terms of tourism development and community well-being. As custodians of the park, their support for this initiative needs to be dismantled so that the management of the national park can still be implemented holistically and does not jeopardise the current UNESCO status of WHS. WHS is UNESCO-designated area of cultural, historical, scientific, or other significant value, legally safeguarded by international agreements, and preserved for the benefit of future generations due to its exceptional value to humanity. Thus, this study aims to examine challenges

The Emerald Handbook of Tourism Economics and Sustainable Development, 99–118

Published under exclusive licence by Emerald Publishing Limited
doi:10.1108/978-1-83753-708-220241006

and socioeconomic impacts of the proposed road among key informants' perspectives at Long Terawan Village, Long Iman Village, Batu Bungan Village and GMNP Headquarters. The study was conducted via in-depth interviews with 10 community members residing in the settlement, including a tribal chief, boatman, lodging operator, park guide and farmer, all of whom are professionals and representatives of the local community. They were designated as key informants on account of their extensive engagement and development within the community. Through a thematic analysis, their perspectives on the proposed roads to be built in the area were elucidated. The study offers pragmatic understanding of socioeconomic impact assessment, which could inform strategic decision-making by incorporating information regarding potential benefits and challenges of the proposed road construction to an isolated protected area.

Keywords: Road development; Mulu; Orang Ulu; Borneo; livelihood; Heart of Borneo

1. Introduction

UNESCO designates Gunung Mulu National Park (GMNP) as a World Heritage Site (WHS) on the basis of its fulfilment of all four World Heritage criteria (UNESCO World Heritage Centre, 2019). Nonetheless, maintaining the GMNP status presents challenges, as any compromise in its criteria – resulting from a failure to underscore the significance of the location – could compromise the possibility of retaining the designation. Both internal and external factors, including but not limited to natural disasters, pollution, poaching, uncontrolled urbanization human development and park management, have the potential to endanger a UNESCO site. Threats that have the potential to cause damage to a site may be imminent dangers that have been proven to be imminent threats or potential dangers (World Heritage in Danger: Compendium II, 2009). A concrete example of this can be seen in the GMNP communities of Penan and Berawan, whose way of life is intricately tied to the forest and who are frequently in conflict with the state government over deforestation. Evaluating the tensions that emerge between stakeholders and nature as a result of human development is vital for the sustainability of biodiversity (Ibrahim, Assim, Johari, Mohammad, Afandi, & Hassan, 2023). Human development, defined as the process of increasing the rights and opportunities of individuals while also enhancing their standard of living, is also susceptible to criticism. With this understanding, the significance of empathy becomes more prominent as the central objective in enabling communities to develop resilience during times of crisis and mending the myriad of factors that rift them apart (Berardi et al., 2020). Effectiveness wise, the communication between the government and the local community that fosters support and comprehension for biodiversity conservation efforts at GMNP must be enhanced. Consequently, to ensure the long-term sustainability of the locals' welfare, it is imperative to tackle the grassroots concerns that form the foundation of the GMNP's effective biodiversity framework.

GMNP is situated in a highly isolated region far from major populated areas; access is restricted to air or water transportation only. In contrast, the government has recently incorporated the GMNP region into the construction proposal for Phase 2of the Sarawak-Sabah Link Road (SSLR), which will link Miri with Limbang and Lawas. Phase 2of the Lawas project, which runs from Simpang Gelugus to Long Lopeng, will span 326 kilometres and cost MYR7.68 billion to connect GMNP to Long Merarap (Tawie, 2022). As part of this 160.5-kilometre single-lane highway, 13 minor bridges, nine major bridges and one passenger jetty terminal, the government is assured that this development project will improve the local economy and society, thereby fostering prosperity for the populace (Kiew, 2021). However, this project might instigate the potential benefits and challenges in terms of tourism development and well-being to the local community. As custodians of the park, their level of support for this initiative needs to be dismantled so that the management of the national park can still be carried out holistically and does not jeopardise the current UNESCO status of WHS. Thus, this study aims to examine challenges and socioeconomic impacts of the proposed road among key informants' perspectives at affected area of Long Terawan village, Long Iman Village, Batu Bungan village, Sungai Melinau village and GMNP headquarters.

2. Literature Review

2.1 Anthropogenic Activities as a Threat to Natural Resources Conservation

Natural resources comprise all substances and elements that can be extracted from any region of the cosmos for human utilization. These are the sources of light, gravitation, precipitation and mineral deposits in the outer environment. This includes both wild and domesticated flora and animals within the ecological domain, as well as human resources (Cook et al., 2013). The primary challenge is the 7.9–10.3 billion global human population that is projected to increase by 2050, according to the United Nations (IPCC, 2014). Consequently, urbanization ensues, exacerbating the strains on natural resources, generating substantial quantities of water contamination, depriving of hygienic water and benefiting electricity in an unjust manner (Medek et al., 2017) and food poverty in metropolitan and villages areas persists (Howes et al., 2017). Both developed and developing countries encounter difficulties concerning the equilibrium between safeguarding the global environment and promoting economic progress (Hassan et al., 2019). Charfeddine (2017) posits that environmental quality is diminished via the ecological footprint as a result of trade liberalization electricity consumption, financial development and urban growth. In accordance with the ecological footprint's dimension, distinct forms of globalization exert distinct impacts on the ecological footprint process. In accordance with the ecological footprint's dimension, distinct forms of globalization exert distinct impacts on the ecological footprint process. There are significant global difficulties that impede the effective management of water resources. Urban residents consume a significantly greater quantity of food, electricity and renewable resources compared to individuals residing in rural areas (Mondal & Palit, 2021).

A number of anthropogenic activities threaten the sustainability of natural environments, including deforestation, soil erosion, water scarcity, over-exploitation resulting in the spread of disease and the introduction and eradication of invasive species (Merlyn Keziah & Subathra Devi, 2017). Additionally, the emergence of many types of contaminants has been found to have detrimental impacts on the climate (Rockström et al., 2009). The rate of climate change is experiencing a significant acceleration as a result of heightened human activity, such as industrialization and modern farming methods (Mahato, 2014). It can also be argued that, in terms of the adverse effects of human activities, global climate changes are at the forefront and must be addressed immediately in order to maintain equilibrium; in this regard, biological climate changes can have a significant impact on the fight against climate change. A decline in ecological diversity has led to a concerning depletion of both terrestrial and aquatic habitats, encompassing both plant and animal life, thereby affecting the overall stability of the ecosystem (Arora et al., 2018). Regrettably, according to the WWF (2014), a significant proportion of global biodiversity, amounting to 52%, has been eradicated. An estimated 30% (interval of uncertainty: 16–50%) of species have been threatened with or driven to extinction on a global scale since 1500, according to biodiversity experts (Isbell et al., 2023). Loss of biological diversity is attributable to a wide variety of human activities. (Bitencourt et al., 2016). Sustainable development is a viable approach to mitigating the ongoing loss of biodiversity, even though its implementation requires consideration of the intricate facets of human welfare.

2.2 Leveraging Digital Tourism to Promote Rural Tourism

The extensive range of choices provided by digital technologies – web pages, search engines, virtual worlds, social media, the Internet, e-mail, smart phones and more – influences the behaviour of consumers (Opute et al., 2020). While the concept of the Internet of Things (IoT) embodies a paradigm in which commonplace objects utilize technology to detect, network and process information in order to establish efficient connections with other devices and services through the internet in order to accomplish objectives (Balaji & Roy, 2017). In the current era, both are crucial in the context of sustainable tourism, which encompasses the entire industry from small- and medium-sized businesses to corporations. It is indisputable that sustainable tourism contributes to the economic development of a nation (Rather et al., 2019). New businesses and non-governmental organisations (NGOs) are powered by the growth of new tourist destinations (Ali, 2023). By leveraging digital competencies to establish consumer confidence in the developing products, the community must adjust to increase the visibility of tourism enterprises. Notwithstanding the governmental initiatives implemented, it remains evident that further effective and coordinated efforts are required to reinstate tourists' trust, ensure the sustainability of the tourism industry, foster the growth of domestic tourism, furnish tourists and businesses with unambiguous information, fortify international and intra-country cooperation and construct a more resilient and sustainable tourism sector (Barykin et al., 2021). According to Camilleri (2018), adventure and rural tourism are the predominant types. Adventure tourism consists of tourists participating in or observing winter sports or other similar activities, frequently in rural locales. In contrast, rural tourism encompasses a variety of pursuits, such as visits to farms (fruit harvesting, for instance), excursions to lakes and mountains, stays in rural retreats and parks and explorations of the countryside. However, tourism is not restricted to rural regions; digitalization can also be used to emphasise the distinctive cultural aspects of the local community (Arora & Sharma, 2020). This can potentially enhance the allure for tourists to visit the region and engage with the indigenous community.

2.3 Political Governance Influence on Sustainable Development

Divergent perspectives prevail regarding policy recommendations, conceptualization of the subject matter and theoretical formulations of governance, despite its universal recognition as an essential concept (Kardos, 2012). Consequently, no single entity can lay claim to the definitive definition of governance. Transparency, participation, accountability, efficacy and coherence are fundamental components that are essential for a comprehensive comprehension of governance, according to the Commission of the European Communities (2001) in its definition of good governance principles. Strategically positioned on the global agenda, sustainable development has been the subject of extensive deliberation, censure and accusations of scientific irresponsibility and political manipulation over the past several decades. Its influence has been both substantial and enduring, resonating at the local and global levels. As Strandeneas (2011) notes, governance systems are required at all levels to direct the implementation of all sustainable development-related concepts. The most common definition of sustainable development is 'development that satisfies the needs of the present without compromising the ability of future generations to meet their own needs' (United Nations, 1988). Governance is present in every component of the strategy, and according to Kemp et al. (2005), the greatest challenge is to ensure that multi-player governance regimes provide a reasonable coherence of vision and commitment and embody the capacity for sustainability-oriented coordination, direction and re-direction. Given that sustainable development extends far beyond environmental management and sectoral policies are insufficient, efforts to strengthen policy integration are urgently required. Decentralization measures aimed at enhancing efficiency are necessary in this regard. Additionally, objectives and actions must be more effectively integrated both 'vertically' (across subnational and international levels) and 'horizontally' (among the entities of a specific stratum of government). Due primarily to inadequate governance of human activities, environmental resource management causes inefficiencies and productivity issues in the majority of developing nations. In order to attain sustainable development goals and promote environmental quality, adequate legislative measures are crucial (Azam et al., 2021). Hence, the efficacy of sustainable development heavily relies on the involvement of institutions, encompassing both formal entities like government agencies and legal frameworks, as well as informal factors like cultural norms and social practises (Hunjra et al., 2023).

3. Methodology

3.1 Study Area

GMNP is a UNESCO World Heritage Site-designated national park situated in the Marudi Division of Sarawak, Malaysia (Fig. 6.1). This approximately 52,864-hectare GMNP area is a mountainous region in northern Sarawak that is isolated from other developing areas. It is situated between the headwaters of the Medalam River, a tributary of the Limbang River and the Tutuh River. It serves as a corridor through the Heart of Borneo, linking Brunei's forests to the remainder of the island. It is situated along the Brunei-Sabah-Sarawak-North Kalimantan Transboundary Landscape, which the WWF (2017) designates as one of Borneo's six priority landscapes with protected areas. The preponderance of GMNP's indigenous Penan and Berawan communities reside in neighbouring villages and GMNP. In comparison to the 18,632 visitors that visited GMNP in 2015, the number of visitors in 2019 was 21,022 (Sarawak Forestry Corporation, 2020). The data indicate that the quantity of international travellers is nearly twofold that of domestic visitors. However, since the pandemic struck in 2020, the management has been unable to disclose the data due to the slow progress of the tourism industry.

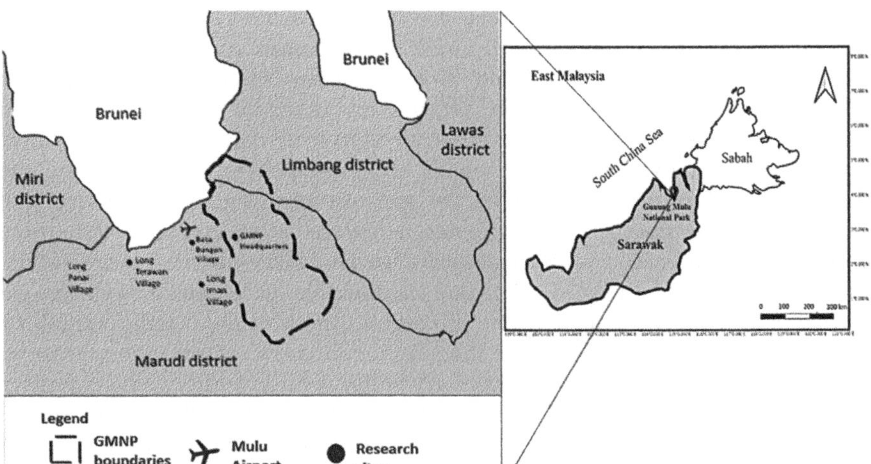

Fig. 6.1. The Distribution of Research Sites in GMNP, Sarawak, Malaysia.

3.2 Research Design

The present study employed a qualitative approach to investigate the subject matter. Specifically, data were collected through personal interviews with 10 individuals who held various roles within the local community, including professionals, tribal chiefs, boatmen, homestay operators, park guides and farmers. These participants were selected from the Long Terawan Village, Long Iman Village, Batu Bungan Village, and Sungai Melinau Village, utilising a snowball sampling technique. They were asked on perceptions towards the road proposal in the area that might instigates benefits and challenges to their well-being. Therefore, these individuals were designated as key informants based on their direct experiential knowledge and active engagement within the community. The narratives presented by the authors offer a qualitative dimension that encompasses meaning, relevance and an in-depth understanding (Tashakkori et al., 2020). The utilization of a sample size of 10 key informants is deemed appropriate for the qualitative research methodology, given its inherent characteristics that necessitate a limited number of participants. Hammarberg et al. (2016) argue against the use of a large sample size due to the potential for increased data difficulties, which may compromise the ability to conduct rigorous analysis with a comprehensive scope. The utilization of key informant voices serves to attain data saturation, enhance external validity and/or ensure information redundancy (Onwuegbuzie & Leech, 2007). The selection of this qualitative longitudinal research approach for the present study was based on its appropriateness for investigating phenomena associated with pandemics or disasters. Such studies often involve dynamic and distinctive contexts that require in-depth and thorough portrayals of the human experience. Consequently, it is a valuable method for investigating the adaptation and impact of the global COVID-19 pandemic (Terzis et al., 2022). Finally, the qualitative data trustworthiness was evaluated where the authenticity and exhaustivity of data was concerned and a comprehensive look at the spatiotemporal context of an experience was taken. It should be consistent with the objective, epistemological stance and design of the study, as well as the extent to which threats of research reactivity and bias are adequately managed (Lietz & Zayas, 2010). Table 6.1 presented to show the characteristics of 10 key informants interviewed in different sessions.

In this study, the reflexivity method was utilized to document memos or field notes during the data collection process. This approach aimed to enhance the understanding of the importance of each interview and observation session within the study area. Additionally, it sought to capture the emotional aspects that key informants sought to convey regarding the domain of local community well-being. The option of reflexivity as the strategy to improve the 'trustworthiness' of the study has allowed social researchers to express their experiences, biases and emotions throughout the research process, as well as inspired them to contribute the article by examining the emotional impact on the study, personal self and professional (Yong et al., 2019).

Table 6.1. List of Key Informants Involved in In-depth Interviews in Different Sessions (n = 10).

Key Informant	Type	Study Site	Gender	Occupation
K1	Professional	GMNP Headquarters	Female	Research liaison officer
K2	Professional	GMNP Headquarters	Male	Wildlife officer
K3	Local community	Batu Bungan village	Male	Tribal chief
K4	Local community	Long Terawan village	Male	Former tribal chief
K5	Local community	Sungai Melinau village	Male	Homestay operator
K6	Local community	Sungai Melinau village	Female	Homestay operator
K7	Local community	Long Terawan village	Male	Park guide (freelance)
K8	Local community	Long Iman village	Male	Boatman
K9	Local community	Batu Bungan village	Male	Boatman
K10	Local community	Sungai Melinau village	Male	Farmer

Note: Authors own source.

4. Data Analysis

The data collected from the narratives were subjected to thematic analysis in order to get insight into the perspectives of key informants about road project initiatives in the GMNP. The data underwent transcription, coding, thematic analysis, validation and interpretation following the methodology outlined by Talib (2019). The utilization of computer-aided qualitative data analysis, more notably ATLAS.ti 8 Windows, played a crucial role in facilitating the process.

5. Result and Discussion

5.1 Community Perspectives on Proposed Road Construction

The proposed High Impact Infrastructure project by the Sarawak government involves the establishment of road connections between Miri-Marudi, Marudi-Mulu

(Kuala Melinau) and Long Panai-Long Lama. The implementation of this project will enhance the level of accessibility between Mulu and surrounding regions. This project facilitates convenient access for residents of Miri to essential amenities, including hospitals, schools and grocery stores, at all times. Despite this, a considerable number of individuals oppose the project on account of the challenges that will emerge from its establishment.

It is sufficient, in my opinion, as it is. There is no need for a road. (K9)

We don't need roads. (K8)

Some residents of the study area believe that this road development project will improve accessibility and make it simpler for them to purchase necessities in the city of Miri, eliminating the need to rely on the extremely expensive goods available in Mulu. Consequently, the road's construction will concurrently reduce their reliance on air and water transportation. It is evident that the local populace would benefit from this endeavour. Local communities frequently benefit from road construction by means of reducing travel time, increasing population, enhancing service accessibility or stimulating economic activity in underdeveloped regions (Ortega-Reig et al., 2023).

Both K5 and K6 hold the identical viewpoint that the road construction has a positive impact on their socioeconomic conditions. This could potentially be attributed to the surge in tourist arrivals, which has enabled locals to augment their monthly earnings via the accommodation sector. Nevertheless, their perception of the potential for business competition involving entrepreneurs from outside the local community does not significantly burden them. This demonstrates their lack of understanding regarding the carrying capacity of tourism, which must be determined based on the following criteria: physical context – the maximum number of tourist arrivals that a particular tourism area can accommodate; economic context – the level of acceptable change between tourist arrivals and the local economy; social context – the point at which visitor satisfaction and the crime rate are equivalent and biophysical context – the capacity of the natural environment to withstand tourist interference (Attallah, 2015). Over tourism and exceeding carrying capacity may result from an excessive number of visitors; therefore, tourism stakeholders should watch out for the need to strike a balance (Wall, 2020). Upon examining their perspectives, it becomes evident that relevant entities must educate tourism stakeholders in this GMNP regarding the practical implementation of sustainable tourism and economic principles that foster a place attachment to this UNESCO site.

5.2 Socioeconomic Impacts From the Development

Boatmen, K8 and K9 share the conviction that the proposed road construction does not warrant their highest priority. This may be because their livelihoods as river boat operators are highly reliant on tourists requiring their services in this tourist area. Their means of subsistence will be adversely affected by the road's construction, as future tourists will be able to access certain regions of Mulu via four-wheel drive vehicles, owing to the bridges that will be built as part of the infrastructure project. Nonetheless, the boatman community in Bako National Park (BNP), Sarawak, continues to provide transportation services for tourists departing from the terminal jetty situated in Bako Village and entering the national park vicinity (Mohamad Junaidi & Hashim, 2023). This demonstrates that while the state government has expressed interest in constructing a road to connect the village area to the national park, it continues to rely on water conveyance to ensure the boatman's livelihood. Thus, the situation at BNP demonstrates the significance of the boating community to the viability of sustainable tourism in this protected area. While the boat service will continue to be utilized by tourists, its availability will be restricted to specific locations, particularly in the upstream region of GMNP where climbers are pursuing the Summit and Pinnacles packages. Concurrently, their income will decrease in comparison to the period preceding the road's construction.

It's not that we don't want to, but it's also not that we really want to. Indeed, there has been a trail or unpaved road connecting this Melinau area for a long time. It begins with Long Iman-Long Lama-Long Bedian-Miri. I don't want a paved road from here (Melinau) to Miri. Our current land is insufficient. It has already been conquered, and any developments will simply make us more stuck. (K3)

The economic well-being of Long Terawan village residents will also be impacted by this proposed road, in addition to the strain it will place on the inhabitants of Batu Bungan and Long Iman Village. This is due to the fact that the road will impede water transportation operations from Miri to GMNP. Long Terawan Village and Marudi will be linked via the Highland Development Agency-proposed road, which is also intended to link the inhabitants of GMNP and Mount Buda with Miri (Louis, 2022). Since before, the Long Terawan community has sustained itself financially through the operation of small rural enterprises that have garnered the attention of foreign visitors. The village serves as an intermediary destination for tourists to visit longhouses and interact with locals prior to continuing their journey to main destination of GMNP. Communities are thus able to sell them locally produced goods, including preserved fish, rice wine (an alcoholic beverage) and other agricultural products. Despite Long Terawan Village's advantages as a transit point, the village's characteristics as an independent tourism attraction present a greater obstacle (Tang et al., 2017). Based on the existing information, it can be inferred that the cultural aspects of the Berawan ethnic group inhabiting in this transit area have not been included in any tour packages offered by the travel agency. Furthermore, the capacity of travel agencies to effectively support the implementation of this cultural product is in question. Therefore, Long Terawan village is perceived as devoid of resilience on account of its inadequate capacity to support tourism via hospitality operations and agricultural commodities. Any substantial alterations to the communities of Long Terawan would merely place them under undue strain in the absence of a thorough examination and suitable guidance to enable them to adapt to imminent infrastructure development, such as the construction of roads and bridges. As an illustration, the monthly earnings of boat operators, who formerly served as the primary transporters for fresh food and retail merchandise, as well as tourists, will be impacted. Regarding the potential socioeconomic impacts of the construction of this road, the population of the area faces a number of major challenges that must be surmounted. The Penan and Berawan communities have undergone a significant life transformation ever since the region they inhabit was designated a national park until it received UNESCO World Heritage Site status. The improvements to the lives of indigenous peoples and their descendants that result from corporate transformations should be in accordance with their rights with regard to traditional lands and resources (Flemmer & Schilling-Vacaflor, 2016). In light of the potential consequences of the proposed road construction, it is preferable to maintain air travel as the primary mode of access to this GMNP, where residents are, in fact, provided with a reasonable subsidy for their airfare from Mulu to Miri Airport.

On the contrary, K4, a former ethnic leader in Long Terawan, holds a contrasting perspective regarding the suggested road development.

I agree if this area of Mulu is connected with a road. It is very difficult to get to Miri because there is no road to get there. Based on the map shown to us, from here there will be a road to Marudi then to Long Panai and straight to the top (Miri). This project should take four to five years. The project is a project by the Chief Minister so that we in this village will benefit if there is such a road. (K4)

Consistent with the aforementioned viewpoint, K7, a native of Long Terawan Village, supports the development of a road that would link their community to the urban centre of Miri. Presently, he is married and resides in Miri City, where occupational prospects are more assured. He is a freelance park guide. While the road construction project does have its drawbacks, it is evident that it provides some local residents with positive aspects. In order to ensure the welfare of the populace by enacting policies that are deemed to strike a balance between economic, social and environmental requirements, however, it reverts to the role and power of the governing authority as an intermediary (Zinda, 2017).

5.3 Digital Initiatives as the Key for Community Resiliency

When contemplating the feasibility of this development, it is imperative that nearby inhabitants be ready to confront any eventuality, including the likelihood that a road will be constructed in their vicinity. Consequently, they must be resilient in the face of the forthcoming tourism challenges. Presently, they are perceived to be somewhat less resilient whereby, during the COVID-19 pandemic, most of them were financially destitute forcing them to depend on their own crops and the river as their primary food source. Prior to the pandemic, their reliance was primarily on tourism-related businesses, which were perceived to be more lucrative than agriculture. They have learnt from the pandemic that certain skills, such as e-business marketing for agricultural products and handicrafts, require further development. At present, tourism has recovered; however, Long Terawan village remains less well-known than GMNP itself, despite serving as a transit point for visitors arriving by water transport from Miri to GMNP. This is due to the fact that the majority of visitors to GMNP depart from Miri via direct flights, which gives them approximately 30 minutes of time compared to the hours-long boat ride. As a result of widespread disinterest regarding natural and cultural heritage values, a considerable number of visitors are hesitant to utilize water transportation. The issue at hand pertains to the inadequate promotion of Long Terawan village's ethnic cultural heritage of Berawan, as well as the aesthetic view that encompasses the mountainous region, including Mount Api and Mount Benarat. The provision of complimentary Wi-Fi in this long house by the state government Sarawak Rural Broadband Network (2021) presents an advantage for the inhabitants of Long Terawan village relative to other villages. Consequently, they ought to make the most of this benefit in a variety of ways. In contrast to Long Iman village, Batu Bungan village and Sungai Melinau village, which are perceived to have a greater emphasis on forest products, this community offers a greater variety of agricultural products. To effectively market their products in the current tourism industry, they must acquire proficiency in these digital competencies. Rural cultural heritage presents remarkable prospects for the integration of digital components and the collaborative development of innovative visitor experiences, thereby fostering a positive feedback loop that encourages repeat visits and further experience (Singh et al., 2023).

Most of the inhabitants of Long Iman Village and Batu Bungan Village, who are Penan and reside in closest proximity to GMNP, will experience the same pressure: a sense of insecurity stemming from the potential influx of individuals whose numbers may surpass the villages' capacity to accommodate. Regarding digital requirements, however, this is regarded as the responsibility of the government and not the individual. The aspiration for digitalization has had a substantial influence on the economic and tourism sustainability of GMNP, despite the fact that the majority of its inhabitants are semi-nomads. As residents in close proximity to GMNP who are consequently frequented by tourists, the government must give precedence to the digitalization of their cultural assets, specifically their intangible heritage, which is experiencing an increasingly rapid decline due to lifestyle changes. The primary menace, exemplified by the Oroo' secret sign language employed within the forest, has practically disappeared and is presently regulated solely by the elderly (Ibrahim et al., 2021). The implication is that promotion and immediate implementation of this digital documentation initiative are required. The government presently promotes primarily the natural heritage of GMNP; efforts to preserve cultural heritage, including the Oroo' language, are limited. The Oroo' language, which symbolises the Penan people's forest navigational wisdom, ought to be leveraged as a selling point for this World Heritage site's tourism opportunities. Moreover, game applications for smartphones can be developed to pique the interest of tourists, particularly the younger and middle-aged demographics, in interactive language learning, in addition to imparting information to them. Based on Mwangonde et al. (2021), the cultural awareness of the participants increased as a result of playing the game, demonstrating that games can aid in the preservation of culture. Poor communities will acquire new opportunities as a result of the preservation and development of their cultural assets, which in turn encourages tourists to visit GMNP. Empirical evidence demonstrates that enthusiastic and serious video game engagement enhances the cognitive, motivational, affective and social competencies and conduct of the player, thereby bolstering the case for video game usage as a method of cultural preservation (DaCosta & Kinsell, 2023).

It is apparent that to attain sustainable development in this tourism sector, it is critical to prioritise community resilience via a digital strategy. This endeavour necessitates the cooperation of multiple stakeholders, including the government, local communities and tourism operators. The specified mediation is the construction of this road, which will significantly alter the way of life of the local populace. There must be an appropriate platform for the community to engage with the government to find the optimal resolution to long-standing problems in this GMNP.

5.4 Challenges for Environmental Degradation

The implementation of the road infrastructure is anticipated to exacerbate the environmental degradation. The presence of roads connecting Mulu to other regions will also result in a rise in population. There will also be pressure on the usage of natural resources in Mulu, resulting in encroachment, poaching, pollution and so on. (K2) What is the guarantee that the companies involved in the road's construction would not take timber while carrying out the project? Make certain that the construction job is done without harming the current environment, which could endanger us. (K10)

While the park and wildlife are being protected, the road connecting to Miri and the rest of the area is a bad idea. It will primarily threaten biodiversity because it will make it easier for outsiders (wildlife traders) to enter the forest and hire locals to gather and transport species directly. The wildlife trade is a major issue for us. They are not involved in the wildlife trade, despite the fact that the local population hunts and the park is mostly for subtenants. Most local communities are also opposed to the proposed route because they are concerned about competition from outsiders. They now have a monopoly on transport in this area. Although some prefer the road to Mulu, it does not immediately lead to here (consider at some points). It should be since it has a direct impact on how people arrive. One solution to this difficulty is to have the road finish at the Tutoh River. The remaining distance to Mulu will be travelled by boat. This has been extensively addressed in the Berawan community. Many homeowners support the road, although many are concerned about the park's encroachment. Because they are active in tourism, they are aware of its effects on their livelihood. (K1)

In spite of objections towards the suggested road development, it is the governing bodies that ascertained the most efficient strategies for managing the situation in order to safeguard the community's welfare. The authority should be aware that locals might have tendency to refuse to cooperate or participate in any conservation activities, which it could pose a direct threat to the protected area (Holmes, 2013). Future biodiversity degradation costs incurred as a result of the road's construction in this area are likely to be the most significant in comparison to road construction expenses. From an economic standpoint, the government should consider this cost. Connecting the mega faunas that are natives to GMNP, habitat fragmentation caused by this construction must be studied in greater detail by relevant parties. As a result, the government should consider increasing funding for habitat corridor initiatives. The implementation of habitat corridors is observed in numerous developed nations, including the United States, with the aim of connecting national parks in North America. This initiative has the potential to substantially mitigate the rate of species extinction, particularly among mammals (Newmark et al., 2023). Nonetheless, the ethnography exhibits variations as a result of several factors, one of which is the community's distinct consciousness regarding the conservation of biodiversity. The Ministry of Natural Resources and Environment of Malaysia (2016) reports that there remains a significant lack of public awareness regarding the conservation of biodiversity in Malaysia. Hence, in the event that a habitat corridor is constructed in the affected area despite the significant concern, the issue of assurance regarding the absence of wildlife mortality resulting from human-wildlife conflicts and exploitation arises.

GMNP is widely recognized as a component of the Heart of Borneo Initiatives, which link the national parks of Sabah, Brunei, Sarawak and Kalimantan (Indonesia) (WWF, 2017). The area is characterized by its transboundary nature, serving as an ecological corridor that links various species of wildlife that flourish in the region, including threatened one such as Borneo orangutans, Borneo pygmy elephants, hornbills and Muller's gibbon, among others (Keong & Onuma, 2021). The potential disruption of biodiversity in GMNP could have a detrimental impact on its UNESCO World Heritage Site status, potentially diminishing the appeal of the destination to international visitors (Ibrahim, Assim, Johari, Mohammad, & Hassan, 2023). When considering road development, there is considerable potential for multiple stakeholders; however, ultimately, it will revert to the image of the GMNP, which is associated with tourists' motivation. It is inconsequential whether the populace has convenient road access; however, the decline in tourism can be attributed to a scarcity of visitors, which impacts the average income of individuals employed in the tourism industry.

While the development of this road may be perceived as detrimental to the environment, it also contributes positively in terms of carbon footprint reduction. Connecting the districts of Limbang and Lawas to Miri via road without requiring a second traversal of Brunei, this certainly presents a tremendous opportunity for country. Consequently, this can reduce reliance on air travel which leave a substantial carbon footprint annually. A comparable circumstance has arisen in which Bruneians are no longer obligated to traverse the Limbang district of Sarawak, Malaysia; instead, they are connected to the Bruneian mainland and its semi-exclave of Temburong via the Temburong Bridge, which spans Brunei Bay, and it reduces vehicle carbon emissions and travel time compared to the previous route. (Kiew, 2022).

In order to regulate the carbon footprint in GMNP, which is projected to increase dramatically, if the government wishes to construct this road as well, it should enact policies mandating the use of hydrogen fuel cell vehicles or electric vehicles for commercial vehicles such as buses and trucks in specific zones. Additionally, subsidies are utilized, as is the case in China (Wu et al., 2017). Sarawak's capability to accommodate an automotive assembly facility that specialises in the manufacturing of hydrogen-powered fuel cell electric vehicles contributes to the feasibility of implementing this transition. This development aligns with the sustainable energy policy of the Sarawak government, which aims to encourage the automotive sector to transition from fossil fuels to clean and renewable sources (Wong, 2022). The primary obstacle is that the government must allocate substantial financial resources towards the construction of infrastructure supporting the operation of these electric vehicles, including electricity sources, fuel storage and transportation, and energy fuel infrastructure (Zhang et al., 2023). Thus, it is crucial to engage with policymakers to ensure that decisions related to the road construction align with the goals of preserving the ecological integrity while promoting the well-being of local communities.

6. Conclusion

This study demonstrates that professionals have a profound understanding of the long-term holistic administration of the national park and its environs. Concerns regarding land encroachment, business pressure and accessibility opportunities are a few of the additional factors that influence the divergent perspectives of the remaining lay key informants. The varied perspectives expressed by these community representatives emphasise the criticality of considering the complex ramifications of the infrastructure endeavour on the local populace. The active participation of community members in the research underscores the importance of continuous engagement with individuals directly impacted, thereby emphasising the necessity for decision-making processes that are inclusive. The application of analysis to narratives facilitates a systematic comprehension of critical matters, thereby providing decision-makers with invaluable knowledge to prioritise issues and customise interventions.

This chapter promotes the implementation of a thorough socioeconomic impact assessment, which would utilize knowledge of prospective advantages and obstacles to guide strategic decision-making. Moreover, by incorporating key informants' stances, a comprehensive approach to park management is emphasized, acknowledging the mutually beneficial connection that exists between ecological conservation and the welfare of the community. In terms of economic, this endeavour is subjective from an economic standpoint and requires further examination of its potential environmental degradation versus road construction costs, among other factors. In the pursuit of reconciling conservation and development objectives, policymakers can rely on these pragmatic understandings to support evidence-driven decision-making. For the benefit of present and future generations, this approach ensures that infrastructure development and the preservation of GMNP's natural and cultural heritage continue to coexist in a sustainable manner while also meeting the diverse needs of tourism, community welfare and environmental conservation.

References

Ali, P. S. (2023). The role of nongovernmental organizations in rural development. *International Journal of Social Science and Humanities, 5*(3), 13–16.

Arora, N. K., Fatima, T., Mishra, I., Verma, M., Mishra, J., & Mishra, V. (2018). Environmental sustainability: Challenges and viable solutions. *Environmental Sustainability, 1*(4), 309–340. https://doi.org/10.1007/s42398-018-00038-w

Arora, S., & Sharma, A. (2020). Digital marketing for religious event of India for tourism sustainability and promotion. https://api.semanticscholar.org/CorpusID:229392741

Attallah, N. F. (2015). The estimation of physical and real carrying capacity with application on Egypt's tourist sites. *Journal of Tourism Research, 12*, 67.

Azam, M., Hunjra, A. I., Bouri, E., Tan, Y., & Al-Faryan, M. A. S. (2021). Impact of institutional quality on sustainable development: Evidence from developing countries. *Journal of Environmental Management, 298*, 113465. https://api.semanticscholar.org/CorpusID:236968252

Balaji, M. S., & Roy, S. K. (2017). Value co-creation with Internet of things technology in the retail industry. *Journal of Marketing Management, 33*(1–2), 7–31. https://doi.org/10.1080/0267257X.2016.1217914

Barykin, S. E., de la Poza, E., Khalid, B., Kapustina, I. V., Kalinina, O. V., & Iqbal, K. M. J. (2021, July). Tourism industry: Digital transformation. In *Handbook of research on future opportunities for technology management education* (pp. 414–434). https://doi.org/10.4018/978-1-7998-8327-2.ch025

Berardi, M. K., White, A. M., Winters, D., Thorn, K., Brennan, M., & Dolan, P. (2020). Rebuilding communities with empathy. *Local Development & Society, 1*(1), 57–67. https://doi.org/10.1080/26883597.2020.1794761

Bitencourt, C., Rapini, A., Santos Damascena, L., & De Marco Junior, P. (2016). The worrying future of the endemic flora of a tropical mountain range under climate change. *Flora – Morphology, Distribution, Functional Ecology of Plants, 218*, 1–10. https://doi.org/10.1016/j.flora.2015.11.001

Camilleri, M. A. (2018). The tourism industry: An overview. In *Tourism, hospitality and event management*. (Issue September 2017). https://doi.org/10.1007/978-3-319-49849-2_1

Charfeddine, L. (2017). The impact of energy consumption and economic development on ecological footprint and CO_2 emissions: Evidence from a Markov switching equilibrium correction model. *Energy Economics, 65*, 355–374. https://doi.org/10.1016/j.eneco.2017.05.009

Commission of the European Communities. (2001). European governance: A white paper, COM(2001) 428 final. *Official Journal of the European Communities*, 1–29. http://eur-lex.europa.eu/LexUriServ/site/en/com/2001/com2001_0428en01.pdf

Cook, J., Nuccitelli, D., Green, S. A., Richardson, M., Winkler, B., Painting, R., Way, R., Jacobs, P., & Skuce, A. (2013). Quantifying the consensus on anthropogenic global warming in the scientific literature. *Environmental Research Letters, 8*(2). https://doi.org/10.1088/1748-9326/8/2/024024

DaCosta, B., & Kinsell, C. (2023). Serious games in cultural heritage: A review of practices and considerations in the design of location-based games. *Education Sciences, 13*(1). https://doi.org/10.3390/educsci13010047

Flemmer, R., & Schilling-Vacaflor, A. (2016). Unfulfilled promises of the consultation approach: The limits to effective indigenous participation in Bolivia's and Peru's extractive industries. *Third World Quarterly*, *37*(1), 172–188. https://doi.org/10.1080/01436597.2015.1092867

Hammarberg, K., Kirkman, M., & Lacey, S. De (2016). Qualitative research methods: When to use them and how to judge them. *Human Reproduction*, *31*(3), 498–501. https://doi.org/10.1093/humrep/dev334

Hassan, S. T., Xia, E., Khan, N. H., & Shah, S. M. A. (2019). Economic growth, natural resources, and ecological footprints: Evidence from Pakistan. *Environmental Science and Pollution Research*, *26*(3), 2929–2938. https://doi.org/10.1007/s11356-018-3803-3

Holmes, D. R. (2013). *Economy of words: Communicative imperatives in central banks*. University of Chicago Press.

Howes, M., Wortley, L., Potts, R., Dedekorkut-Howes, A., Serrao-Neumann, S., Davidson, J., Smith, T., & Nunn, P. (2017). Environmental sustainability: A case of policy implementation failure? *Sustainability*, *9*(2), 1–17. https://doi.org/10.3390/su9020165

Hunjra, A. I., Azam, M., Bruna, M. G., & Bouri, E. (2023). A cross-regional investigation of institutional quality and sustainable development. *Journal of International Financial Markets, Institutions and Money*, *84*, 101758. https://doi.org/10.1016/j.intfin.2023.101758

Ibrahim, M. S. N., Assim, M. I. S. A., Johari, S., Mohammad, S. K. W., & Hassan, S. (2021). Perception on Penan's Oroo' language among local community at Batu Bungan, Mulu, Sarawak, Malaysia. *Iasaýı Ýnıversıtetiniń Habarshysy*, *122*(4), 27–36. https://doi.org/10.47526/2021-4/2664-0686.03

Ibrahim, M. S. N., Assim, M. I. S. A., Johari, S., Mohammad, S. K. W., Afandi, S. H. M., & Hassan, S. (2023). Public awareness on biodiversity conservation and well-being: Case of Gunung Mulu National Park, Sarawak. *Geojournal*, *88*(3), 3471–3496. https://doi.org/10.1007/s10708-022-10818-x

Ibrahim, M. S. N., Assim, M. I. S. A., Johari, S., Mohammad, S. K. W., & Hassan, S. (2023). Tourists' knowledge on the status of UNESCO world heritage site for Gunung Mulu National Park. *Malaysian Forester*, *86*(2), 325–338.

IPCC, I. (2014). *Climate change 2014: Synthesis report. Contribution of working groups I, II and III to the fifth assessment report of the intergovernmental panel on climate change*. IPCC. https://epic.awi.de/id/eprint/37530/

Isbell, F., Balvanera, P., Mori, A. S., He, J. S., Bullock, J. M., Regmi, G. R., Seabloom, E. W., Ferrier, S., Sala, O. E., Guerrero-Ramírez, N. R., Tavella, J., Larkin, D. J., Schmid, B., Outhwaite, C. L., Pramual, P., Borer, E. T., Loreau, M., Omotoriogun, T. C., Obura, D. O., & Palmer, M. S. (2023). Expert perspectives on global biodiversity loss and its drivers and impacts on people. *Frontiers in Ecology and the Environment*, *21*(2), 94–103. https://doi.org/10.1002/fee.2536

Kardos, M. (2012). The reflection of good governance in sustainable development strategies. *Procedia – Social and Behavioral Sciences*, *58*, 1166–1173. https://doi.org/10.1016/j.sbspro.2012.09.1098

Kemp, R., Parto, S., & Gibson, R. (2005). Governance for sustainable development: Moving from theory to practice. *International Journal of Sustainable Development*, *8*. https://doi.org/10.1504/IJSD.2005.007372

Keong, C. Y., & Onuma, A. (2021). Transboundary ecological conservation, environmental value, and environmental sustainability: Lessons from the Heart of Borneo. *Sustainability*, *13*(9727), 21. https://doi.org/10.3390/su13179727

Kiew, P. (2021, March 23). Value assessment lab held for proposed road network. *The Borneo Post*. https://www.theborneopost.com/2021/03/23/value-assessment-lab-held-for-proposed-road-network/

Kiew, P. (2022, September 14). Brunei opens Asean's longest bridge to foreign vehicles but limits weight. *Borneo Post Online Sdn Bhd*. https://www.theborneopost.com/2022/09/14/brunei-opens-aseans-longest-bridge-to-foreign-vehicles-but-limits-weight/

Lietz, C. A., & Zayas, L. E. (2010). Evaluating qualitative research for social work practitioners. *Advances in Social Work*, *11*(2), 188–202. https://doi.org/10.18060/589

Louis, N. A. (2022, February 20). Limbang–Miri route to connect Mount Buda community to Greater Miri. *Dayak Daily*. https://dayakdaily.com/limbang-miri-route-to-connect-mount-buda-community-to-greater-miri/

Mahato, A. K. (2014). Ultra low frequency CMOS ring oscillator design. In *2014 Recent Advances in Engineering and Computational Sciences (RAECS)* (pp. 1–5). https://doi.org/10.1109/RAECS.2014.6799627

Medek, D. E., Schwartz, J., & Myers, S. S. (2017). Estimated effects of future atmospheric CO_2 concentrations on protein intake and the risk of protein deficiency by country and region. *Environmental Health Perspectives*, *125*(8), 1–8. https://doi.org/10.1289/EHP41

Merlyn Keziah, S., & Subathra Devi, C. (2017). Essentials of conservation biotechnology: A mini review. *IOP Conference Series: Materials Science and Engineering*, *263*(2). https://doi.org/10.1088/1757-899X/263/2/022047

Ministry of Natural Resources and Environment of Malaysia. (2016). *National policy on biological diversity 2016–2025*. Ministry of Natural Resources and Environment.

Mohamad Junaidi, N. A., & Hashim, H. (2023). Covid-19: Impacts on tourism in Bako National Park and survival strategies of local tourism operators. *Malaysian Journal of Social Sciences and Humanities*, *8*(9), 1–16. https://doi.org/10.47405/mjssh.v8i9.2534

Mondal, S., & Palit, D. (2021). Challenges in natural resource management for ecological sustainability. In *Natural resources conservation and advances for sustainability*. (Issue October). https://doi.org/10.1016/B978-0-12-822976-7.00004-1

Mwangonde, K., Ntinda, M., & Hasheela-Mufeti, V. (2021). A game-based approach to revive cultural heritage amongst the youth. In *2021 IST-Africa conference, IST-Africa 2021* (pp. 1–8). https://ieeexplore.ieee.org/abstract/document/9577035

Newmark, W. D., Halley, J. M., Beier, P., Cushman, S. A., McNeally, P. B., & Soulé, M. E. (2023). Enhanced regional connectivity between western North American national parks will increase persistence of mammal species diversity. *Scientific Reports*, *13*(1), 1–12. https://doi.org/10.1038/s41598-022-26428-z

Onwuegbuzie, A. J., & Leech, N. L. (2007). Sampling designs in qualitative research: Making the sampling process more public. *Qualitative Report*, *12*(2), 19–20.

Opute, A. P., Irene, B. O., & Iwu, C. G. (2020). Tourism service and digital technologies: A value creation perspective. *African Journal of Hospitality, Tourism and Leisure*, *9*(2), 1–18.

Ortega-Reig, M., Schürmann, C., Ferrandis Martínez, A., & Copus, A. (2023). Measuring access to services of general interest as a diagnostic tool to identify well-being disparities between rural areas in Europe. *Land, 12*(5). https://doi.org/10.3390/land12051049

Rather, R. A., Hollebeek, L. D., & Islam, J. U. (2019). Tourism-based customer engagement: The construct, antecedents, and consequences. *Service Industries Journal, 39*(7–8), 519–540. https://doi.org/10.1080/02642069.2019.1570154

Rockström, J., Steffen, W., Noone, K., Persson, Å., Chapin, F. S., Lambin, E. F., Lenton, T. M., Scheffer, M., Folke, C., Schellnhuber, H. J., Nykvist, B., de Wit, C. A., Hughes, T., van der Leeuw, S., Rodhe, H., Sörlin, S., Snyder, P. K., Costanza, R., Svedin, U., & Foley, J. A. (2009). A safe operating space for humanity. *Nature, 461*(7263), 472–475. https://doi.org/10.1038/461472a

Sarawak Forestry Corporation. (2020). *Visitors statistic from 2015–2019*. https://www.sarawakforestry.com/visitor-statistics/

Sarawak Rural Broadband Network. (2021). The launching of MySRBN in Long Terawan. *Miri, Sarawak. MySRBN*. https://mysrbn.sarawak.digital/the-launching-of-mysrbn-in-long-terawan-miri-sarawak/

Singh, A. S., Parahoo, S. K., Ayyagari, M., & Juwaheer, T. D. (2023). Conclusion: How could rural tourism provide better support for well-being and socioeconomic development? *Worldwide Hospitality and Tourism Themes, 15*(1), 84–93. https://doi.org/10.1108/WHATT-09-2022-0108

Strandeneas, J. G. (2011). *Rio+20: Towards achieving sustainable development by greening the economy and improving governance—Council conclusions* (Vol. 32, Issue 0). https://www.consilium.europa.eu/uedocs/cms_data/docs/pressdata/en/envir/125027.pdf

Talib, O. (2019). *Analisis data kualitatif dengan Atlas.ti* (1st ed.). Penerbit Universiti Putra Malaysia.

Tang, C., Weaver, D., & Lawton, L. (2017). Can stopovers be induced to revisit transit hubs as stayovers? A new perspective on the relationship between air transportation and tourism. *Journal of Air Transport Management, 62*, 54–64. https://doi.org/10.1016/j.jairtraman.2017.02.008

Tashakkori, A., Johnson, R. B., & Teddlie, C. (2020). *Foundations of mixed methods research: Integrating quantitative and qualitative approaches in the social and behavioral sciences* (2nd ed.). SAGE Publications, Inc.

Tawie, S. (2022, August 7). PM: Construction of phase two of Sarawak-Sabah Link Road to begin early next year. *MalayMail*. https://www.malaymail.com/news/malaysia/2022/08/07/pm-construction-of-phase-two-of-sarawak-sabah-link-road-to-begin-early-next-year/21550

Terzis, L. D., Saltzman, L. Y., Logan, D. A., Blakey, J. M., & Hansel, T. C. (2022). Utilizing a matrix approach to analyze qualitative longitudinal research: A case example during the COVID-19 pandemic. *International Journal of Qualitative Methods, 21*, 1–10. https://doi.org/10.1177/16094069221123723

UNESCO World Heritage Centre. (2019). *World heritage list: The criteria for selection*. UNESCO World Heritage Centre. https://whc.unesco.org/en/criteria/

United Nations, World Commission on Environment and Development. (1988). *The Brundtland report: "Our common future"* (Vol. 4, Issue 1). United Nations, World Commission on Environment and Development. https://doi.org/10.1080/07488008808408783

Wall, G. (2020). From carrying capacity to overtourism: A perspective article. *Tourism Review*, *75*(1), 212–215. https://doi.org/10.1108/TR-08-2019-0356

Wong, J. (2022, August 1). Sarawak gets into hydrogen-powered fuel cell electric vehicles. *Star*. https://www.thestar.com.my/business/business-news/2022/08/01/sarawak-gets-into-hydrogen-powered-fuel-cell-electric-vehicles

World Heritage in Danger: Compendium II, 59. (2009). https://doi.org/10.1371/journal. pone.0221947

Wu, Y., Zhang, S., Hao, J., Liu, H., Wu, X., Hu, J., Walsh, M. P., Wallington, T. J., Zhang, K. M., & Stevanovic, S. (2017). On-road vehicle emissions and their control in China: A review and outlook. *Science of the Total Environment*, *574*, 332–349. https://doi.org/10.1016/j.scitotenv.2016.09.040

WWF. (2014). Living planet report 2014: Species and spaces, people and places. https://www.wwf.or.jp/activities/data/WWF_LPR_2014.pdf

WWF. (2017). *Updates on Heart of Borneo Programme*.

Yong, P. L., Mas'ud, F., & Ling, H. K. (2019). Reflexivity: Doing research with women in a mental health care facility. *Asian Social Work Journal*, *4*(4), 19–29. https://doi.org/10.47405/aswj.v4i4.106

Zhang, W., Fang, X., & Sun, C. (2023). The alternative path for fossil oil: Electric vehicles or hydrogen fuel cell vehicles? *Journal of Environmental Management*, *341*(February), 118019. https://doi.org/10.1016/j.jenvman.2023.118019

Zinda, J. A. (2017). Tourism dynamos: Selective commodification and developmental conservation in China's protected areas. *Geoforum*, *78*, 141–152. https://doi.org/10.1016/j.geoforum.2015.08.004

Chapter 7

The Role of ICT and Infrastructure in Tourism Development: New Evidence From Asian Economies

Saira Arsh[a], *Samia Nasreen*[a] *and Xuan-Hoa Nghiem*[b]

[a]Lahore College for Women University, Pakistan
[b]Vietnam National University, Vietnam

Abstract

The adoption and usage of information and communication technology (ICT) has introduced transformation in the tourism arena with ICT applications extensively used in tourism industry. In addition to ICT, an advanced infrastructure is essential for the development of tourism industry. Thus, the goal of present research is to probe the impact of ICT and infrastructure on tourism development (TD) in 28 Asian economies using method of moments panel quantile regression (MM-QR) model introduced by Machado and Silva (2019) applied to a panel data from 2008 to 2020. Empirical findings demonstrate that there is an asymmetric non-linear effect of ICT and infrastructure through all quantile range. This indicates that ICT has negative effect on TD in poor countries while positive impact in rich countries. Negative impact in poor countries may be due to higher establishment cost and information technology (IT) productivity paradox. However, results confirm the importance of ICT and infrastructure in endorsing the development of tourism sector in Asian nations by lessening time and money costs and facilitating travelers.

Keywords: ICT; infrastructure; tourism development; panel quantile regression; Asia

The Emerald Handbook of Tourism Economics and Sustainable Development, 119–145

Published under exclusive licence by Emerald Publishing Limited
doi:10.1108/978-1-83753-708-220241007

1. Introduction

The travel and tourism (T&T) have become the predominant and extensively varied industry worldwide in the past few decades. Many nations depend on this vibrant industry as a chief source to generate revenues, accelerate private sector growth and facilitate employment and infrastructure development. Tourism industry is rapidly growing, especially in developing countries where other means of development such as the progression of manufacturing sector or the export of natural resources are not commercially feasible (WTO, 1997). Tourism is a major global economic activity and considered to be an important component of growth and development because it has beneficial effects on employment generation and infrastructure provision, among others (Eilat & Einav, 2004; Turayev & Sharma, 2021).

The socio-economic growth of a touristic country is heavily reliant on the tourism business. This industry generates foreign cash, improves the external balance of trade and creates job related opportunities. Mobility-created money is believed to have a multiplier effect on the economy and aids in the growth of other industries (Rhaman, 2016).

Nowadays, tourism has become a significant and frequently used tool in developing nations for long-term development and is highly well-liked there. Tourism may bring about numerous benefits such as to provide job opportunities, reduce poverty, help the poor construct their own prospects and improve quality of life. Additionally, tourism is also believed to offer non-economic effects on livelihoods, such as increased socio-cultural status, opportunities for training, education and skill building, as well as improvements in health and other aspects of well-being (Rhaman, 2016).

At the United Nations Commission (UNC) on Statistics (April, 1968) and the Conference on Tourism and International Travel, held in Rome in 1963, the United Nations recognized this basic concept of tourism (Vellas et al., 1995).[1] Several modifications have been made in the definitions of tourism in order to modernize them in the conference of World Tourism Organization (WTO) at Ottawa in June 1991. One of the definitions offered was as tourism encompasses the activities of individuals travelling to and staying in places external to their usual environment for one consecutive year and not more than this period for business, leisure or other purposes (UNWTO, 2010).

Economic benefits provided by (international) tourism are being proved to be among crucial factors for international relations and growth of economy for several nations. According to El Meriague (2014), tourism accounts for 8.7% of total employment and 10% of GDP worldwide. Moreover, with the increase in global networking, strengthened tourism sector is regarded as a symbol of the social expansion, progression as well as the evolution of a country (El Meriague, 2014). With more than half a billion international tourist arrivals at present, the tourism industry has been experiencing tremendous growth. So far, it is clear that it is not the same scenario in all regions of the globe. No doubt, the main attribute

---

[1]United Nations Commission.

of international tourism trend is unbalanced allocation of international tourism flows to the diversified regions of the world.

Information and communication technology (ICT) has brought revolution to the people's lives and operations of business organization. Different organizations exploit ICT applications to process, accumulate, distribute and endorse their goods and services worldwide. Over the last several decades, the application of ICT has transformed the tourism industry across the globe (Adeola & Evans, 2020). The term ICT envisages a wide range of technological instruments and resources are employed for the purpose of conveying and generating information, as well as distributing, retaining and organising it (Enakrire & Ocholla, 2017). ICT in general and the Internet particularly have emerged as one of the most effective tools that transform the tourism industry (Navío-Marco et al., 2018). Tourism sector has broadly used ICT tools to minimize costs, saving labour hours, make operations efficient and most importantly improve customer experience and service quality (Law et al., 2009).

There is a vast body of empirical evidence on the link between tourism and economic development (Lee & Chang, 2008) and ICT and economic growth (Jorgenson & Vu, 2016; Rahman et al., 2021). While there are quite a few studies confirming the relationship among ICT, infrastructure and tourism development separately (Adeola et al., 2018; Haini, 2019; Law et al., 2018; Tsokota et al., 2019), few of them explored the effects of ICT and infrastructure on TD simultaneously (Adeola & Evans, 2020) and the effect of COVID-19 on tourism industry (Arora & Sharma, 2021). Therefore, this panel study is intended to bridge this gap by elucidating the role played by ICT and infrastructure on TD using a dataset of 28 Asian economies over the time period of 2008–2020.

This research contributes to the literature in several ways: (i) First and foremost, to the authors' best knowledge, there has been no study in the case of Asian economies that simultaneously scrutinized the impact of both ICT and infrastructure on TD. (ii) Second, in this chapter, we generate an ICT index by using five indicators namely; fixed broadband subscription, mobile cellular subscriptions, secured Internet servers, individuals using the Internet and computers, communications and other services. We believe that this is a novel contribution of our research as the majority of studies usually use one or two indicators. (iii) Third, we employ the latest estimation techniques, namely the panel quantile regression (PQR) proposed by Powell (2022) and Machado and Silva (2019) that can curtail the problem of outliers and heterogeneity more suitably. This is an advantage as compared to other traditional regression techniques which ignore the issues related to outliers and heterogeneity and provide biased and inconsistent results.

The subsequent sections of our manuscript will be outlined as follows: The second section shall furnish a comprehensive literature review. The third section will be dedicated to delineating the data, sources and model employed. In Section 4, the empirical methods employed will be expounded upon. The fifth section will showcase the estimated results obtained. Lastly, Section 6 will serve as the concluding segment of the chapter.

2. Literature Review

2.1 ICT and Tourism

This section directs some attention to a number of empirical findings from tourism-related literature, especially the ICT-tourism nexus. Zhang et al. (2019) noted that thanks to rapid advances in technology and its penetration into nearly all facets of the universe, the tourism sector has employed a wide range of technologies to assist travelling activities, thus enhancing improve tourists' experiences at their destinations. Tcheng et al. (2007) unveiled the positive influence of ICT as it is associated with other utilities including electricity, transportation and water availability. Andrianaivo and Kpodar (2011) stated that as fixed investments prove to be constructive for the growth of economy, investments in general infrastructures especially in ICT may boost the overall development of a country. Waverman et al. (2005) stated that since transaction expenses are reduced due to communication utilities, making investments in ICT infrastructure cost-saving. ICT can advance information flows, which, in turn, allow markets to function in better ways and improve price discovery by facilitating procedures. Furthermore, ICT may also assist in regulating supply and demand by cutting the costs of retrieving information (Andrianaivo & Kpodar, 2011).

There are quite a few studies (e.g. Bethapudi, 2013) lending support to the significant role that ICT plays in developing the tourism sector. However, while ICT is believed to make major contributions to the expansion of tourism industry, many previous studies appear to overlook the most significant connection linking the tourism and ICT with the exception of Ma et al. (2003) and Firoiu and Croitoru (2013).

Information is usually considered the most important component of the tourism industry and thus, as according to Poon (1993) a whole system of ICT is spreading rapidly across the tourism industry, and no player will escape its effects. Poon (1993) posited that the adoption of advanced technology such as ICT can make a sharp difference between the 'traditional tourism sector' and 'new tourism industry' since greater number of customers adopt and adapt the latest processes, products and system of hospitality and tourism delivery by the blessing of advanced technologies. Moreover, the adoption of advanced technologies by destination has continued this trend at diverse phases of TD across the globe (Li & Suomi, 2008).

Buhalis and Law (2008) opined that ICT adoption has provided the ease and the opportunity for tourists to get latest information regarding their destination and tour as a whole. Berne et al. (2012) employed structural equation modeling with survey data from Spain to demonstrate that increased ICT tool utilization in the tourism industry enhances business performance and promotes industry growth.

Ramos and Rodrigues (2012) applied fixed effect model to a sample of 18 Western European countries from 1993 to 2007 to conclude that Internet usage has significant positive impact on TD in Western Europe. Bethapudi (2013) argued that ICT has an important role to play in the development of tourism sector and has inevitably become the main contributor to the expansion of tourism industry.

Ollo-López and Aramendía-Muneta (2012) conducted a study focusing upon the tourism utilizing a sample of 2,665 organizations in 26 European countries. Using probit model, the study concluded that ICT adoption has modest impact on competition level and on raising the productive capacity of tourism industry.

Feshari (2017) investigated short- and long-run relationship between ICT indices and the ratio of government expenditures on ICT to GDP in Iran over the time span between 1994 and 2015. Using fully modified ordinary least square (FMOLS) co-integrating method, the study concluded increased application of ICT attracts greater number of tourist arrival in Iran. Additionally, with the development of supporting sectors such as telecoms and transportation, there is an increase in the potentials of tourism sector as a medium of revenue generation for an economy (Min et al., 2016). In his study, Hoonsawat (2016) employed a gravity model to analyze bilateral tourism flow and concluded that, with the aid of ICT, tourists are able to acquire necessary information regarding their preferred destinations, thereby increasing the demand for tourism services.

Wahab (2017) investigated the effect of ICT on TD in India from 1975 to 2013 and concluded that ICT provided ease in reducing the cost incurred in the propagation of knowledge. Bekteshi and Bekteshi (2017) analyzed how ICT index affects the level of demand for tourism in Albania during 2015. By analyzing the data collected through survey and Internet searches, Bekteshi and Bekteshi (2017) concluded that higher quality of ICT and improved infrastructure would lead to increase in tourism demand in Albania. Furthermore, to attain resourceful tourism sector management, immense computer system usage for tourist guides, reservation and accommodation as well as the enterprise management are highly recommended. Ukpabi and Karjaluoto (2017), Tan et al. (2017) and Kim and Lee (2017) all showed that ICT adoption has become an indispensable tool for tourists to obtain information about the tourist sites, food and accommodation. Kotiloglu et al. (2017) showed that mobile phones and Internet access may assist tourists in making friends with local citizens, local guides and find tourist sites in four cities of United States of America (USA).

Moreover, there is further evidence on the performance of ICT tools in tourism. Ilic and Nikolic (2018) tried to explore the role of new technology in the Serbian tourist business and examined how it contributes to increased competency. By employing the rundown relied upon methodology of World Economic Forum (WEF) methodologies to reflect the competitiveness of the T&T sector, including the T&T Competitiveness Index, they concluded that ICT positively impacts the development of tourism industry.

Tsaurai and Chimbo (2019) studied the effect of ICT on the growth of tourism industry in emerging markets over the period 1996–2016. By applying fixed effects and generalized method of moments (GMM), they conclude that the impact of ICT on tourism growth is positive. Moreover, both Law et al. (2018) and Haini (2019) found that internet services have become more important for tourism sector. Tsokota et al. (2019) explored the effects of ICT on tourism sector in Zimbabwe and concluded that ICT has both positive impacts on tourism sector if utilized and handled efficiently. However, if ICT is not handled adequately, the impact of ICT on tourism could be reversed.

Adeola and Evans (2019) examined the effects of internet usage and mobile phones on tourism sector in Africa for the period of 1996–2017. They used system GMM and concluded that Internet usage and mobile penetration encourage tourism, which subsequently enhances the usage of internet even more in these economies. Adeola and Evans (2019) argue that both Internet usage and mobile penetration affected tourism negatively whereas the square of Internet usage and square of mobile penetration affected tourism positively in Africa. Nadeem et al. (2020) accomplished a study in order to investigate the impact of ICT infrastructures on Pakistan's tourism economy between 2002 and 2016. Using the autoregressive distributed lag (ARDL) model, the study demonstrated that ICT infrastructure has considerable negative effect on TD. Mohamed Aslam (2020) investigated the long-term association between ICT and TD for Sri Lankan tourism business during time span from 1990 to 2018. Using bivariate cointegration approach, his research concluded that ICT triggers the growth of Sri Lankan tourism business.

Various other studies also reaffirm the importance of ICT for TD (Lopez Cordova, 2020; Choudhary et al., 2020). Lopez Cordova (2020) found that countries that are less developed in digital adoption also lag behind the other counties in terms of tourism demand. Lopez Cordova (2020) also concluded that countries in the Middle East, North Africa and Sub-Saharan Africa could stimulate economic activity in the T&T industry by exploiting such technologies. Thus, the study concluded that digital adoption enhances international tourism.

Similarly, Kumar and Kumar (2020) applied pooled mean group (PMG) on a sample of nine major tourist's spots to come to the conclusion that broadband and mobile subscriptions increased the international visitor's arrival by 0.11% and 0.04%, respectively. Taking a sample of 40 African countries over the period 1996–2015 and using GMM, Adeola and Evans (2020) found that ICT is positively related to TD.

Sharma and Arora (2022) argued that virtual reality and travel blogging have the potential to revolutionize the tourism industry. Virtual reality enhances the tourist experience by creating a parallel universe, while travel blogs offer insights and reflections on destinations.

2.2 Infrastructure and Tourism

Infrastructure is commonly regarded as a fundamental ingredient in the tourism package of a country, a key determinant in attracting tourists in a country and therefore one of the key elements of TD (Arshad et al., 2018; Khan & Rasheed, 2016). Some of the studies have exposed the importance of infrastructure as an essential factor in the international tourism demand (Adeola et al., 2018; Adeola & Evans, 2020; Khadaroo & Seetanah, 2007). Both transport and non-transport infrastructure (including physical infrastructure) constitute key component in TD. Smith (1984) determined the role of service infrastructure in building a product experience. He argued that service infrastructure surrounded by a larger macro-environment of the destination as well as the destination's technology and infrastructure make up critical components that can improve the experience of tourists. Kaul (1985) was the first-ever to acknowledge the significance of transportation infrastructure as an indispensable part of effective development as it encourages and facilitates the creation of new attractions and the growth of existing ones.

While transportation linkage systems may not be able to meet the demand, but ICT advances can definitely transform the information communication infrastructure of the 1990s, thus fostering global tourism expansion (Mowlana & Smith, 1990). In recognition of the value of infrastructure, Rowley (1992) concluded that 'the physical infrastructure of Europe, Americas and Asia all needs to be created, renovated or upgraded' (p. 47).

Prideaux (2000) defined the tourism-related transportation system as 'the process of and communication between, modes of transportation, terminals and ways that give support visitors to travel in and out of destinations; in addition, it provides transport services inside the destinations'. Naudé and Saayman (2005) concluded that tourism infrastructure is the main determinant of tourist arrival in Africa because improved infrastructure attracts greater tourist arrival from other countries. Boopen (2006) investigated the significance of transportation capital in the overall destination's attractiveness. By using Johansen cointegration, Boopen (2006) found that the transport capital stock of the economy is significant and contributing positively to the number of tourist arrival. Khadaroo and Seetanah (2007) found that Asian and European/American tourists are unenthusiastic to spend holidays in Mauritius due to inferior transport infrastructure. Thus, the results of their study showed that improved transport and non-transport infrastructures are necessary for TD. Khadaroo and Seetanah (2008) employed GMM over the period between 1990 and 2000 in order to examine the role of transportation infrastructure in tourism flows in a sample of 28 countries. Their study confirmed that transport infrastructure is top on the list of tourism infrastructure and important determinant of tourism inflow into the destination.

Seetanah et al. (2010) assessed the number of tourist arrivals in South Africa from 38 different origins over the period of 1985–2000. The study employed FMOLS to document a significant and positive coefficient for infrastructure (rooms). Seetanah et al. (2011) assessed the importance of infrastructure in tourism business in Mauritius over the time span of 1985–2006. Using both random effects and fixed effects models, Seetanah et al. (2011) found that infrastructure has a positive impact on tourism industry as more rooms attract greater number of tourist arrival. Muchapondwa and Pimhidzai (2011) employed ARDL modelapproach to cointegration over the period 1998–2005 and concluded that improved international tourism infrastructure builds up taste formation and thus play a crucial role in attracting greater number of tourist's arrival in Zimbabwe. Imikan and Ekpo (2012) using multiple correlation and stepwise regression to conclude that infrastructure is highly significant and positively contributed to the TD in Nigeria. Moreover, Bramo (2013) concluded that railway infrastructure in one of the most important means of TD and has positive impact on tourism industry in Albania. Adeola et al. (2018) conducted a study to identify the determinants of international TD in 44 African economies over time span of 1995–2015. Results of Poisson regression showed that infrastructure is an important driver of international tourism in Africa.

Khoshnevis Yazdi and Khanalizadeh (2017) identified factors influencing international tourism demand in the USA from 1995 to 2014 using ARDL across 14 countries. The results indicated that transport infrastructure and gross domestic product (GDP) positively determine United States (US) tourist arrivals, highlighting the significant impact of infrastructure on shaping traveller preferences. Nadeem et al. (2020) investigated the impact of physical infrastructures on tourism industry in an emerging economy, Pakistan, from 2002 to 2016. Employing ARDL and dynamic OLS (DOLS), they found that physical infrastructure negatively affects tourism in both short and long term. This underscores the importance of upgrading infrastructure in emerging economies to promote tourism (Nadeem et al., 2020).

In summary, considering the preceding literature, it can be concluded that the crucial role of ICT and infrastructure in creating favourable conditions for TD is well-established in the literature. However, studies that have discussed simultaneously the effect of ICT and infrastructure are limited. Hence, the current chapter aims to assess the impact of ICT and infrastructure on TD in Asian economies using quantile panel regression.

3. Data and Methodology

Following Lee et al. (2021), Adeola and Evans (2020), Hunjra et al. (2022) and Hunjra et al. (2023), the general specification form of the model used in this chapter to examine the effect of ICT and infrastructure on tourism industry is described as follows:

$$\text{Tour\_Ind}_{it} = \beta_0 + \beta_1 \text{ICT}_{it} + \beta_2 \text{INFRA}_{it} + \beta_3 \text{K}_{it} + \varphi_i + \nu_{it} \quad (7.1)$$

Where $i = 1, 2, \ldots, 28$ $t = 2008, 2009, \ldots, 2020$

The detail of potential variables is provided below.

3.1 Tourism Industry (Tour_Ind)

Most commonly used proxies for the tourism industry in literature are number of tourist arrivals (Adeola & Evans, 2020; Evans & Kelikume, 2018; Lee et al., 2021) and international tourism receipts (Husein & Kara, 2020; Imikan & Ekpo, 2012; Lee et al., 2021). Both these proxies are used in the present study.

3.2 ICT Indicators (ICT)

ICT is a multi-dimensional concept, so it is not captured by a single variable. For empirical analysis, ICT index made with five indicators, namely (i) fixed broadband subscriptions, (ii) secure Internet servers, (iii) computer, communications and other services, (iv) individuals using the Internet and (V) mobile cellular subscriptions (see e.g., Adeola & Evans, 2020; Enakrire & Ocholla, 2017; Lee et al., 2021).[2]

3.3 Infrastructure (INFRA)

Infrastructure is proxied by gross fixed capital formation (Adeola & Evans, 2020).

3.4 Control Variables

K is the vector of all control variables namely; economic growth which is proxied by income per capita (GDP), foreign direct investment (FDI) and labour force (LAB).

This study has used a panel of 28 Asian countries from 2008 to 2020 with data collected from World Bank's World Development Indicators (WDI, 2021).[3]

3.4.1 Descriptive Analysis

Table 7.1 reports summary statistics. Labour force (LAB) has the highest mean value followed by INFS while GDP per capita has the lowest mean value. The standard deviation of tourism receipts (TR) illustrates more volatility as compared to other variables. GDP per capita is least volatile since it has smallest value of standard deviation.

3.5 Econometric Methods

As there are variety of panel data types, there exist various estimation methods and empirical models examining the impact of explanatory variables on the mean value of explained variable. However, these methods usually ignore the influence for the whole distribution of explanatory variables. As Lin and Xu (2018) argue, outliers and non-normal distribution are the most two common problems inherited with economic variables which may lead to spurious regression, necessitating the use of appropriate estimation methods. Among potential solutions to outliers and non-normal distribution, quantile regression is desirable as it is robust to these problems (Koenker & Bassett, 1978). Lin and Xu (2018) argue that residuals from quantile regression do not have to satisfy classical linear regression model assumptions such as normal distribution, constant variance or zero mean. In quantile regression, the objective function minimizes the absolute value of the

[2]PCA is used for the construction of ICT index. Results of PCA will be available upon request.

[3]Armenia, Azerbaijan, Bahrain, Bangladesh, Cambodia, Georgia, Hong Kong SAR (China), Indonesia, Israel, Japan, Jordan, Kazakhstan, Korea, Lebanon, Malaysia, Maldives, Mongolia, Myanmar, Nepal, Oman, Pakistan, Philippines, Saudi Arabia, Singapore, Sri Lanka, Tajikistan, Thailand and Turkey.

Table 7.1. Data Description, Source of Data and Summary Stats.

Variable	Measurement	Source	Mean	St.D	Minimum	Maximum
TA	International tourism, no. of arrival	WDI	15.19528	1.475973	11.55215	18.05954
TR	International tourism, receipts (% of total exports)	WDI	15.49556	17.23754	0.2864618	86.18721
ICT	ICT index	Authors calculation	11.47204	11.58394	0	100
INFS	Gross fixed capital formation	WDI	25.05761	5.873874	10.00578	48.41233
GDP	GDP per capita (constant 2015 US$)	WDI	8.716918	1.235754	6.51686	11.02148
FDI	Foreign direct investment (US$)	WDI	21.7606	2.355884	0.0312671	25.92203
LAB	Labour force (% of population)	WDI	66.16075	10.78562	40.74758	87.98

Source: WDI (2021) and author's calculation.

Note: WDI = World Development Indicators. TA, GDP and FDI are used in log form.

residuals rather than the sum of squared residuals in the traditional ordinary least squares (OLS) estimator.

In addition to being robust to outliers, quantile regressions can control the effect of unobserved heterogeneity. In this chapter, PQR methodology is employed to examine the effect of ICT and infrastructure on TD at heterogeneous conditional distribution of dependent variable. Under this technique, the conditional distribution of explained variable is divided into multiple quantiles and 50th quantile is called median (Ummalla et al., 2019).

Koenker (2004) suggested a very brief procedure to estimate the coefficients of individual effects but the effect of unrecognized factors is not captured by this method. Canay (2011) put forward the two-step framework of FE PQR by using the technique of Koenker (2004).[4] In first step, additive fixed effects are computed by using expected mean and coefficients of unobserved variables. In second step, difference is estimated between additive fixed effects and original dependent variables and lastly, PQR is used. Nevertheless, in the course of the process, the structure of given model could be changed due to incorporation of additive fixed effects. In order to address this limitation, Powell (2022) presented the PQR, which includes non-additive fixed effect because he noted that the impact of policy variable on the DV is not clearly defined in AFE model. Hence, Powell (2022) technique estimates that are more reliable with smaller period (T).[5,6]

This study goes one more step further by employing the MM-QR introduced by Machado and Silva (2019).[7] Machado and Silva (2019) present a versatile tool to compute PQR, particularly where parameter estimation is challenging or even infeasible, hence generating consistent and robust results that can be applied to policy formulation. The unique advantage of MM-QR technique is that in nonlinear models, it offers simpler calculation especially in case of multiple endogenous variables. The MM-QR methodology is therefore employed to evaluate the impact of ICT factors on TD by intensifying the descriptive statistics by using Ln TA and TR dependent variables. Ike et al. (2020) argue that traditional quantile regression methods do not account for unobserved heterogeneity across individuals. In addition to addressing this problem, the MM-QR also proves helpful in dealing with models where independent variables are endogenous or identifying conditional heterogeneous effects of the independent variables.

The following text specification of the MM-QR is adopted from Ike et al. (2020), Anwar et al. (2021), Padhan et al. (2020),

$$Y_{\text{it}} = \gamma_i + X_{\text{it}}'\beta + \left(Z_{\text{it}}'\alpha + \theta_i\right) U_{\text{it}} \tag{7.2}$$

in which (γ_i, θ_{it}) are individual fixed effects, Z is a k-vector of transformable identified components, X_{it} is independent across time and independently and

---

[4]Fixed effect.

[5]Dependent variable.

[6]Additive fixed effect.

[7]Method of moment quantile regression.

identically distributed (i.i.d) with any individual i, while U_{it} is i.i.d across time t and individual i. X_{it} and U_{it} are assumed to b orthogonal to each other. Finally, the probability $P(Z'_{it}\alpha + \theta_{it} > 0) = 1.(\gamma, \beta', \theta, \alpha')'$ represent parameters that will be estimated.

Eq. (7.2) can be rewritten as

$$Q_Y(\varphi|X_{it}) = (\gamma_i + \theta_i q(\varphi)) + X'_{it}\beta + Z'_{it}\alpha * q(\varphi) \quad (7.3)$$

In Eq. (7.3), $Q_Y(\varphi|X_{it})$ denotes the quantile distribution of the dependent variable Y_{it}. The quantile distribution of Y_{it} depends on where the independent variable X_{it} is located; X'_{it} is a vector of explanatory variables consisting of: ....; $(\gamma_i + \theta_i q(\varphi))$ is scalar coefficient denoting quantile-φ fixed-effect for individual ith. Finally, $q(\varphi)$ indicates the φth sample quantile estimated through solving the optimization exercise:

$$\min_q \sum_i \sum_t \rho_\varphi \left[R_{it} - (Z'_{it}\alpha + \theta_i) q(\varphi)\right]$$

In which $\rho_\varphi(A) = (\varphi - 1)AI\{A \leq 0\} + TAI\{A > 0\}$ is the check function.

4. Findings and Discussion

4.1 Preliminary Analysis

Conditional mean regression requires variables to follow a normal distribution. If they do, it is suitable for assessing ICT and infrastructure's impact on tourism. If not, using it would yield unreliable results, making it unsuitable. In such cases, the PQR model may be better. We initially checked data normality using Q-Q plots indicated in Fig. 7.1. These plots did not form a diagonal line, indicating non-normal distribution. Refer to Table 7.2 for more details. Based on these results, conditional mean regression is not suitable, and we prefer the PQR model for analysis.

Subsequently, as panel data can be subject to cross-sectional dependence (CD) which may make results biased and unreliable, it is necessary to test for the presence of CD (Pesaran, 2007). In this chapter, we employed the CD test by Pesaran (2007).[8] Results of the Pesaran (2007) CD (see Table 7.3) confirm the existence of the CD problem in the data. As a result, we applied Pesaran second-gen panel unit root test to address the CD. Table 7.3 illustrates that all the series are stationary (first difference stationary).

4.2 PQR Results

As previously mentioned, our variables deviate from normal distribution. Thus, we apply Powell (2022) PQR method with 11 quantiles (5th, 10th, 20th, 30th, 40th, 50th, 60th, 70th, 80th, 90th and 95th) for analysis. We also include OLS results (first column in Table 7.4) for comparison with PQR outcomes. The remaining table columns show PQR results. Table 7.4 data clearly reveal the

[8]Cross-sectional dependence.

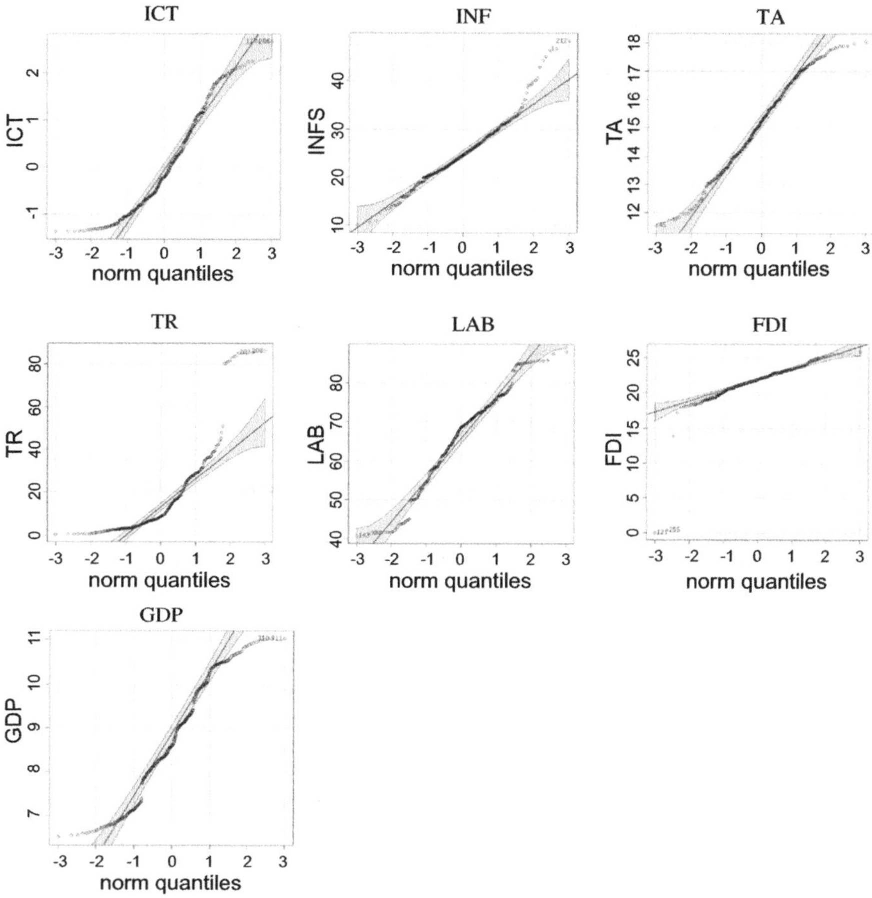

Fig. 7.1. Q-Q Plot. *Source:* Graphs are made using R studio.

Table 7.2. Shapiro-Wilk W Test, Shapiro-Francia W′ and Jarque Bera Tests.

Variable	Shapiro-Wilk W Test	Shapiro-Francia W′ Test	JB Test
TA	4.186	3.931	9.397
	(0.000)	(0.001)	(0.009)
TR	67.548	72.869	888.9
	(0.000)	(0.000)	(0.002)
ICT	74.043	81.04	48.02
	(0.000)	(0.000)	(0.000)
INFS	9.212	10.18	81.99
	(0.000)	(0.000)	(0.000)

Continued

Table 7.2. *(Continued)*

Variable	Shapiro-Wilk W Test	Shapiro-Francia W′ Test	JB Test
GDP	10.35 (0.000)	10.339 (0.000)	16.34 (0.000)
FDI	68.82 (0.000)	76.209 (0.000)	2.210 (0.038)
LAB	8.008 (0.000)	8.070 (0.000)	12.25 (0.002)

Source: Author's calculation.

Note: Probability value in parentheses.

Table 7.3. Cross-Sectional Dependence and Unit Root Tests.

Variable	TA	TR	ICT	INFS	GDP	FDI	LAB
Pesaran CD test							
Statistic	40.99	23.77	21.12	48.54	32.78	1.683	5.854
P value	0.000	0.000	0.033	0.000	0.000	0.092	0.000
CADF unit root test (at level)							
Statistic	−2.653	−2.654	−2.015	−3.192	−2.116	−2.690	−2.595
P value	0.032	0.032	0.078	0.000	0.028	0.021	0.057

Source: Author's calculation.

Note: Probability value in parentheses.

heterogeneous effects of independent variables on TD. Notably, ICT index estimates are significant and negative (positive) at lower (upper) quantiles.

Infrastructure (INFS) has a positive and significant effect on tourist arrival at second-half quantile range and significant but negative at first five quantiles. There is asymmetric impact of INF on tourist arrival, a negative (positive) relation at lower (upper) quantiles.

Concerning control variables, GDP per capita, labour force and FDI are also positively related to TD but results are heterogeneous. GDP exerts a positive impact on tourism.

With a view to checking the robustness of results, we have also used tourism receipts as proxy for dependent variable (see Table 7.5). Results show that ICT has statistically significant and positive impact on international tourist receipts at upper quantile range.

Regarding control variables, GDP is significantly and positively affecting tourism receipts at entire quantile range, implying that countries with high income can afford to invest in building new and maintaining old infrastructures, which in turn increase tourism receipts. FDI has negative as well as insignificant effect on tourism receipts.

Table 7.4. OLS and Powell (2022) Results (Dependent Variable: TA).

Variables	**(1) OLS**	**(2)** **0.05**	**(3)** **0.1**	**(4)** **0.2**	**(5)** **0.3**	**(6)** **0.4**	**(7)** **0.5**	**(8)** **0.6**	**(9)** **0.7**	**(10)** **0.8**	**(11)** **0.9**	**(12)** **0.95**
ICT	0.016 ***	0.023 ***	0.028 ***	0.028***	0.024***	0.032***	0.032***	0.024***	0.024***	0.029***	0.022***	0.023***
	(0.005)	(0.012)	(0.007)	(0.007)	(0.008)	(0.006)	(0.0087)	(0.0064)	(0.002)	(0.0052)	(0.0003)	(0.000)
INFS	0.022***	0.002	−0.002	−0.0002	−0.0007	−0.008	−0.0078	0.0005	0.0006	0.0003	0.002**	0.003***
	(0.009)	(0.006)	(0.006)	(0.005)	(0.005)	(0.005)	(0.0103)	(0.005)	(0.002)	(0.002)	(0.001)	(0.000)
GDP	0.590**	0.603*	0.559**	0.572**	0.601*	0.555**	0.5551***	0.592*	0.5764**	0.588**	0.599***	0.584***
	(0.049)	(0.060)	(0.048)	(0.042)	(0.059)	(0.051)	(0.0406)	(0.057)	(0.016)	(0.018)	(0.008)	(0.003)
FDI	0.150**	0.179*	0.154**	0.156**	0.169**	0.1555**	0.1555***	0.1609**	0.154**	0.1351**	0.138**	0.121***
	(0.025)	(0.063)	(0.016)	(0.027)	(0.037)	(0.0252)	(0.0243)	(0.0275)	(0.022)	(0.010)	(0.013)	(0.0044)
LAB	0.017***	0.012***	0.010***	0.011***	0.012***	0.009***	0.0096**	0.012***	0.011***	0.012***	0.014***	0.012***
	(0.005)	(0.002)	(0.002)	(0.002)	(0.002)	(0.002)	(0.0038)	(0.002)	(0.002)	(0.001)	(0.000)	(0.0009)

Source: Author's calculation.

Note: Parentheses show robust S.E. $^{***}p < 0.01$, $^{**}p < 0.05$, $^{*}p < 0.1$.

Table 7.5. OLS and Powell (2022) Results (Dependent Variable: TR).

Variable	(1) OLS	(4) 0.05	(5) 0.1	(6) 0.2	(7) 0.3	(8) 0.4	(9) 0.5	(10) 0.6	(11) 0.7	(12) 0.8	(13) 0.9	(14) 0.95
ICT	0.4222***	−0.544*	−0.361*	−0.264*	−0.330*	– 0.179***	−0.194	0.222***	0.158***	0.182***	0.175***	0.194***
	(0.076)	(0.250)	(0.197)	(0.110)	(0.172)	(0.042)	(0.154)	(0.080)	(0.042)	(0.022)	(0.026)	(0.018)
INFS	0.276*	0.167	0.219	−0.013	0.032	−0.051	−0.005	−0.064	−0.081	−0.065*	0.090*	0.101*
	(0.140)	(0.241)	(0.333)	(0.179)	(0.173)	(0.099)	(0.257)	(0.118)	(0.066)	(0.026)	(0.051)	(0.047)
GDP	0.586	6.152***	6.023***	3.620**	4.193**	4.406 ***	5.013***	4.040***	4.031***	4.353 ***	4.159***	4.169 ***
	(0.760)	(1.562)	(1.565)	(1.750)	(1.634)	(0.474)	(0.853)	(0.739)	(0.391)	(0.251)	(0.113)	(0.175)
FDI	−1.791***	−4.277	– 6.893***	– 6.654***	– 6.046***	– 8.568***	– 8.929***	– 7.882***	– 8.663***	−8.682 ***	−8.428 ***	– 8.462***
	(0.384)	(3.163)	(2.536)	(2.417)	(2.212)	(0.300)	(1.994)	(0.659)	(0.651)	(0.326)	(0.094)	(0.100)
LAB	−0.094	0.013	0.014*	−0.100 *	−0.080	– 0.076***	−0.075	– 0.087***	– 0.086***	– 0.082***	– 0.083***	−0.087 ***
	(0.078)	(0.078)	(0.089)	(0.058)	(0.074)	(0.023)	(0.047)	(0.026)	(0.017)	(0.013)	(0.010)	(0.013)

Source: Author's calculation.

Note: Parentheses show robust SE. *** indicate $p < 0.01$, ** indicate $p < 0.05$ and * indicates $p < 0.1$.

4.3 MM-QR Results

In order to ensure the robustness of results, we go one step further by employing MM-QR by Machado and Silva (2019) which can account for unobserved heterogeneity as well as endogeneity. Our results are significant when we used tourist arrival as dependent variable (see Table 7.6). ICT is positively related with TD at entire quantile range. Infrastructure negatively affects TD from 5th to 60th quantile, but again it has a positive impact on TD in more developed nation as its coefficients are positive at upper quantile range. Furthermore, GDP, labour force and FDI all exert positive impacts on TD in selected Asian economies. Explanations of these results are similar to those of the results of the Powell (2022) PQR method above. Finally, Table 7.7 produces the results of MM-QR. MM-QR produced mostly insignificant results when tourists' receipt is used as dependent variable (see Table 7.7).

5. The Discussion of Results

ICT has a positive effect on TD in upper quantile range as ICT enables tourists to acquire necessary information regarding their desired, consequently boosting the demand for tourism services. Findings suggest ICT significantly fosters tourism growth in developed Asian countries. Our findings corroborate with Hoonsawat (2016), Adeola and Evans (2020) and Lee et al. (2021). The reason for negative effect of ICT in lower five quantiles may be that the establishment cost of ICT and IT productivity paradox that lowers tourist arrival (Lee et al., 2021).

Infrastructure also has a positive effect on TD. This finding also corroborates the conclusion of Meo et al. (2018) who showed a long-term relationship between infrastructure and tourism. The negative impact of infrastructure on tourism could be attributed to the fact that poor countries do not contain improved infrastructure, which attracts more tourists. Our outcomes are in line with Adeola et al. (2018) and Adeola and Evans (2020) who demonstrated that as better infrastructure may lead to stronger tourism.

Government in high-income nations tend to spend a huge amount of investment in constructing and maintaining infrastructures for strengthening tourism sector (Saha & Yap, 2014). Similarly, upsurge in GDP can also have a direct impact on tourism (Castro-Nuño et al., 2013). Labour force releases a positive impact on tourist arrival. The positive coefficient of labour force participation rate illustrates that Asian economies have a huge amount of useful labour force, which contributes to tourism expansion. This is in line with Barkas et al. (2020) and Haini (2019). FDI is positively related to TD which could be attributed to the fact that FDI is required for establishing the enabling environment for tourism to flourish.

ICT boosts TD in developed nations. It is evident that more developed countries are able to introduce and exploit state-of-the-art technology with lower costs, thus accelerating TD. The negative coefficients of ICT from 5th to 50th quantiles are consistent with findings of Lee et al. (2021) where mobile cellular subscription was negatively related with tourist receipts. This result may be

Table 7.6. MM-QR Regression Results (Dependent Variable: TA).

Variable	**(1)**	**(2)**	**(3)**	**(4)**	**(5)**	**(6)**	**(7)**	**(8)**	**(9)**	**(10)**	**(11)**
	0.05	**0.1**	**0.2**	**0.3**	**0.4**	**0.5**	**0.6**	**0.7**	**0.8**	**0.9**	**0.95**
ICT	0.006**	0.008***	0.010***	0.010***	0.011***	0.012**	0.014 ***	0.015***	0.016***	0.018**	0.019**
	(0.012)	(0.009)	(0.006)	(0.006)	(0.005)	(0.005)	(0.006)	(0.007)	(0.008)	(0.011)	(0.013)
INFS	−0.038**	−0.033**	−0.029**	−0.028**	–	–	−0.020**	−0.017**	0.015**	0.011**	0.008**
	(0.020)	(0.015)	(0.011)	(0.010)	0.026***	0.024***	(0.010)	(0.012)	(0.014)	(0.019)	(0.020)
					(0.009)	(0.009)					
GDP	0.489	0.519 *	0.548*	0.5580*	0.571**	0.584***	0.612*	0.628*	0.643*	0.672*	0.687
	(0.104)	(0.079)	(0.058)	(0.053)	(0.048)	(0.046)	(0.055)	(0.064)	(0.075)	(0.099)	(0.112)
FDI	0.149*	0.150**	0.151**	0.152**	0.152**	0.153***	0.154**	0.155**	0.155**	0.156*	0.156*
	(0.056)	(0.042)	(0.031)	(0.028)	(0.025)	(0.025)	(0.029)	(0.035)	(0.040)	(0.053)	(0.060)
LAB	0.038**	0.032***	0.026***	0.0242***	0.021***	0.018***	0.013***	0.009***	0.006***	0.0009**	−0.002**
	(0.010)	(0.008)	(0.006)	(0.005)	(0.005)	(0.004)	(0.005)	(0.006)	(0.007)	(0.010)	(0.011)

Source: Author's calculation.

Note: Parentheses show robust SE. *** indicate $p < 0.01$, ** indicate $p < 0.05$ and * indicates $p < 0.1$.

Table 7.7. MM-QR Regression Results (Dependent Variable: TR).

Variable	(1) 0.05	(2) 0.1	(3) 0.2	(4) 0.3	(5) 0.4	(6) 0.5	(7) 0.6	(8) 0.7	(9) 0.8	(10) 0.9	(11) 0.95
ICT	−0.338**	−0.356***	−0.398***	−0.430	−0.457***	−0.481***	0.514***	0.555***	0.618**	0.700*	0.837
	(0.176)	(0.152)	(0.105)	*** (0.085)	(0.088)	(0.105)	(0.141)	(0.196)	(0.284)	(0.406)	(0.611)
INFS	−0.145	−0.105	−0.011	0.0600	0.119	0.173	0.247	0.340	0.479	0.664	0.968
	(0.305)	(0.264)	(0.184)	(0.150)	(0.155)	(0.185)	(0.247)	(0.342)	(0.496)	(0.711)	(1.077)
GDP	−3.075	−2.701	−1.834	−1.168	−0.613	−0.117	0.563	1.425	2.719	4.425	7.244
	(2.317)	(2.013)	(1.415)	(1.162)	(1.202)	(1.428)	(1.900)	(2.624)	(3.789)	(5.463)	(8.336)
FDI	1.374	1.061	0.332	−0.225	−0.691	−1.107	−1.679	−2.402	−3.488	−4.919	−7.285
	(2.585)	(2.244)	(0.570)	(1.280)	(1.330)	(1.581)	(2.108)	(2.914)	(4.217)	(6.061)	(9.208)
LAB	0.062	0.046	0.008*	−0.020*	−0.042*	−0.066	−0.096	−0.134	−0.191	−0.266	−0.389
	(0.147)	(0.127)	(0.088)	(0.072)	(0.074)	(0.088)	(0.118)	(0.164)	(0.239)	(0.341)	(0.515)

Source: Author's calculation.

Note: Parentheses show robust S.E. ***$p < 0.01$, **$p < 0.05$, *$p < 0.1$.

explained by the fact that high establishment costs of installing and setting up ICT in less developed countries may lower the tourism receipts. Infrastructure also has insignificant impact at 10th, 20th, 30th, 50th and 60th quantiles and significant positive impact on tourism receipts at 90th and 95th quantiles. It implies that in less developed countries, infrastructure does not play a significant role in developing tourism because these nations do not have better infrastructures. This result also suggests that a higher scale of infrastructural development may be needed to promote the tourism sector, lending support to Lee et al. (2021).

FDI has negative as well as insignificant effect on tourism receipts. This implies that in Asia, poor countries have low FDI inflow and so make no use of FDI in tourism-development related ventures, as per the case of more developed nations, low FDI inflows may be owing to some terrorism-related issues. This suggests that governments in emerging economies should leave no stone unturned to combat and kerb terrorism.

6. Conclusion

Despite the crucial importance of ICT and infrastructure in TD, little attention is paid to these main components of TD in Asian region. Thus, the current chapter aims to examine the role of ICT and infrastructure in TD of Asian region using data from 2008 to 2020. To achieve this goal, our study utilized the PQR estimation method by Powell (2022) and the MM-QR technique by Machado and Silva (2019), offering consistent estimates for short panel data while addressing outlier and heterogeneity challenges efficiently.

Empirical results reveal varying impacts of ICT and infrastructure on TD across quantiles, with higher quantiles showing stronger statistical significance. In the Asian region, ICT and infrastructure significantly enhance TD. Control variables, including income per capita, FDI and the labour force, also play an important role in the tourism sector.

It is evident that a number of policy implications stem from these results on both domestic and international level. First, from the perspective of policy-making, it is recommended that such policies should be devised to boost TD by promoting digital tourism since ICT has a significant and positive impact on international tourism. Moreover, rather than just upgrading multiple ICTs, it is advisable to prioritize TD quartiles and allocate resources to specific ICTs. Our findings have significant implications for enhancing the industry across various tourism factors, aiding effective management strategies in all conditions. Second, the governments of Asian countries should create favourable conditions for distinctive and impressive tourism experiences by raising their efforts to develop strong infrastructure. The governments of Asian nations should promote the policies to build enhanced physical infrastructure so that better prospects can be realized regarding increased tourist arrival as well as increased tourism receipt.

Finally, to facilitate the development of tourism industry, the governments of these Asian economies should carry out policies aimed at boosting economic growth in these economies as economic growth is confirmed to have a significant impact on invigorating investment in new infrastructural projects, which would be vital for healthy tourism activity in economy.

The crux of above policy suggestions is that a policy framework needs to be projected that facilitate the introduction of better developed and modern ICT tools, construction of better and advanced infrastructure and growth of developmental projects. However, there are some limitations to this research work. First, our study focused on just five ICT tools and excludes consideration of other cutting-edge technologies like robots and artificial intelligence (AI), which offer insights into the impact of advanced technology on the T&T industry's development. Second, the ongoing challenges posed by the COVID-19 pandemic to the T&T sector are not addressed in this research. Future studies should explore the benefits of advanced ICT during pandemic periods.

7. Declarations

- Ethical approval: Not applicable.
- Consent to participate: Not applicable.
- Consent to publish: Not applicable.
- Authors contributions: **S. Arsh:** Conceptualization, Data curation, Software, Formal analysis, Roles/Writing – original draft; Writing – review and editing. **S. Nasreen**: Conceptualization, Data curation, Methodology, Software, Formal analysis, Roles/Writing – original draft; Writing – review and editing. **X-H. Nghiem:** Methodology, Roles/Writing – original draft; Writing – review and editing.
- Funding: This research did not receive any specific grant from funding agencies in the public, commercial or not-for-profit sectors.
- Competing interests: None.
- Availability of data and materials: The datasets generated during and/or analysed during the current study are available from the corresponding author on reasonable request.

References

Adeola, O., Boso, N., & Evans, O. (2018). Drivers of international tourism demand in Africa. *Business Economics*, *53*, 25–36. https://doi.org/10.1057/s11369-017-0051-3

Adeola, O., & Evans, O. (2019). Digital tourism: Mobile phones, internet and tourism in Africa. *Tourism Recreation Research*, *44*(2), 190–202. https://doi.org/10.1080/02508281.2018.1562662

Adeola, O., & Evans, O. (2020). ICT, infrastructure, and tourism development in Africa. *Tourism Economics*, *26*(1), 97–114. https://doi.org/10.1177/1354816619827712

Andrianaivo, M., & Kpodar, K. (2011). *ICT, financial inclusion, and growth: Evidence from African countries*. https://burkina-ntic.net/ressources/assets/docP/Document_N013.pdf

Anwar, A., Siddique, M., Dogan, E., & Sharif, A. (2021). The moderating role of renewable and non-renewable energy in environment-income nexus for ASEAN countries: Evidence from Method of Moments Quantile Regression. *Renewable Energy*, *164*, 956–967. https://doi.org/10.1016/j.renene.2020.09.128

Arora, S., & Sharma, A. (2021). Covid-19 impact on overtourism: Diversion from mass tourism to alternative tourism, *Overtourism as destination risk* (pp. 275–283). Emerald Publishing Limited. https://doi.org/10.1108/978-1-83909-706-520211018

Arshad, M. I., Iqbal, M. A., & Shahbaz, M. (2018). Pakistan tourism industry and challenges: A review. *Asia Pacific Journal of Tourism Research*, *23*(2), 121–132. https://doi.org/10.1080/10941665.2017.1410192

Barkas, P., Honeck, D., & Rubio, E. (2020). International trade in travel and tourism services: Economic impact and policy responses during the COVID-19 crisis. https://doi.org/10.30875/3514a476-en

Bekteshi, L., & Bekteshi, J. (2017). Use of ICT and development of tourism sector in Albania. *European Scientific Journal*, *13*(25), 138–149. http://doi.org/10.19044/esj.2017.v13n25p138

Berne, C., Garcia-Gonzalez, M., & Mugica, J. (2012). How ICT shifts the power balance of tourism distribution channels. *Tourism Management*, *33*(1), 205–214. https://doi.org/10.1016/j.tourman.2011.02.004

Bethapudi, A. (2013). The role of ICT in tourism industry. *Journal of Applied Economics and Business*, *1*(4), 67–79. https://www.aebjournal.org/articles/0104/010406.pdf

Boopen, S. (2006). Transport capital as a determinant of tourism development: A time series approach. *Tourismos*, *1*(1), 53–71. https://doi.org/10.26215/tourismos.v1i1.9

Bramo, F. (2013). The role of railway infrastructure on the tourism development (case Albania). *Albanian Journal of Agricultural Sciences*, *12*(2), 297–300.

Buhalis, D., & Law, R. (2008). Progress in information technology and tourism management: 20 years on and 10 years after the internet – The state of eTourism research. *Tourism Management*, *29*(4), 609–623. https://doi.org/10.1016/j.tourman.2008.01.005

Canay, I. A. (2011). A simple approach to quantile regression for panel data. *The Econometrics Journal*, *14*(3), 368–386. https://doi.org/10.1111/j.1368-423X.2011.00349.x

Castro-Nuño, M., Molina-Toucedo, J. A., & Pablo-Romero, M. P. (2013). Tourism and GDP: A meta-analysis of panel data studies. *Journal of Travel Research*, *52*(6), 745–758. https://doi.org/10.1177/0047287513478500

Choudhary, S. A., Khan, M. A., Sheikh, A. Z., Jabor, M. K., bin Nordin, M. S., Nassani, A. A., Alotaibi, S. M., Abro, M. M. Q., Vo, X. V., & Zaman, K. (2020). Role of information and communication technologies on the war against terrorism and on the development of tourism: Evidence from a panel of 28 countries. *Technology in Society*, *62*, 101296. https://doi.org/10.1016/j.techsoc.2020.101296

Eilat, Y., & Einav, L. (2004). Determinants of international tourism: A three-dimensional panel data analysis. *Applied Economics*, *36*(12), 1315–1327. https://doi.org/10.1080/0003684042000180897

El Meriague, Q. (2014). The role of ICT in the tourism industry. *Morocco News*.

Enakrire, R. T., & Ocholla, D. N. (2017). Information and communication technologies for knowledge management in academic libraries in Nigeria and South

Africa. *South African Journal of Information Management*, *19*(1), 1–9. https://hdl.handle.net/10520/EJC-8c673bb95

Evans, O., & Kelikume, I. (2018). The effects of foreign direct investment, trade, aid, remittances and tourism on welfare under terrorism and militancy. *International Journal of Management, Economics and Social Sciences*, *7*(3), 206–232. https://papers.ssrn.com/sol3/papers.cfm?abstract_id=3287759

Feshari, M. (2017). The role of ICT indices in tourism demand of Iran (the FMOLS co-integrating approach). *Iranian Journal of Economic Studies*, *5*(2), 209–221. https://doi.org/10.22099/ijes.2017.22823.1280

Firoiu, D., & Croitoru, A.-G. (2013). Tourism and tourism infrastructure from the perspective of technological changes. *Romanian Economic and Business Review*, *8*(2), 93–103. http://www.rebe.rau.ro/REBE%208%202.pdf#page=93

Haini, H. (2019). Internet penetration, human capital and economic growth in the ASEAN economies: Evidence from a translog production function. *Applied Economics Letters*, *26*(21), 1774–1778. https://doi.org/10.1080/13504851.2019.1597250

Hoonsawat, R. (2016). Information searching: The case of tourism promoted through the internet. *Global Economy Journal*, *16*(1), 33–47. https://doi.org/10.1515/gej-2014-0056

Hunjra, A. I., Azam, M., Bruna, M. G., & Taskin, D. (2022). Role of financial development for sustainable economic development in low middle income countries. *Finance Research Letters*, *47*, 102793. https://doi.org/10.1016/j.frl.2022.102793

Hunjra, A. I., Hassan, M. K., Zaied, Y. B., & Managi, S. (2023). Nexus between green finance, environmental degradation, and sustainable development: Evidence from developing countries. *Resources Policy*, *81*, 103371. https://doi.org/10.1016/j.resourpol.2023.103371

Husein, J., & Kara, S. M. (2020). Nonlinear ARDL estimation of tourism demand for Puerto Rico from the USA. *Tourism Management*, *77*, 103998. https://doi.org/10.1016/j.tourman.2019.103998

Ike, G. N., Usman, O., & Sarkodie, S. A. (2020). Testing the role of oil production in the environmental Kuznets curve of oil producing countries: New insights from Method of Moments Quantile Regression. *Science of the Total Environment*, *711*, 135208. https://doi.org/10.1016/j.scitotenv.2019.135208

Ilic, I., & Nikolic, A. (2018). Implications of modern technology development for the tourism sector of the Republic of Serbia. *Ekonomika*, *64*(1), 37–52.

Imikan, A., & Ekpo, K. (2012). Infrastructure and Tourism Development in Nigeria: A case study of rivers state. *International Journal of Economic Development Research and Investment*, *3*(2), 53–60.

Jorgenson, D. W., & Vu, K. M. (2016). The ICT revolution, world economic growth, and policy issues. *Telecommunications Policy*, *40*(5), 383–397. https://doi.org/10.1016/j.telpol.2016.01.002

Kaul, R. N. (1985). *Dynamics of tourism: A trilogy. Transportation and marketing*. Sterling Publishers.

Khadaroo, J., & Seetanah, B. (2007). Transport infrastructure and tourism development. *Annals of Tourism Research*, *34*(4), 1021–1032. https://doi.org/10.1016/j.annals.2007.05.010

Khadaroo, J., & Seetanah, B. (2008). The role of transport infrastructure in international tourism development: A gravity model approach. *Tourism Management*, *29*(5), 831–840. https://doi.org/10.1016/j.tourman.2007.09.005

Khan, R. E. A., & Rasheed, M. K. (2016). Political economy of tourism in Pakistan: The role of terrorism and infrastructure development. *Asian Development Policy Review*, *4*(2), 42–50.

Khoshnevis Yazdi, S., & Khanalizadeh, B. (2017). Tourism demand: A panel data approach. *Current Issues in Tourism*, *20*(8), 787–800. https://doi.org/10.1080/13683500.2016.1170772

Kim, J., & Lee, C.-K. (2017). Role of tourism price in attracting international tourists: The case of Japanese inbound tourism from South Korea. *Journal of Destination Marketing & Management*, *6*(1), 76–83. https://doi.org/10.1016/j.jdmm.2016.03.002

Koenker, R. (2004). Quantile regression for longitudinal data. *Journal of Multivariate Analysis*, *91*(1), 74–89. https://doi.org/10.1016/j.jmva.2004.05.006

Koenker, R., & Bassett, Jr, G. (1978). Regression quantiles. *Econometrica: Journal of the Econometric Society*, 33–50. https://doi.org/10.2307/1913643

Kotiloglu, S., Lappas, T., Pelechrinis, K., & Repoussis, P. (2017). Personalized multi-period tour recommendations. *Tourism Management*, *62*, 76–88. https://doi.org/10.1016/j.tourman.2017.03.005

Kumar, N., & Kumar, R. R. (2020). Relationship between ICT and international tourism demand: A study of major tourist destinations. *Tourism Economics*, *26*(6), 908–925. https://doi.org/10.1177/1354816619858004

Law, R., Chan, I. C. C., & Wang, L. (2018). A comprehensive review of mobile technology use in hospitality and tourism. *Journal of Hospitality Marketing & Management*, *27*(6), 626–648. https://doi.org/10.1080/19368623.2018.1423251

Law, R., Leung, R., & Buhalis, D. (2009). Information technology applications in hospitality and tourism: A review of publications from 2005 to 2007. *Journal of Travel & Tourism Marketing*, *26*(5–6), 599–623. https://doi.org/10.1080/10548400903163160

Lee, C.-C., & Chang, C.-P. (2008). Tourism development and economic growth: A closer look at panels. *Tourism Management*, *29*(1), 180–192. https://doi.org/10.1016/j.tourman.2007.02.013

Lee, C.-C., Chen, M.-P., Wu, W., & Xing, W. (2021). The impacts of ICTs on tourism development: International evidence based on a panel quantile approach. *Information Technology & Tourism*, *23*, 509–547.

Li, H., & Suomi, R. (2008). Internet adoption in tourism industry in China. In Towards sustainable society on ubiquitous networks: The 8th IFIP conference on e-Business, e-Services, and e-Society (I3E 2008), 24–26 September 2008, Tokyo, Japan. https://doi.org/10.1007/978-0-387-85691-9_17

Lin, B., & Xu, B. (2018). Factors affecting CO_2 emissions in China's agriculture sector: A quantile regression. *Renewable and Sustainable Energy Reviews*, *94*, 15–27. https://doi.org/10.1016/j.rser.2018.05.065

Lopez Cordova, J. E. (2020). *Digital platforms and the demand for international tourism services*. World Bank Policy Research Working Chapter, 9147, 36. https://ssrn.com/abstract=3537945

Ma, J. X., Buhalis, D., & Song, H. (2003). ICTs and Internet adoption in China's tourism industry. *International Journal of Information Management*, *23*(6), 451–467. https://doi.org/10.1016/j.ijinfomgt.2003.09.002

Machado, J. A., & Silva, J. S. (2019). Quantiles via moments. *Journal of Econometrics, 213*(1), 145–173. https://doi.org/10.1016/j.jeconom.2019.04.009

Meo, M. S., Chowdhury, M. A. F., Shaikh, G. M., Ali, M., & Masood Sheikh, S. (2018). Asymmetric impact of oil prices, exchange rate, and inflation on tourism demand in Pakistan: New evidence from nonlinear ARDL. *Asia Pacific Journal of Tourism Research, 23*(4), 408–422. https://doi.org/10.1080/10941665.2018.1445652

Min, C.-K., Roh, T.-S., & Bak, S. (2016). Growth effects of leisure tourism and the level of economic development. *Applied Economics, 48*(1), 7–17. https://doi.org/10.1080/00036846.2015.1073838

Mohamed Aslam, A. L. (2020). Long-run relationship between ICT and tourism demand in Sri Lank. *Journal of Information Systems & Information Technology, 5*, 68–74. http://ir.lib.seu.ac.lk/handle/123456789/5435

Mowlana, H., & Smith, G. (1990). Tourism, telecommunications and transnational banking: A framework for policy analysis. *Tourism Management, 11*(4), 315–324. https://doi.org/10.1016/0261-5177(90)90066-I

Muchapondwa, E., & Pimhidzai, O. (2011). Modelling international tourism demand for Zimbabwe. *International Journal of Business and Social Science, 2*(2), 71–81.

Nadeem, M. A., Liu, Z., Xu, Y., Nawaz, K., Malik, M. Y., & Younis, A. (2020). Impacts of terrorism, governance structure, military expenditures and infrastructures upon tourism: Empirical evidence from an emerging economy. *Eurasian Business Review, 10*, 185–206. https://doi.org/10.1007/s40821-020-00152-y

Naudé, W. A., & Saayman, A. (2005). Determinants of tourist arrivals in Africa: A panel data regression analysis. *Tourism Economics, 11*(3), 365–391. https://doi.org/10.5367/000000005774352962

Navío-Marco, J., Ruiz-Gómez, L. M., & Sevilla-Sevilla, C. (2018). Progress in information technology and tourism management: 30 years on and 20 years after the internet-Revisiting Buhalis & Law's landmark study about eTourism. *Tourism Management, 69*, 460–470. https://doi.org/10.1016/j.tourman.2018.06.002

Ollo-López, A., & Aramendía-Muneta, M. E. (2012). ICT impact on competitiveness, innovation and environment. *Telematics and Informatics, 29*(2), 204–210. https://doi.org/10.1016/j.tele.2011.08.002

Padhan, H., Padhang, P. C., Tiwari, A. K., Ahmed, R., & Hammoudeh, S. (2020). Renewable energy consumption and robust globalization(s) in OECD countries: Do oil, carbon emissions and economic activity matter? *Energy Strategy Reviews, 32*, 100535. https://doi.org/10.1016/j.esr.2020.100535

Pesaran, M. H. (2007). A simple panel unit root test in the presence of cross-section dependence. *Journal of Applied Econometrics, 22*(2), 265–312. https://doi.org/10.1002/jae.951

Poon, A. (1993). *Tourism, technology and competitive strategies.* CAB international. https://www.cabdirect.org/cabdirect/abstract/19931857161

Powell, D. (2022). Quantile regression with nonadditive fixed effects. *Empirical Economics, 63*(5), 2675–2691. https://doi.org/10.1007/s00181-022-02216-6

Prideaux, B. (2000). The role of the transport system in destination development. *Tourism Management, 21*(1), 53–63. https://doi.org/10.1016/S0261-5177(99)00079-5

Rahman, H. U., Ali, G., Zaman, U., & Pugnetti, C. (2021). Role of ICT investment and diffusion in the economic growth: A threshold approach for the empirical evidence from Pakistan. *International Journal of Financial Studies, 9*(1), 14. https://doi.org/10.3390/ijfs9010014

Ramos, C. M., & Rodrigues, P. M. (2012). The importance of ICT for tourism demand: A dynamic panel data analysis. In *Quantitative methods in tourism economics* (pp. 97–111). Springer. https://doi.org/10.1007/978-3-7908-2879-5_6

Rhaman, M. R. (2016). *The socio-economic importance of tourism and its impact on the livelihood in south Asia: Case Rangamati, Bangladesh.* https://urn.fi/URN:NBN:fi:amk-2016103115572

Rowley, A. (1992). The way ahead: Real demand for infrastructure spending will fuel 1990s lending. *Far Eastern Economic Review*, *155*, 47–48.

Saha, S., & Yap, G. (2014). The moderation effects of political instability and terrorism on tourism development: A cross-country panel analysis. *Journal of Travel Research*, *53*(4), 509–521. https://doi.org/10.1177/0047287513496472

Seetanah, B., Durbarry, R., & Ragodoo, J. N. (2010). Using the panel cointegration approach to analyse the determinants of tourism demand in South Africa. *Tourism Economics*, *16*(3), 715–729. https://doi.org/10.5367/000000010792278437

Seetanah, B., Juwaheer, T. D., Lamport, M. J., Rojid, S., Sannassee, R. V., & Subadar, A. U. (2011). Does infrastructure matter in tourism development? *University of Mauritius Research Journal*, *17*, 89–108. https://doi.org/10.4314/umrj.v17i1.70731

Sharma, A., & Arora, S. (2022). Virtual reality and blogging in tourism: Approached towards creating better tourist experience. *Journal of Tourism Insights*, *12*(1), 6. https://doi.org/10.9707/2328-0824.1252

Smith, S. L. (1984). A method for estimating the distance equivalence of international boundaries. *Journal of Travel Research*, *22*(3), 37–39. https://doi.org/10.1177/004728758402200305

Tan, G. W.-H., Lee, V. H., Lin, B., & Ooi, K.-B. (2017). Mobile applications in tourism: The future of the tourism industry? *Industrial Management & Data Systems*, *117*(3), 560–581. https://doi.org/10.1108/IMDS-12-2015-0490

Tcheng, H., Huet, J. M., Viennois, I., & Romdhane, M. (2007). Telecoms and development in Africa: The chicken or the egg. *Convergence Letter*, *8*(16).

Tsaurai, K., & Chimbo, B. (2019). Information and communication technology (ICT) led tourism growth nexus in transitional markets. *Acta Universitatis Danubius – Œconomica*, *15*(5), 36–53. https://www.ceeol.com/search/article-detail?id=855777

Tsokota, T., von Solms, R., & van Greunen, D. (2019). The reticent effect of ICT on tourism: A case study of Zimbabwe. *8*(3), 1–17.

Turayev, B., & Sharma, A. (2021). Development of the tourism industry before and after the covid pandemic. *Procedia of Social Sciences and Humanities*, *1*, 178–192. https://doi.org/10.21070/pssh.v1i.40

Ukpabi, D. C., & Karjaluoto, H. (2017). Consumers' acceptance of information and communications technology in tourism: A review. *Telematics and Informatics*, *34*(5), 618–644. https://doi.org/10.1016/j.tele.2016.12.002

Ummalla, M., Samal, A., & Goyari, P. (2019). Nexus among the hydropower energy consumption, economic growth, and CO_2 emissions: Evidence from BRICS countries. *Environmental Science and Pollution Research*, *26*, 35010–35022. https://doi.org/10.1007/s11356-019-06638-1

UNWTO. (2010). *Definition of tourism (UNWTO definition of tourism)/what is tourism?* https://www.tugberkugurlu.com/archive/definintion-of-tourism-unwto-definition-of-tourism-what-is-tourism. Accessed on 4/1/2021.

Vellas, F., Bécherel, L., Vellas, F., & Bécherel, L. (1995). The theoretical economic determinants of international tourism. *International Tourism: An Economic Perspective*, 63–95. https://doi.org/10.1007/s11356-019-06638-1

Wahab, I. N. (2017). Role of information technology in tourism industry: Impact and growth. *International Journal of Innovative Research in Computer and Communication Engineering*, *5*(2), 260–263.

Waverman, L., Meschi, M., & Fuss, M. (2005). The impact of telecoms on economic growth in developing countries. *The Vodafone Policy Chapter Series*, *2*(03), 10–24.

WDI. (2021). *World Development Indicators*. https://databank.worldbank.org/source/world-development-indicators. Accessed on 1/1/2021.

WTO. (1997). *International tourism: A global perspective* (English version). World Tourism Organization.

Zhang, T., Wei, W., Fu, X., Hua, N., & Wang, Y. (2019). Exploring the roles of technology, people, and organization in building a tourism destination experience: Insights from the 2nd USA-China tourism research summit and Industry Dialogue. *Journal of Destination Marketing & Management*, *12*, 130–135. https://doi.org/10.1016/j.jdmm.2019.03.001

Chapter 8

The Tourism–Climate Nexus: A Conceptual Review of Their Interactions

Xuan-Hoa Nghiem[a], *Huong Trang Pham*[b], *Thu Giang Nguyen*[c] *and Thi Kim Duyen Nguyen*[a,*]

[a]International School, Vietnam National University, Vietnam
[b]The Business School, RMIT University, Vietnam
[c]Foreign Trade University, Vietnam

Abstract

Climate change has been universally recognized as a major threat to human well-being, necessitating a comprehensive transformation of people's activities. Various measures have been proposed to contain climate change among which the green transformation grabs special attention, thanks to its desirable properties. Within the green transformation process, green tourism comes to prominence with huge potential. As one of the largest carbon emitters, the transition towards green tourism may offer substantial benefits not only for tourism companies but also for the whole economy. Yet, most studies tend to focus on the adverse effects of tourism on climate change while overlooking the potential impact of climate change on tourism. This chapter clarifies the feedback relationship between climate change and tourism and makes some recommendations.

Keywords: Climate change; tourism impacts; tourism stakeholders; sustainable tourism; green tourism

1. Introduction

Climate change is widely acknowledged as one of the most pressing issues of the 21st century. It presents formidable challenges to governments and policymakers

*Corresponding Author

The Emerald Handbook of Tourism Economics and Sustainable Development, 147–167

Published under exclusive licence by Emerald Publishing Limited
doi:10.1108/978-1-83753-708-220241008

globally. According to the United Nations (UN), consequences caused by climate change, to name but a few, include ever-higher global temperatures, more frequent and devastating storms, destruction of natural habitats and species, harsher drought, food insecurity, human poverty and displacement, and health-related problems. In response to the severity of these consequences, the UN in 2015 adopted the so-called 17 Sustainable Development Goals (SDGs).[1] These goals aim to eliminate poverty, safeguard the planet, promote peace and prosperity by the year 2030.[2] Among these SDGs, SDG 13 is exclusively aimed at tackling climate change, while others (including SDGs 6, 7 and 11) contribute significantly to mitigating it.

The tourism sector is not exempt from the extensive and profound impacts of climate change. Fang et al. (2018) suggest that the tourism industry is particularly susceptible to climate change, owing to its strong dependence on and connection with the climate and natural environment. The UN identifies key factors contributing to climate change as energy use, transportation, the production and consumption of foods and goods as well as power plant developments. These factors are, to some degree, associated with tourism activities. A 2008 report by the United Nations World Tourism Organisation (UNWTO), the United Nations Environment Programme (UNEP) and the World Meteorological Organization (WMO) showed that in 2005, travel and tourism's greenhouse gas (GHG) emissions made up about 5% of global emissions. The bulk of these emissions came from transportation (75%) and accommodation (21%), with tourism-related transport contributing to 18% of all transport emissions. Lenzen et al. (2018) revealed a notable rise in the global carbon footprint of tourism between 2009 and 2013. This surge accounted for approximately 8% of worldwide greenhouse gas emissions. The breakdown of emissions within the sector was as follows: transportation was responsible for 49%, retail accounted for 12%, food and beverage services contributed 10% and accommodations made up 6%. Further, a collaborative study in 2019 by the UNWTO and the International Transport Forum (ITF) highlighted that emissions related to both international and domestic tourism transportation made up 5% of all man-made emissions and 22% of total transportation emissions as of 2016. Turayev and Sharma (2021) noted occurrences of environmental rejuvenation during the pandemic, attributed to the decrease in tourist activities. This led to outcomes like more pristine beaches and a reduction in pollution levels. Although this number could have decreased owing to the impact of the COVID-19 pandemic, a solid global economic recovery indicates that this number will rise again. Pang et al. (2013) posit that the tourism sector is distinct in its dual role related to climate change: it is not only affected by climate change but also contributes to it. Despite the intricate interplay and mutual dependence between tourism and climate change, research exploring this

---

[1] https://www.un.org/en/climatechange/science/causes-effects-climate-change. Access date 01 November 2022.

[2] https://www.undp.org/sustainable-development-goals?utm_source=EN&utm_medium=GSR&utm_content=US_UNDP_PaidSearch_Brand_English&utm_campaign=CENTRAL&c_src=CENTRAL&c_src2=GSR&gclid=CjwKCAiA_vKeBhAdEiwAFb_nraYI7Q-nXOK1BorDa9WoRnIornvTmM0KMHjnXTf2Oh_7PaIS3TgCMhoCQysQAvD_BwE

specific nexus seems to be less prevalent compared to other areas of climate change studies. Furthermore, the majority of prior research has concentrated on either the influence of climate change on tourism or vice versa without systemically reviewing and proposing solutions that stakeholders can implement to make the tourism industry become resilient to climate change. Sharma and Yedla (1970) underscore the disequilibrium stemming from the excessive crowding of famed tourist spots, detrimentally affecting both the areas and their local communities. Notably, there is a discernible lack of initiative by government bodies, private sectors and media outlets in bringing lesser-known travel destinations to the fore. The tourists' overconsumption and poor handling of natural resources have significantly depleted what is available for the local inhabitants, catalysing a mounting opposition to the further expansion of tourism. Moreover, there is a notable absence of studies that examine tourism and climate change within the framework of the SDGs. This chapter is aimed at filling this void in the literature.

This conceptual review chapter presents an analysis of the tourism–climate change nexus, structured into five sections: The introductory section frames the interlinkages. Section 2 delves into the reciprocal impacts between climate change and tourism. It explores how tourism influences climate change through facets like transportation, accommodations and food services, as well as how climate change adversely affects the natural resources fundamental to tourism. Section 3 details climate change adaptation responses, highlighting the importance of tourism stakeholders, sustainability frameworks and green tourism. Section 4 discusses implications and critical perspectives on sustainable tourism measures. Finally, the conclusion synthesizes key insights on navigating tourism industry decarbonization and resilience strategies in light of substantial projected growth and already locked-in climate alterations.

This chapter makes significant contributions to understanding how tourism and climate change interact. First, it offers an up-to-date and thorough conceptual review that looks at various aspects of tourism, such as transportation, accommodations and food systems, to assess the industry's total emissions. While many studies focus on specific areas like air travel or hotels, this chapter adopts a comprehensive approach to the entire industry. Additionally, the chapter highlights the substantial current and future increase in emissions from tourism. It emphasizes the urgent need for measures to reduce this growth, as many studies downplay the impact of tourism on climate change. Despite perceptions of modest effects, the chapter contends that the industry's rapid expansion could have significant consequences without a good sustainability planning.

2. The Tourism–Climate Change Nexus

2.1 The Impact of Tourism on Climate Change

Tourism's relationship with climate change is twofold: it not only experiences the effects of climate change but also plays a role in exacerbating it, thereby contributing to the degradation of natural tourist destinations. Notably, tourism is responsible for 8% of global carbon emissions (Lenzen et al., 2018). With the annual increase in travel, these GHG emissions are becoming increasingly unsustainable.

Further emphasising this point, Lenzen et al. (2018) found that in 2013, activities associated with tourism, including travel, accommodation, dining and other related actions, constituted 8% of all human-caused GHG emissions. The study showed that tourism generated 4.5 GtCO2e of carbon dioxide emissions in 2013, an increase from 3.9 GtCO2e in 2009. The COVID-19 pandemic caused a 7% reduction in global GHG emissions in 2020, primarily due to decreased economic activity and travel restrictions, according to research by the Global Carbon Project (Le Quéré et al., 2020). In order to align with the objectives set by the Paris Agreement, the Intergovernmental Panel on Climate Change (IPCC, 2018) calculates that a yearly reduction in emissions of approximately 7% will be necessary over the coming decade. However, based on current trends, the tourism sector is predicted to emit 6.5 GtCO2e by 2025, making it one of the fastest-growing sources of GHG emissions (UNEP, 2017).

Transport plays a critical role in the GHG emissions of the tourism sector. It is estimated that global tourism accounts for about 5% of worldwide carbon dioxide emissions, with transportation constituting the majority (75%) of these tourism-related emissions. Specifically, air travel is responsible for half of the total emissions from transportation, particularly in high-income countries (Lenzen et al., 2018). The significance of travel during arrival and departure phases cannot be overlooked, as demonstrated by the research of Markus et al. (2019). They conducted a study on the mobility-related CO_2 balance in Alpbach, a Tyrolean destination and found that around 80% of the destination's mobility-related CO_2 emissions were attributable to guests. The majority of these emissions were from car travel for arrivals and departures (approximately 60% of guest emissions) and air travel (about 30%).

Abegg et al. (2019) emphasized that the energy and emission intensities in tourism are highly variable, influenced by factors such as different markets (e.g. domestic, short-haul and long-haul), types of tourism products (e.g. bicycle vs cruise tourism), travel distances (often linked to preferred transportation modes) and choices of accommodation and activities (Chenoweth, 2009; Higham et al., 2013). This variability underscores the complexity of assessing and addressing the environmental impact of tourism-related transportation.

Lodging is responsible for 6% of emissions, primarily due to the energy consumed for heating, cooling, lighting and operating appliances and other services (Lenzen et al., 2018). Similar to the way residential homes produce emissions through energy use, hotels, homestays and rental properties during vacations also generate emissions. A significant portion of these emissions comes from heating and air conditioning systems used in accommodations to maintain comfortable temperatures in varying climates. Additionally, water heating for showers, pools and spas, as well as electricity for lighting, televisions, refrigerators, laundry machines and other equipment, contribute to CO_2 emissions. This is particularly true in places with outdated or inefficient energy systems. Implementing infrastructure upgrades and integrating renewable energy sources can significantly reduce the climate impact of these lodging facilities.

Food and beverages is responsible for approximately 10% of global emissions (Lenzen et al., 2018). The acquisition of food for tourists involves a chain of carbon-emitting processes, including the cultivation, processing, transportation, storage and preparation of these items. Additionally, the tendency for overconsumption while on vacation further escalates these emissions (Gössling et al., 2011). A significant issue arises with the importation of food, a phenomenon particularly noticeable in tropical island states. Here, hospitality venues frequently serve high-quality, protein-rich dishes designed to cater to the tastes of affluent tourists. These discerning visitors typically anticipate finding familiar food items from their home regions. In such locales, a considerable portion of the food supply depends largely on air transportation, which includes a range of products such as soft drinks, dairy and vegetables. This reliance on imported goods, often transported by air, contributes substantially to the environmental footprint of these destinations (Gössling & Schumacher, 2010). It is important to highlight that in regions like Europe, the USA or Australia, the transportation of food items can significantly impact GHG emissions. This underlines the need to acknowledge and address the environmental consequences associated with the movement of food across these areas (Gössling et al., 2011; Gössling & Schumacher, 2010). Substantial *food waste* also increases emissions. Hotel buffets and large portions lead to excess waste, negating emissions from production.

Tourists' shopping activities are another matter of concern. When travelling, many tourists indulge in purchasing keepsakes such as magnets and various trinkets as a memory of their journeys. However, their shopping extends beyond mere souvenirs, encompassing everything from quaint local stalls to high-end designer stores. These products bear an environmental cost due to their production, manufacturing and transportation processes. While the charm of locally crafted items is undeniable, the reality often involves mass-produced goods originating from distant locations. Take, for example, a typical t-shirt bought in New York: the cotton might be sourced from China, manufactured in Vietnam and finally transported to New York for sale. Such consumer patterns among tourists tend to amplify the environmental impact compared to local shopping practices, predominantly through increased production emissions.

Destruction of natural attractions by tourism-induced climate change: The impact of tourism-induced climate change is a significant threat to natural attractions worldwide. This change endangers every form of life and the very destinations that draw tourists. From the verdant tropics to the pristine snow-capped mountains, many beloved travel destinations face imminent risks. As global temperatures rise, phenomena like escalating sea levels, extreme weather events and increasing temperatures are affecting ecosystems and communities globally. This results in diminishing beaches, bleaching of coral reefs and alpine resorts losing their snow cover. The link between tourism, particularly nature-based tourism, and biodiversity is profound, relying heavily on the allure of diverse and rich environments. However, this relationship is strained when tourism leads to biodiversity loss. Excessive use of land and resources, impacting flora and fauna and stressing various ecosystems, including marine, mountainous and coastal areas, along with water resources, can lead to their degradation. The introduction of exotic species by tourists and suppliers – ranging from insects and plants to diseases – can disrupt local ecosystems, sometimes leading to their destruction. Moreover, the development of tourism often increases resource consumption in areas where they are already scarce, adding further pressure to natural resources. This complex interplay of factors underscores the challenges in balancing tourism with the preservation of natural environments and resources.

Fresh water stands as a critical natural resource, yet its excessive consumption in the tourism sector is a growing concern. Resorts, hotels, swimming pools, restaurants and golf courses, along with the personal usage of visitors, contribute to this overuse. This pattern often leads to water shortages, the deterioration of water sources and an increase in wastewater production. In hot and dry regions, such as the Mediterranean, the scarcity of water poses a significant challenge. Additionally, tourists tend to consume more water while on vacation. Factors such as the hot climate contribute to tourists using up to 440 litres of water per day, nearly double the average daily usage of a typical resident in a Spanish city (UNEP, 1995).

Local Resources: Tourism can significantly strain vital local resources, including food and energy, which might be scarce. This increased demand often exacerbates the physical toll on these resources, leading to a need for their further extraction and relocation. Additionally, due to the seasonal nature of tourism, some destinations experience a dramatic fluctuation in population, often increasing tenfold between peak and off-peak seasons. This surge necessitates an increased supply of resources to support additional tourist amenities, such as heating and hot water (Cohen et al., 2013).

Land degradation is another concern. Essential natural resources such as fossil fuels, minerals, forests, fertile land, wetlands and wildlife face increasing pressure from tourism and its related activities. The development of recreational facilities and enhanced infrastructure attracts more people to these areas, leading to heightened utilization and enjoyment of these natural resources and scenic landscapes (UNWTO & UNEP, 2011). The provision of visitor accommodations and amenities directly impacts both renewable and non-renewable natural resources. These land uses, mainly gathering firewood and clearing land, have the worst impact on forest resources, according to (Mohamed et al., 2019). For instance, trees have been cut down because of trekking in the Himalayas, Sikkim and Assam.

The consequences of climate change mitigation efforts on tourists' freedom of movement and spending patterns will become apparent as the tourism sector adapts to the phenomenon. The long-term viability of the tourism sector may be hampered by efforts to mitigate the consequences of climate change. Products for tourism may only be available for a shorter period, prices may increase due to increased operating expenses and there may be fewer stunning natural destinations to visit. Therefore, mitigation is crucial in the tourism industry. However, mitigation measures must consider various aspects, including maintaining a stable global climate, people's right to leisure, and achieving the UN's SDGs (Cohen et al., 2021).

2.2 The Impact of Climate Change on Tourism

The widespread negative impacts of climate change are expected to affect numerous economic sectors, and the tourism industry is no exception to this trend. Fang et al. (2018) contend that due to its intimate connection with the environment, the tourism industry is particularly susceptible to the ramifications of climate change. They also identify several aspects through which climate change could influence tourism, including tourist safety, thermal comfort and weather conditions.

Tourism, increasingly subjected to the diverse and adverse effects of climate change, is identified as one of the five industries most vulnerable to various climate hazards.[3] These hazards include changes in warming temperatures, heatwaves, fires, drought, precipitation patterns, floods, alterations in natural land cover, rising sea levels and shifts in ocean chemistry (Mora et al., 2018). Pang et al. (2013) also emphasize that tourism is not immune to climate change's impacts. They further highlight the unique position of the tourism industry, as it not only significantly contributes to GHG emissions – a primary cause of climate change – but is also likely to suffer from its consequences.

Given the broad-ranging effects of climate change on the tourism sector, this section aims to provide a comprehensive enumeration of these impacts. Mora et al. (2018) point out that among various income-generating activities, tourism is especially susceptible to climate change. Yet, research exploring the intersection between climate change and the tourism industry seems to be less advanced compared to other climate-related fields, with initial studies emerging only in the 1990s (Giles & Perry, 1998; Harrison et al., 1999; Smith, 1990; Wall, 1998). Consequently, there is a scarcity of systemic evaluations of climate change's potential impacts on tourism. This section intends to address this gap in the literature.

As the old saying 'every coin has two sides' goes, the climate–tourism nexus should not be an exception. In addition to undesirable effects, climate change's impact may also bring unexpected side benefits to a proportion of people and firms engaging in the tourism industry. Therefore, it should be noted that while most players in the tourism industry are expected to be adversely affected, some winners will benefit from the situation.

To the best of the authors' understanding, the study by Scott et al. (2008) stands as the most extensive work detailing the adverse effects of climate change on tourism. In contrast to Scott et al. (2008), which differentiates the implications of climate change into direct and indirect impacts, this study opts to categorize the effects of climate change on tourism based on their beneficial and detrimental nature.

2.2.1 Detrimental Impacts

Lower Number of Tourists: Climate change is believed to negatively impact the global economy. According to the Swiss Re Institute (2021), if left unchecked, in the worst-case scenario, climate change may wipe off 18% of global gross domestic product (GDP) in 2050. In another study, Kahn et al. (2021) confirm that climate change significantly and negatively impacts real GDP in 174 countries from 1960 to 2014. They also estimate that if climate change is left unaddressed, the loss in global

---

[3]Although they may have different meanings, in this chapter, the two words 'tourism sector' and 'tourism industry' are considered as having the same meaning and will be used interchangeably.

GDP in 2100 may reach 7%. Lower income, in turn, will likely cause potential tourists to rethink, reduce, delay or even abandon their travel plans, resulting in fewer tourists.

Decreasing Competitiveness and Attractiveness of Tourism Destinations: It is widely acknowledged that climate change has detrimental impacts on the global environment, and it stands to reason that tourism destinations would also be subject to these negative influences. Specifically, consequences of climate change include loss of biodiversity, rising sea levels, frequent natural catastrophes, water pollution and unavailability, alteration of traditional production activities, erosion of coastal areas, destructions of infrastructures, etc. These consequences will likely make places of interest less attractive and less competitive, leading to losses.

Mora et al. (2018) confirm that tourism is one of the few sectors subject to the full impacts of climate change. For example, in their study, Buckley and Foushee (2012) estimate that visitors cancel their plans to go to US national parks due to higher temperatures. Similarly, Liu (2014) affirms the similar impact of storms on reducing the number of tourists in Taiwan. Moreover, the effects of higher temperatures on tourism activities in the winter and related to the ocean could even be more severe. For example, Pickering et al. (2010) and Scott and Mcboyle (2007) confirm the negative influence of higher temperatures on skiing activities in Australia and the Alps, respectively. In addition, Wilkinson (1996) and Meynecke et al. (2017) conclude that variations in ocean chemistry components may cause coral reefs' degradation, resulting in lower profits for tourism activities in Thailand and Australia, respectively.

2.2.2 Beneficial Impacts

On the one hand, climate change may bring about disastrous consequences. But, on the other hand, it can also invite some unexpected beneficial effects as follows.

Redistribution of Tourists: One of the unexpected benefits of climate change is the redistribution of tourists from climate-sensitive destinations to climate-resilient ones. As tourism destinations may suffer severe consequences from climate change, people planning to visit these beauty spots will likely reconsider their plans and choose a different place of interest. So there will be a 'loser-winner' situation created by climate change: some destinations may be hurt while others may benefit from climate change.

Higher Level of Awareness and Adaptation: Lower revenues or even losses incurred by tourism destinations and tourism companies at the micro-level and lower contributions to GDP at the macro-level may prompt relevant authorities and stakeholders to rethink their business strategies and pursue more sustainable approaches. Local governing bodies are expected to implement strategies for adapting to climate change, which could subsequently enhance environmental performance in these areas.

In conclusion, the influence of climate change on tourism is multifaceted, presenting both potential benefits and drawbacks. Conversely, tourism is recognized as a substantial contributor to climate change.

3. Tourism Adaptive Responses to Climate Change

3.1 Engaging Tourism Stakeholders for Climate Change Adaptation

Adaptation to climate change is necessary and has received attention from policymakers and tourism service providers. However, given the high dependency of tourism on climate factors, the core of adaptation activities is the cooperation of stakeholders. Tourism can be characterized as a dynamic interplay, encompassing the processes, activities and results that emerge from interactions among various stakeholders. These include tourists, providers of tourism services, host governments, local communities and the environments in which they operate, all playing a role in attracting and accommodating visitors (Goeldner & Ritchie, 2007). Thus, the interactions and cooperation among stakeholders are critical for tourism development towards sustainability. Freeman (1984, 2010) describes stakeholders as any individual or group that has the potential to influence or be influenced by the accomplishment of an organization's goals. The stakeholder-based theory, as elaborated by Phillips et al. (2003), is a functional approach that emphasizes ethics and values as central to organizational management. Based on this theory, one organization's past, present and future state to its stakeholders can be described. Goeldner and Ritchie (2007) integrated the stakeholder theory into their analysis, identifying four key stakeholders in the tourism industry: tourists, service providers, the host community and the host government. These stakeholders play a vital role in shaping the historical, current and future conditions of an organization in which they are involved. Consequently, they are essential in the process of adapting to climate change. Essentially, human activities in tourism are central to climate change adaptation efforts. The level of awareness among these stakeholders about climate change and its effects on tourism is crucial in directing their responsible actions. Each stakeholder might be involved in one way or another, primarily due to the fragmented feature of the tourism industry. As a result, the government may take the leading role in decision-making, thus, having broader impacts and many tools to promote adaptation activities towards climate change compared to other stakeholders.

The stakeholder theory identifies three aspects: the defining aspect, the instrumental aspect and the normative aspect. The descriptive aspect focuses on describing some characteristics and behaviour that affect the organization's past, present and future state. The instrumental aspect identifies the connection in a good or bad way between stakeholders, which affects the achievement of an organization's goals. Finally, the normative aspect illustrates the function, moral and philosophical guidelines for managing that organization (Donaldson & Preston, 1995). Thus, the normative aspect is considered the most vital aspect of the theory.

Linked to tourism, this theory emphasizes how the primary stakeholders in tourism are managed to ensure climate change adaptation. Tourists travel to new destinations far from home for business or professional purposes. Their actions and behaviour in different places outside of their usual environment left behind their footprints which lead to social, cultural and economic issues in both positive and negative ways. With good awareness and knowledge about climate change and its impact on tourism, their behaviour will be more responsible.

Service providers in the tourism sector often view tourism activities as avenues for profit generation through the provision and sale of goods and services that meet tourists' needs. Their understanding of how climate change affects tourism influences their strategies for adaptation and mitigation. In recent years, there has been a noticeable shift towards more sustainable practices in the operations of tourism destinations and enterprises. These changes, aimed at adapting to climate change, include measures such as reducing energy consumption and enhancing energy efficiency through the use of renewable and eco-friendly resources (Pandy, 2017). Specifically, in regions severely impacted by climate change, like ski resorts experiencing reduced snowfall, innovative adaptations have been implemented. These adaptations range from the utilization of snow-making machines and the construction of snow pools to the introduction of alternative tourism offerings (Brouder & Lundmark, 2011; Pandy, 2017).

Local authorities view tourism as a GDP generation factor for the economy and are an essential stakeholder in policy formulation and implementation (Amoako et al., 2022). Although other stakeholders are involved in these processes in one way or another, their power in decision-making or participation is unequal (Byrd, 2007). The current view of climate change adaptation highlights local authorities' critical role in developing national, regional, and local strategies. Consequently, various strategies, methods, and recommendations have been developed for public bodies, companies, and individuals. These resources aim to enhance understanding of climate change and foster readiness for both adaptation and mitigation measures (Baker et al., 2012; Lawrence et al., 2015). The approaches for climate change policy at local levels are based on a quasi-legal framework, emergency management focussing on immediate responses (Dilling et al., 2017; Lawrence et al., 2015). As the tourism industry is fragmented, UNEP (2005) suggested that local government should lead in climate change adaptation. Despite the growing recognition of climate change impacts that requires immediate and long-term strategies, international and national authorities are being relatively slow in their responses (Lesnikowski et al., 2021).

Local communities view tourism as cultural exchange and job creation (Amoako et al., 2022). However, local people are vulnerable to climate change as they depend on land and resources within the ecosystems, particularly prone to climate change impacts. In response, local communities are critical for taking action for adaptation. In tourism, indigenous people protect biological diversity, forests and natural resources. As a result, a wide range of community-led activities has been implemented worldwide (Ramos-Castillo et al., 2017). Tourism involves interactions of different stakeholders, in various roles, from other places outside their zones resulting in social, cultural and economic movements; it is essential to understand the stakeholder theory and their engagement in climate change adaptation.

3.2 Sustainable Tourism: An Adaptive Response to Climate Change

The evolution of sustainable tourism has recently become a pivotal topic in the tourism industry. Defined by the World Tourism Organization, sustainable tourism is an approach that thoroughly considers its impacts – both present and future – across economic, social and environmental spheres, while fulfiling the needs of visitors, the industry, the environment and host communities. Alternatively, it can be viewed as a harmonious interplay of three dimensions: environmental, economic, and social-cultural.

From 2000 to 2019, the tourism of Association of South East Asian Nations (ASEAN) witnessed robust growth, averaging a yearly rate of 6.7% (Hong & Giang, 2022). Despite this, the equilibrium among these three sustainability factors has not been effectively maintained. The rapid expansion of the tourism sector has often overshadowed environmental concerns. Environmental degradation poses a significant threat to the achievements in other areas; a polluted destination loses its allure, resulting in decreased tourist numbers and a direct impact on the income of local communities. Moreover, restoring the environment can be financially burdensome and is often complicated by economic considerations, particularly when tourist areas need to close for environmental rejuvenation.

This scenario is not uncommon in ASEAN countries. For example, Maya Bay in Thailand saw its coral reefs decimated due to a tourism boom that began in the early 2000s. The bay's visitor count soared nearly 3000-fold in two decades, enriching the local economy but devastating its marine life. Activities like snorkelling and diving, compounded by the harmful effects of sunscreen, led to significant coral bleaching and death by 2017. In response, Thai authorities closed the bay from 2018 to 2021 for ecological restoration, highlighting the critical balance between tourism and environmental preservation (Rajendra et al., 2021). Similarly, the Philippines' Boracay Island was temporarily shut down in 2018 (Al Jazeera, 2018). In Vietnam, pollution along a 300-kilometre stretch of beach resorts has led to visitor dissatisfaction. Reflecting these environmental challenges, the World Economic Forum's Travel and Tourism Competitiveness Report 2019 ranked several ASEAN countries low on environmental sustainability.

The surge in tourist arrivals has stressed infrastructure and administrative management in these destinations, leading to overcrowding and significant pollution. However, overcrowding is not the sole culprit. For instance, Bali's beaches remained littered with plastic waste during the COVID-19 pandemic due to single-use plastics, inadequate waste management and lack of marine environmental regulations.

The environmental neglect in ASEAN tourism may be attributed to insufficient funding and capacity. However, the unequal distribution of tourism profits is also a significant factor. Studies indicate high tourism leakage in the accommodation industry in Bali (51%) and Siem Reap (around 40%), where profits primarily benefit external investors rather than local reinvestment for tourism preservation.

This disparity in benefits may drive residents to further exploit resources for economic gain, diminishing their interest in sustainable practices. This short-term mindset, neglecting local heritage conservation, could increase the vulnerability of local communities in the long run. The COVID-19 pandemic, despite impacting tourism revenues, provided an environmental respite and a chance for countries to better prepare for tourists' return. The gradual reopening to a controlled number of visitors offers an opportunity to refine sustainable tourism development strategies, involving key stakeholders, especially local communities. In Bali, a collaborative action plan involving the government, international organizations and locals has been initiated to combat plastic waste, demonstrating the importance of collective responsibility in sustainable tourism development.

3.3 Tourism as a Solution: Responsible Tourism and Green Tourism

Tourism is not just a victim of climate change but has contributed to at least 8% of carbon emissions (Campos Herrero et al., 2022). Transforming tourism for climate actions and sustainable development promotes a new tourism trend named responsible tourism. Responsible tourism necessitates that every stakeholder acknowledges and assumes responsibility for the environmental integrity, the local populace and the cultural heritage of the destination. In doing so, responsible tourism aims to reduce adverse social, economic and environmental effects, simultaneously fostering advantages that contribute to the welfare of the local community (Shuxin et al., 2022). As a result, responsible tourism creates a shared value and vision among stakeholders (Camilleri, 2016). Many intergovernmental organizations, activists and scholars have contributed to the concept and development of responsible and sustainability (Camilleri, 2016; Goodwin, 2007). In addition, responsible tourism works on alternatives for reducing climate change and scarce resource management (Camilleri, 2016).

One primary goal to combat climate change is reducing tourism's carbon footprint, in which local government and tourism enterprises are taking the leading roles. Approaches towards low-carbon travel and hotels have been adopted towards this goal. Pieri et al. (2014) measure and reduce the tourist carbon footprint by strategically mapping the existing hotel stock. Many alternatives to reduce food waste, energy and solid waste have been researched and provided to the hotel and accommodation industry (Chan, 2021; Hu et al., 2015).

The UNWTO, along with numerous environmental activists, have formulated a set of guidelines known as the 'Global Code of Ethics for Tourism', aimed at informing 'responsible tourists'. This guidance was encapsulated in a leaflet titled 'The Responsible Tourist and Traveller', which was released in 2005 by the World Committee on Tourism Ethics. This guideline consists of seven prominent tips for a responsible traveller focussing on respect and converse environment and culture of the destination visiting. In this direction, responsible tourism made tourists an undeniable part of the local community for a shared value and shared goal of promoting, protecting the culture, and enhancing the economic well-being of the local community without compromising the environment.

In conclusion, responsible tourism is centered on leveraging tourism as a means to enhance destinations, making them superior places for both residents and visitors. This concept necessitates strategic efforts from various stakeholders (Goodwin, 2007). While it is challenging for tourism activities to be completely devoid of negative impacts, the goal of responsible tourism is to minimize these adverse effects and amplify the positive outcomes.

4. Implications of Sustainable Tourism Towards Climate Change

Developing tourism into a key economic sector is an essential strategic direction for many countries' development. Tourism development can promote the development of other related industries. Many countries aim for sustainable tourism. Developing tourism and preserving the cultural heritage at the same time. One country cannot develop tourism at all costs. Tourism development also means protection of the natural environment. At the same time, sustainable tourism also creates jobs and improves social welfare.

Tourism development must be associated with environmental protection, ensuring sustainable tourism development. Tourism development leading to environmental damage is a problem many countries with developed tourism industries in the region and the world have faced. Unfortunately, this problem is becoming more severe in many countries, threatening sustainable tourism development.

Tourism development is leading to increasing environmental pollution, including pollution of water sources, waste and air, and threatening the biodiversity of the ecological environment. The COVID-19 epidemic left valuable lessons on how people exploit the natural environment, promoting changes in more environmentally friendly consumer behaviour. The trend of sustainable tourism will increase in the coming time. Tourists are increasingly interested in trips that bring sustainability, close to both the environment and life at the destination. Tourists will consider choosing airlines with measures to reduce carbon emissions, prioritising choosing eco-friendly travel services and accommodation establishments, applying eco-friendly standards, saving energy and recycling waste.

Relevant authorities are advised to strengthen propaganda to raise awareness of sustainable tourism development for society as a whole; tourism development must attach importance to the protection of tourism to meet the new trends of tourism in the world. The environment should be treated as a core factor for tourism development. The authorities need to improve the legal system on environmental protection; strengthen sanctions for environmental violations; step up the work of environmental inspection, inspection and supervision at tourist routes and spots; to enhance the role and responsibility of local authorities in protecting the tourism environment; to invest in improving professional and professional capacity for staffs in charge of the environment, invest in equipment and modern technical means.

Policymakers should strengthen communication activities and raise awareness of environmental protection and sustainable tourism development. Tourism businesses must enhance their corporate responsibility for effective and sustainable tourism development, build a sustainable and appropriate management system to control environmental, cultural and social issues and prioritize public health and tourism safety, building sustainable competitiveness. Tourism staff needs to be regularly trained to improve their knowledge about the role of individuals in protecting the environment, contributing to socio-cultural development and preserving heritages and national tourism resources.

Businesses need to determine sustainability by conserving natural and cultural resources, ensuring tourism contributes to socio-economic development on a large-scale, building products, responsible tourism products and responsible businesses and building a green working environment. Unfortunately, the COVID-19 epidemic poses many challenges for the world economy in general and international tourism. Still, from another angle, this is also the time for governments and tourism businesses to move towards a new and more sustainable future in an increasingly volatile global economic picture.

International tourists prefer eco-friendly tourism products closer to nature; products with specific and unique characteristics will be more and more popular than mass tourism, mainly for sightseeing purposes. Therefore, Vietnam must develop new tourism products with specialization and individualization to meet new requirements. Consequently, creating high-quality, diverse, different and high-value tourism products is necessary to increase the tourists' experience.

Thematic tourism products are diverse but must be unique, associated with national identity, cultural traditions, history, art, customs and habits of the country, even each region, to create a competitive advantage. Once unique tourism products have been formed, it is necessary to have a market growth strategy, constantly improve the quality, and continuously innovate tourism products to meet the increasing needs of tourists.

It is vital to develop models of resorts, entertainment and shopping complexes to diversify tourism products, decisive to success in attracting customers and increasing revenue. In countries like Thailand, Malaysia and especially Singapore, the tourism complex model has been invested and developed in parallel and synchronously with the tourism market from 20 years ago up to 30 years. Large-scale and modern tourist, entertainment, shopping and resort centres, such as Genting (Malaysia), Pattaya and Phuket (Thailand), are destinations that annually attract tens of millions of international visitors, generating revenue mainly for the tourism industry of these countries.

Especially for Singapore, a small island nation in both area and population, natural tourism resources are minimal, but this country has taken human creativity as a 'resource' for travel. Artificial works such as the Merlion statue, Esplanade shopping entertainment centre, Marina Bay Sands, Garden by the Bay, and Sentosa tourist island complex can meet all the high-class needs of visitors during their extended stay. Rich, unique and modern tourism products have helped Singapore become the leading entertainment tourism centre in the continent and the world. Vietnam should work on policies to build large-scale, modern tourism and resort centres offering rich tourism products. In this way, the destination can satisfy visitors' needs, attract the high-end tourist segment and prolong their stay, thereby increasing revenue and encouraging a higher rate of return visitors.

It is suggested that the government should allocate a more reasonable budget for the national tourism promotion fund, creating favourable conditions to promote the public–private cooperation as a model to effectively manage and use tourism promotion funds. It is necessary to provide comprehensive and valuable information to visitors. Tourism businesses need to actively coordinate in implementing campaigns emphasising the safe tourism criteria that ensuring the health and safety of tourists.

Creating and promoting a safe national image requires close and synchronous coordination between ministries, branches and localities nationwide while encouraging all sectors' active and creative participation. The reputation of the destination's safety is desirable for international tourists in the current context and in the coming years. Recognized as a safe country, a country will undoubtedly have great opportunities to attract tourists, international events and foreign investment. Tourism promotion should shift from supporting promotion at trade fairs and exhibitions to e-marketing to increase investment and develop websites that provide both tourist information while receiving customer feedback and connecting with search systems and social networks that are popular in the world today. Thus, appropriate policies to improve the transportation and infrastructure of tourist areas and tourist resorts and to invest in road stops and breaks are crucial.

When conditions are exceeded, tourism overload damages tourism resources, thus threatening sustainable development. Many countries worldwide have gone through this phase and are now dealing with the negative economic, political and social consequences. In recent years, environmental pollution and destruction due to overheating tourism growth has forced some famous East Asia and Pacific destinations to take extreme measures to deal with the consequences.

The economic loss in forgone tourism revenue and the cost of remedial measures are warning examples of the consequences that a country will face if it does not proactively and effectively deal with the risks. A country will likely miss opportunities caused by continued rapid tourism growth because redirecting and mitigating damage after it is over will often be more expensive and complicated than preventive measures. For example, the famous tourist island of Boracay in the Philippines, due to heavy pollution, 2018 had to be temporarily closed for six months. As a result of the closure, over 36,000 jobs are estimated to be impacted, and the financial losses are estimated to be approximately $1 billion, based on the information from Boracay Foundation Inc (Al Jazeera, 2018). Similarly, the tourist island of Koh Phi Phi of Thailand, 2018 also had to close to overcome environmental pollution, which has hurt the annual revenue around 400 million baht generated by tourism (Ellis-Petersen, 2018). To prevent and limit tourism overload, in addition to traditional measures, a country needs to study and apply some management measures that some countries have implemented and initially brought effectively.

The number of visitors will decrease shortly, but in the long term, it will gradually increase again. Therefore, the government should take the initiative in developing reasonable measures to deal with the overcrowding of tourists in destinations with really effective management measures to limit the number of tourists gathered together. For example, tiered pricing can be introduced by charging higher fees for access to destinations at a high risk of overcrowding. In addition, limiting the maximum number of guests allowed to visit certain locations at a specific time is possible. Around the world, many famous tourist destinations have applied robust management solutions to preserve the core values of the destination while maintaining a reasonable number of visitors. Some measures to limit tourist overload in some countries In Spain, at famous monuments, to prevent the deterioration of monuments, the government sells tickets with limited time to visit.

Barcelona's city government has closed new hotels in the historical downtown area to limit the number of boats that dock at the port, avoiding adverse environmental impacts. Amsterdam, the Netherlands – a city of less than a million people but welcoming more than 19 million visitors annually, has reduced visitor numbers by introducing a 6% increase in tax on all hotels and, at the same time, limiting the operation of short-term house rental services on Airbnb – an online booking and rental system for rooms and apartments. In New Zealand, to avoid overcrowding, the National Conservation Service has introduced a mandatory reservation system for visitors who wish to join several day trips to natural attractions. In Bhutan, to control the number of tourists reasonably, the government has required visitors to buy a package trip from domestic tour operators and, simultaneously, ensure each visitor has to spend at least 200–250 USD per day, depending on the time of year.

5. Conclusion

International tourism has played a significant role in the global economy, contributing substantially to the socio-economic development of many countries, particularly those that are developing. Recently international tourism has grown rapidly. It leads to the increase in revenue and changes in the structure of the international tourism market. Numerous countries have successfully exerted efforts to attract international tourists. The COVID-19 pandemic has inflicted severe damage on tourism in almost all countries. With the growing demand, tourism is forecasted to recover and maintain its crucial role in the global economy. Timely and appropriate solutions and policies by the government will be vital for the swift recovery of the tourism sector in this new context. Tourism development must be linked with environmental protection. In other words, tourism development must be sustainable. The development of tourism will cause an increase in environmental pollution such as water and air pollution. It is a threat to the balance and biodiversity of the ecosystem.

The pandemic has taught us valuable lessons about how humans should interact with nature. It promotes changes in consumer behaviour towards environmentally friendly practices. The trend of sustainable tourism is expected to become increasingly popular and important in the future. Tourists are likely to become more concerned about the environmental consequences of their travel, such as carbon emissions from airlines.

References

Abegg, B., Jänicke, L., Unger, R., & Mailer, M. (2019). Alpine winter tourists' view on climate change and travel mobility. In *In winter tourism: Trends and challenges* (pp. 82–91). CABI. https://doi.org/10.1079/9781786395207.0082

Al Jazeera. (2018, April 5). Philippines to shut down tourist island Boracay for six months. *Al Jazeera*. https://www.aljazeera.com/economy/2018/4/5/philippines-to-shut-down-tourist-island-boracay-for-six-months

Amoako, G. K., Obuobisa-Darko, T., & Ohene Marfo, S. (2022). Stakeholder role in tourism sustainability: The case of Kwame Nkrumah Mausoleum and centre for art and culture in Ghana. *International Hospitality Review*, *36*(1), 25–44. https://doi.org/10.1108/IHR-09-2020-0057

Baker, I., Peterson, A., Brown, G., & McAlpine, C. (2012). Local government response to the impacts of climate change: An evaluation of local climate adaptation plans. *Landscape and Urban Planning*, *107*(2), 127–136. https://doi.org/10.1016/j.landurbplan.2012.05.009

Brouder, P., & Lundmark, L. (2011). Climate change in Northern Sweden: Intra-regional perceptions of vulnerability among winter-oriented tourism businesses. *Journal of Sustainable Tourism*, *19*(8), 919–933. https://doi.org/10.1080/09669582.2011.573073

Buckley, l. B., & Foushee, M. S. (2012). Footprints of climate change in US national park visitation. *International Journal of Biometeorology*, *56*, 1173–1177. https://doi.org/10.1007/s00484-011-0508-4

Byrd, E. T. (2007). Stakeholders in sustainable tourism development and their roles: Applying stakeholder theory to sustainable tourism development. *Tourism Review*, *62*(2), 6–13. https://doi.org/10.1108/16605370780000309

Camilleri, M. A. (2016). Responsible tourism that creates shared value among stakeholders. *Tourism Planning & Development*, *13*(2), 219–235. https://doi.org/10.1080/21568316.2015.1074100

Campos Herrero, C., Laso, J., Cristobal Garcia, J., Albertí, J., Bala, A., Fullana, M., Fullana-i-Palmer, P., Margallo, M., & Aldaco, R. (2022). Towards more sustainable tourism under a carbon footprint approach: The Camino Lebaniego case study. *Journal of Cleaner Production*, *369*, 133222. https://doi.org/10.1016/j.jclepro.2022.133222

Chan, E. S. (2021). Influencing stakeholders to reduce carbon footprints: Hotel managers' perspective. *International Journal of Hospitality Management*, *94*, 102807. https://doi.org/10.1016/j.ijhm.2020.102807

Chenoweth, J. (2009). Is tourism with a low impact on climate possible? *Worldwide Hospitality and Tourism Themes*, *1*(3), 274–287. https://doi.org/10.1108/17554210910980611

Cohen, B., Cowie, A., Babiker, M., Leip, A., & Smith, P. (2021). Co-benefits and trade-offs of climate change mitigation actions and the Sustainable Development Goals. *Sustainable Production and Consumption*, *26*, 805–813. https://doi.org/10.1016/j.spc.2020.12.034

Cohen, S. A., Higham, J. E., & Reis, A. C. (2013). Sociological barriers to developing sustainable discretionary air travel behaviour. *Journal of Sustainable Tourism*, *21*(7), 982–998. https://doi.org/10.1080/09669582.2013.809092

Dilling, L., Pizzi, E., Berggren, J., Ravikumar, A., & Andersson, K. (2017). Drivers of adaptation: Responses to weather-and climate-related hazards in 60 local governments in the Intermountain Western US. *Environment and Planning A: Economy and Space*, *49*(11), 2628–2648. https://doi.org/10.1177/0308518X16688686

Donaldson, T., & Preston, L. E. (1995). The stakeholder theory of the corporation: Concepts, evidence, and implications. *Academy of Management Review*, *20*, 65–91. https://doi.org/10.5465/amr.1995.9503271992

Ellis-Petersen, H. (2018, October 3). Thailand bay made famous by 'The Beach' closed indefinitely. *The Guardian*. https://www.theguardian.com/world/2018/oct/03/thailand-bay-made-famous-by-the-beach-closed-indefinitely

Fang, Y., Yin, J., & Wu, B. (2018). Climate change and tourism: A scientometric analysis using CiteSpace. *Journal of Sustainable Tourism*, *26*, 108–126. https://doi.org/10.1080/09669582.2017.1329310

Freeman, R. E. (1984). *Strategic management: A stakeholder approach*. Pitman. https://doi.org/10.1017/CBO9781139192675

Freeman, R. E. (2010). *Strategic management: A stakeholder approach*. Cambridge University Press. https://doi.org/10.1017/CBO9781139192675

Giles, A. R., & Perry, A. H. (1998). The use of a temporal analogue to investigate the possible impact of projected global warming on the UK tourist industry. *Tourism Management*, *19*, 75–80. https://doi.org/10.1016/S0261-5177(97)00093-9

Goeldner, C. R., & Ritchie, J. B. (2007). *Tourism principles, practices, philosophies*. John Wiley and Sons.

Goodwin, H. (2007). Responsible tourism in destinations. Accessed on 11 December 2013.

Gössling, S., Garrod, B., Aall, C., Hille, J., & Peeters, P. (2011). Food management in tourism: Reducing tourism's carbon 'foodprint'. *Tourism Management*, *32*, 534–543. https://doi.org/10.1016/j.tourman.2010.04.006

Gössling, S., & Schumacher, K. (2010). Implementing carbon neutral destination policies: Issues from the Seychelles. *Journal of Sustainable Tourism*, *18*, 377–391. https://doi.org/10.1080/09669580903147944

Harrison, S., Winterbottom, S. J., & Sheppard, C. (1999). The potential effects of climate change on the Scottish tourist industry. *Tourism Management*, *20*, 203–211. https://doi.org/10.1016/S0261-5177(98)00072-7

Higham, J., Cohen, S. A., Peeters, P., & Gössling, S. (2013). Psychological and behavioural approaches to understanding and governing sustainable mobility. *Journal of Sustainable Tourism*, *21*(7), 949–967. https://doi.org/10.1080/09669582.2013.828733

Hong, P. M., & Giang, N. T. (2022). The reopening of ASEAN: Trade, tourism and travel. ISEAS-Yusof Ishak Institute. https://www.iseas.edu.sg/wp-content/uploads/2022/09/ASEANFocus-Mar-2022.pdf

Hu, A. H., Huang, C.-Y., Chen, C.-F., Kuo, C.-H., & Hsu, C.-W. (2015). Assessing carbon footprint in the life cycle of accommodation services: The case of an international tourist hotel. *International Journal of Sustainable Development & World Ecology*, *22*(4), 313–323. https://doi.org/10.1080/13504509.2015.1049674

IPCC. (2018). Global warming of 1.5°C. https://www.ipcc.ch/site/assets/uploads/sites/2/2019/06/SR15_Full_Report_High_Res.pdf. Accessed on October 1, 2023.

Kahn, M. E., Mohaddes, K., NG, R. N., Pesaran, M. H., Raissi, M., & Yang, J.-C. (2021). Long-term macroeconomic effects of climate change: A cross-country analysis. *Energy Economics*, *104*, 105624. https://doi.org/10.1016/j.eneco.2021.105624

Lawrence, J., Sullivan, F., Lash, A., Ide, G., Cameron, C., & McGlinchey, L. (2015). Adapting to changing climate risk by local government in New Zealand: Institutional practice barriers and enablers. *Local Environment*, *20*(3), 298–320. https://doi.org/10.1080/13549839.2013.839643

Le Quéré, C., Jackson, R. B., Jones, M. W., Smith, A. J., Abernethy, S., Andrew, R. M., & Canadell, J. G. (2020). Temporary reduction in daily global CO_2 emissions during the COVID-19 forced confinement. *Nature Climate Change*, *10*(7), 647–653. https://doi.org/10.1038/s41558-020-0797-x

Lenzen, M., Sun, Y.-Y., Faturay, F., Ting, Y.-P., Geschke, A., & Malik, A. (2018). The carbon footprint of global tourism. *Nature Climate Change*, *8*(6), 522–528. https://doi.org/10.1038/s41558-018-0141-x

Lesnikowski, A., Biesbroek, R., Ford, J. D., & Berrang-Ford, L. (2021). Policy implementation styles and local governments: The case of climate change adaptation. *Environmental Politics*, *30*(5), 753–790. https://doi.org/10.1080/09644016.2020.1814045

Liu, T.-M. (2014). Analysis of the economic impact of meteorological disasters on tourism: The case of typhoon Morakot's impact on the Maolin National Scenic Area in Taiwan. *Tourism Economics*, *20*, 143–156. https://doi.org/10.5367/te.2013.0258

Markus, M., Bruno, A., Leandra, J., & Bartosz, B. (2019). Mobilitätsbedingte Klimawirkung einer alpinen Tourismusdestination: CO_2-Bilanz und Einschätzung durch Touristen, Bewohner und Beschäftigte. *Zeitschrift für Tourismuswissenschaft*, *11*(2), 211–236. https://doi.org/10.1515/tw-2019-0013

Meynecke, J.-O., Richards, R., & Sahin, O. (2017). Whale watch or no watch: The Australian whale watching tourism industry and climate change. *Regional Environmental Change*, *17*, 477–488. https://doi.org/10.1007/s10113-016-1034-z

Mohamed, E., Belal, A.-A., Ali, R., Saleh, A., & Hendawy, E. A. (2019). Land degradation. *The Soils of Egypt*, 159–174. https://doi.org/10.1007/978-3-319-95516-2_9

Mora, C., Spirandelli, D., Franklin, E. C., Lynham, J., Kantar, M. B., Miles, W., Smith, C. Z., Freel, K., Moy, J., & Louis, L. V. (2018). Broad threat to humanity from cumulative climate hazards intensified by greenhouse gas emissions. *Nature Climate Change*, *8*, 1062–1071. https://doi.org/10.1038/s41558-018-0315-6

Pandy, W. (2017). Tourism enterprises and climate change: Some research imperatives. *African Journal of Hospitality, Tourism and Leisure*, *6*, 1–18. https://doi.org/10.4324/9781351022545-8

Pang, S. F., Mckercher, B., & Prideaux, B. (2013). Climate change and tourism: An overview. *Asia Pacific Journal of Tourism Research*, *18*, 4–20. https://doi.org/10.1080/10941665.2012.688509

Phillips, R., Freeman, R. E., & Wicks, A. C. (2003). What stakeholder theory is not. *Business Ethics Quarterly*, *13*(4), 479–502. http://www.jstor.org/stable/3857968

Pickering, C. M., Castley, J. G., & Burtt, M. (2010). Skiing less often in a warmer world: Attitudes of tourists to climate change in an Australian ski resort. *Geographical Research*, *48*, 137–147. https://doi.org/10.1111/j.1745-5871.2009.00614.x

Pieri, S., Stamos, A., & Tzouvadakis, I. (2014). Reducing tourist carbon footprint through strategic mapping of the existing hotel stock – Attica. *International Journal of Sustainable Energy*, *35*, 1–12. https://doi.org/10.1080/14786451.2014.943757

Rajendra, K., Boonyaroj, V., & Garcia-Hernandez, J. (2021). Paradox of over-tourism, income opportunities, and coral degradation: A case of Maya Bay, Thailand. In *Proceedings of the 13th AUN/SEED-Net Regional Conference on Chemical Engineering 2020 (RCChE-2020) jointly held with the 5th international symposium on conservation and management of tropical lakes*, February 4–5, 2021, Phnom Penh, Cambodia. https://www.apn-gcr.org/wp-content/uploads/2021/02/MayaBay.pdf

Ramos-Castillo, A., Castellanos, E. J., & Galloway McLean, K. (2017). Indigenous peoples, local communities and climate change mitigation. *Climatic Change*, *140*(1), 1–4. https://doi.org/10.1007/s10584-016-1873-0

Scott, D., Amelung, B., Becken, S., Ceron, J.-P., Dubois, G., Gössling, S., Peeters, P., & Simpson, M. (2008). Climate change and tourism: Responding to global challenges. *World Tourism Organization*, *230*, 1–38. https://doi.org/10.18111/9789284412341

Scott, D., & Mcboyle, G. (2007). Climate change adaptation in the ski industry. *Mitigation and Adaptation Strategies for Global Change*, *12*, 1411–1431. https://doi.org/10.1007/s11027-006-9071-4

Sharma, A., & Yedla, V. R. (1970, January 1). *Facing overtourism: Social media to advise tourists to visit unexplored destinations*. SpringerLink. https://link.springer.com/chapter/10.1007/978-981-16-2474-2_3

Shuxin, W., Chunping, T., Jiankuo, D., Zi, T., Chenyan, L., & Yarong, W. (2022). A responsible tourism system at glacier tourism sites: Reducing the impacts of tourism activities on glaciers. *Journal of Resources and Ecology*, *13*(4), 697–707. https://doi.org/10.5814/j.issn.1674-764x.2022.04.015

Smith, K. (1990). Tourism and climate change. *Land Use Policy*, *7*, 176–180.

Swiss Re Institute. (2021). *This is how climate change could impact the global economy*. https://www.weforum.org/agenda/2021/06/impact-climate-change-global-gdp/

Turayev, B., & Sharma, A. (2021). Development of the tourism industry before and after the covid pandemic. *Procedia of Social Sciences and Humanities*, *1*, 178–192. https://doi.org/10.21070/pssh.v1i.40

UNEP. (1995). *Blue plan for the Mediterranean: "Tourism and environment in the Mediterranean"*. United Nations Environment Programme.

UNEP. (2005). *Making tourism more sustainable: A guide for policy makers*. United Nations Environment Programme, Division of Technology, Industry and Economics. https://wedocs.unep.org/20.500.11822/8741

UNEP. (2017). *Emissions Gap Report 2017*. United Nations Environment Programme. https://www.unep.org/resources/emissions-gap-report-2017. Accessed on October 1, 2023.

UNWTO & UNEP. (2011). *Towards a green economy: Pathways to sustainable development and poverty eradication* (pp. 409–447). United Nations Environment Programme. https://www.unep.org/resources/report/towards-green-economy-pathways-sustainable-development-and-poverty-eradication-10. Accessed on October 1, 2023.

United Nations Environment Programme, United Nations World Tourism Organization World Meteorological Organization. (2008). *Climate change and tourism – Responding to global challenges*. https://www.e-unwto.org/doi/epdf/10.18111/9789284412341. Accessed on 1 October 2023.

United Nations World Tourism Organization and International Transport Forum. (2019). *Transport-related CO_2 emissions of the tourism sector*. https://www.e-unwto.org/doi/pdf/10.18111/9789284416660?download=true. Accessed on 1 October 2023.

Wall, G. (1998). Implications of global climate change for tourism and recreation in wetland areas. *Climatic Change*, *40*, 371–389. https://doi.org/10.1023/A:1005493625658

Wilkinson, C. R. (1996). Global change and coral reefs: Impacts on reefs, economies and human cultures. *Global Change Biology*, *2*, 547–558. https://doi.org/10.1111/j.1365-2486.1996.tb00066.x

Chapter 9

Tourists' Behaviour Post-COVID-19 Pandemic: Do Education, Perceived Knowledge and Perceived Risk Matter?

Do Hai Yen[a], *Truong Thi Xuan Dao*[b,*], *Huong Trang Pham*[c], *Jackie Lei Tin Ong*[c] *and Phuong Mai Nguyen*[d,*]

[a]East Asia University of Technology, Vietnam
[b]Hong Bang International University, Vietnam
[c]RMIT University Vietnam
[d]International School, Vietnam National University Hanoi, Vietnam

Abstract

This study combines perceived knowledge, perceived psychologic risk with Theory of Planned Behaviour (TPB) to examine the influence on tourists' intention to choose a safe destination and willingness to pay (WTP) more for safety measures. An online survey was conducted in 2021 to approach tourists globally. After 10 weeks, we received 365 valid responses. SmartPLS software version 3.3 was applied to run structural equation modelling to test the proposed hypotheses. Research results reveal that intention to choose a safe destination mediates the relationship between perceived knowledge of COVID-19, perceived psychological risk and the WTP more for safety measures while moderating role of educational level is also addressed. In turn, perceived psychological risk mediates the relationship between perceived knowledge of COVID-19 and intention to choose a safe destination. As a result, this study implies that destination managers should take actions to promote their WTP more for safety measures.

Keywords: Educational level; perceived knowledge; perceived psychological risk; Theory of Planned Behaviour; willingness to pay; tourism

*Corresponding author.

The Emerald Handbook of Tourism Economics and Sustainable Development, 169–187

doi:10.1108/978-1-83753-708-220241009

1. Introduction

Tourism is one of the most vulnerable industries to natural and artificial crises. Its recovery time is generally longer than that of other industries, especially regarding potential visitors' health and personal safety (Cró & Martins, 2017; Estevão & Costa, 2020). Since the first outbreak in December 2019, the coronavirus pandemic known as COVID-19 pandemic has spread worldwide. It has far-reaching consequences of a worldwide economic crisis, with COVID-19 causing havoc on almost all businesses. The tourism business is most affected by COVID-19 pandemic in terms of income and employment (del Rio-Chanona et al., 2020).

Economic analyses strongly correlate the progression of the COVID-19 outbreak with declining tourism demand (Yang et al., 2021). Compared to prior epidemics like Severe Acute Respiratory Syndrome (SARS), Ebola, Middle East Respiratory Syndrome (MERS) and H1N1, the projected tourism impacts of COVID-19 pandemic are unparalleled in scale (Global Rescue, World Travel & Tourism Council, 2019). While Severe Acute Respiratory Syndrome (SARS) led to $7.7 billion in losses in China and Asia, Ebola $152 million in Africa, Middle East Respiratory Syndrome (MERS) $2.4 billion in Saudi Arabia and H1N1 $45–55 billion globally, COVID-19 tourism losses are forecast to drop these figures dramatically 2020 estimates indicate over 75 million tourism job losses and up to a $2.1 trillion tourism GDP reduction, dwarfing previous crises (World Travel and Tourism Council, 2020). Effective and urgent policy interventions are critical to mitigate these severe economic and employment ramifications as the pandemic continues unfolding.

Recognising the downturn in tourism industry, nations are carrying out a wide range of measures to minimize the influence of COVID-19 pandemic and encourage tourists to visit their destinations. Thus, it is critical to understand the determinants of tourists' intention to choose a safe destination and needs when travelling post the COVID-19 pandemic (Prelims, 2022).

Various theories and models have been used to examine consumers' behaviours in tourism. Nevertheless, the Theory of Planned Behaviour (TPB) has been widely used in tourism literature to explain tourists' decision-making processes and behaviours. Prior studies have employed this theory to explain tourist behaviours in a variety of tourism contexts, mainly for the intention to visit a destination (Chaulagain et al., 2021) or to measure how risk and uncertainty affect tourists' decision-making about their trips (Quintal et al., 2010) or the tourists' environmentally responsible behaviour towards public environmental facilities (Wang et al., 2020). COVID-19 pandemic has become one of the world's most serious crises, affecting nearly every country because of its complicated characteristics and rapid expansion. However, studies to predict tourists' behaviours during and post a health crisis like the COVID-19 pandemic is still limited in tourism literature. As things are stabilising with this pandemic, there is still a scarcity of studies on travel intention and environmental pandemic risk (Wen et al., 2021).

Studies of perceived knowledge of COVID-19 consider that the educational level is not consistent and clear in the extant literature (Ciancio et al., 2020; Harapan et al., 2020). As tourists' knowledge of COVID-19 increases, they may be willing to pay more for additional safety measures. Therefore, our paper will fill these gaps by expanding the TPB model in the context of tourism post-COVID-19 pandemic, analysing tourists' perceived risk and perceived knowledge in the relationship between intention to choose a safe destination and willingness to pay (WTP) for safety measures at the destination. Particularly, the educational level of respondents is also validated in the model as prime importance in influencing perceived risk and perceived knowledge. These factors are critical in facilitating the decision-making process to recover tourism demand after COVID-19 pandemic. For the above arguments, this present study expects to shed light on determinants of global tourists' behaviours while addressing both the mediating and moderating factors in an extended TPB framework.

At present, there is a conspicuous gap in the literature examining travel intentions within the context of post-pandemic risk (Arora & Sharma, 2021; Wen et al., 2020). Addressing this analytical void is imperative for evidence-based policymaking aimed at revitalising tourism demand. The current study seeks to fulfil this need, adopting the well-established TPB (Ajzen, 1985), we investigate the influence of perceived risk on intentions to travel during the post-pandemic period. Moreover, we explore risk as a moderator of TPB antecedents underlying travel intent. Notably, our model further incorporates the effect of such intent on WTP for additional safety precautions at destinations. Empirically, insights from this inquiry can inform more effective strategies for tourism recovery and guide the development of enhanced security measures in the tourism sector. Thus, the present work makes both a substantive and methodological contribution in elucidating decision-making processes during global crises with implications for scholars and practitioners.

The rest of this paper is structured as follows. Section 2 explains the literature review and proposed hypotheses. Section 3 discusses research methods and measurements. Section 4 presents the findings. Section 5 discusses our research results. Finally, Section 6 mentions the limitations of our study and suggests future research directions.

2. Literature Review and Hypotheses

2.1 Tourists' Behaviours in Post-COVID-19 Era

The significance of the setting, specifically the COVID-19 pandemic, is of comparable importance because to its widespread influence on the tourism industry at a global scale. While past health crises, such as SARS and MERS, have exerted an influence on the tourist industry, the impact of COVID-19 has been the most profound, leading to a prolonged downturn in global tourism growth (Gössling et al., 2020).

The enduring effects of the pandemic on travel and tourism, particularly the heightened apprehension associated with post-COVID-19 journeys, give rise to significant apprehensions about the future trajectory of the sector. Epidemic occurrences markedly amplify perceptions of risk and vulnerability in the context of tourism activities and transportation. The 'COVID-19 tourist Recovery Technical Assistance Package' provided by the World Tourism Organization aims to offer guidance and technical support to stakeholders in the tourist sector. This package focuses on areas such as impact assessment and defining roles and responsibilities. However, it should be noted that the recommendations lack enforceability. The implementation of healthcare systems in different countries can vary due to budgetary limitations, disparities in healthcare standards and a scarcity of essential resources like water and waste management. In the context of tourist rivalry, countries with higher economic prosperity, well-developed healthcare systems and robust sanitary standards are likely to exhibit a comparative advantage. In such a context, it is argued that there may be a link between perceived risk and tourist travel intentions in the post-COVID-19 global tourism market. However, this link is understandably yet to be established. Furthermore, post-health crises in tourist behaviour are understudied. Understanding travel motivators and factors influencing tourists' behaviour and intention in choosing a travel destination is beneficial to tourism planning and marketing (Lam & Hsu, 2006). In the post-COVID-19 context, tourists' attitude towards health and hygienic conditions is regarded as an important factor for increasing destination competitiveness and attracting tourists (Williams & Baláž, 2013; Frost et al., 2019). Tourists expect a safe destination with clean drinking water, good sanitary conditions, hygiene of accommodation, environment and in restaurants (Liu et al., 2014). Inadequate cleaning may be problematic because studies have suggested that infectious diseases may be transmitted during air travel. Thus, this present study focuses on the intention to choose a safe destination as one of the changes in the behaviours of tourists in post-COVID-19 era.

Furthermore, the COVID-19 outbreaks have changed our usual perceptions and behaviours regarding safety and hygiene measures, such as applying the sanitising gel, maintaining safety distances, disinfecting furniture objects, etc. When tourist service providers can offer additional safety and hygienic measures to their consumers in response to the COVID-19 epidemic, it helps to increase service quality and customer satisfaction. Furthermore, it can be argued that part of the cost for these measures would be passed on to tourists through the price, who would perceive it with the advantage of a decreased risk of infection. In other words, tourists aware of the effects of COVID-19 will be more likely to travel to a safe destination and be ready to pay more for safety measures. For the above arguments, in this study, WTP more for extraordinary hygiene and safety measures at the destination can be used in the broaden research framework based on the TPB model as a dependent variable.

Ajzen and Peterson (1988) proposed the idea of 'WTP' as an individual's intention to allocate a specific monetary value in order to participate in a recreational activity or obtain a public utility (Ajzen & Peterson, 1988). In their study, Ajzen and Driver (1992) conceptualized WTP for public goods as a behavioural intention. They employed the TPB to investigate the underlying significance of contingent value metrics (Ajzen & Driver, 1992). In other words, WTP refers to the monetary worth of a change in the present state of an environment. According to Ramdas and Mohamed (2014), this notion can be defined as the monetary amount or cost that an individual is willing to pay for a higher or better quality of a product or service (Ramdas & Mohamed, 2014). WTP can be a suitable tool for evaluating goods or services with no clear economic or market value. However, this metric is subject to change as individuals gain more knowledge and additional information (Nowacki, 2013).

2.2 Theory of Planned Behaviour (TPB) and Tourists' Behaviours

The TPB is one of the most widely used sociopsychological theories for predicting human behaviours and decisions (Chaulagain et al., 2021). The TPB was first proposed by Ajzen (1985) as an extension of the theory of reasoned action (TRA) to confirm that individuals behaving in a certain way would produce a particular outcome that they value (Ajzen, 1985). Thus, individuals' intentions and behaviours can be explained through volitional and nonvolitional processes based on the TPB. According to the TPB, behavioural intention is the closest determinant to actual behaviour. This intention is formed based on attitudes towards the action, subjective norms, and perceived behavioural control (Guggenheim et al., 2020). Attitude towards behaviour (ATB) is considered a prominent determinant of tourists' intention because it indicates the general judgements of individuals regarding whether a particular behaviour is rated favourably or unfavourably (Ajzen, 1985; Lam & Hsu, 2006). If visitors have a favourable attitude towards the behaviour, they are more likely to engage positively in that behaviour.

Subjective norms (SUN) is an individual's perception of a particular behaviour, which is influenced by the judgement of others about that behaviour. In other words, subjective norms are an individual's perception of social pressure to perform or refrain from performing a particular action (Ajzen & Kruglanski, 2019).

Thus, we developed the hypotheses as follows:

H1: Intention to choose a safe destination (INT) mediates the relationship between attitude (ATB) and willingness to pay more for safety measures (WTP).

H2: Intention to choose a safe destination (INT) mediates the relationship between subjective norms (SUN) and the willingness to pay more for safety measures (WTP).

2.3 Perceived Knowledge, Perceived Psychological Risks, and Tourists' Behaviours

Even though TPB's competence has been demonstrated in diverse environments, prior studies suggest that its predictability of human behaviours or decisions still needs to be improved by including contextual factors (Kim & Hwang, 2020). Prior studies, especially in the tourism field, have confirmed the critical importance of perceived knowledge and perceived psychological risk to clearly understand customers' decision-making towards safe/reliable or risky travel products. Furthermore, studies in consumer behaviour and social psychology also highlighted the essential effects of these concepts in intention formation (Chan et al., 2014).

In the tourism sector, perceived knowledge indicates the ability to know and understand tourism-related issues and problems, so it is considered an important concept to explain tourists' behaviours. When tourists' perceived knowledge is insufficient, they tend to avoid any uncertain situation or try to reduce uncertainty by not taking any action. Their high level of perceived knowledge can impact their intention or behaviour to practice any action (Han et al., 2020). In this study, the perceived knowledge of COVID-19 refers to the cognitive assessment of their ability to comprehend the pandemic and its dangerous effects on humans and its significance on the tourism industry.

The existing body of literature suggests that there is a notable inclination among tourists to engage in travel activities both during and following the COVID-19 pandemic. This inclination is particularly evident after individuals have experienced the impact of lockdown measures, quarantines, and the closure of national borders (Li et al., 2020; Matiza, 2020). The findings of these studies also align with the literature to indicate that if their perceived knowledge about COVID-19 is insufficient, they will minimize any uncertainty by refusing to engage in specific tourist behaviour. They also tend to choose safe destinations or tourist providers who provide good hygiene and safety measures. Thus, we formulated the following hypothesis.

H3: Intention to choose a safe destination (INT) mediates the relationship between perceived knowledge of COVID-19 (PKC) and the willingness to pay more for safety measures (WTP).

Perceived psychological risk is a potential subjective loss generated from uncertainty to which certain probabilities can be assigned. Although risk construct has gained more attention in research, its conceptualization depends on the service or research context (Bruwer & Cohen, 2019). Perceived psychological risk in tourism refers to situations in which it predicts the decision to avoid particular places or destinations for any potential dangers related to the trip, such as terrorism, political instability, or health concerns (Fuchs & Reichel, 2006). In other words, concern for the possibility of any fatal incidents during their trip raises tourists' psychological risk perception. In this regard, tourists tend to choose safer tourism destinations to avoid any potential threats or may change their travel plans if their perceived risk exceeds an acceptable level. Thus, we developed these hypotheses.

H4: Intention to choose a safe destination (INT) mediates the relationship between perceived psychological risk (PSR) and the willingness to pay more for safety measures (WTP).

Furthermore, it is argued that the higher perceived knowledge about COVID-19 that the tourists have, the less the probability of psychological risks they encounter. Therefore, we hypothesized as follows.

H5: Perceived knowledge of COVID-19 (PKC) negatively impacts the perceived psychological risk (PSR).

H6: Perceived psychological risk (PSR) mediates the relationship between perceived knowledge of COVID-19 (PKC) and willingness to pay more for safety measures (WTP).

Prior studies on consumers' behaviour during the COVID-19 pandemic have looked at perceived knowledge and perceived psychological risk, but most of them have mostly neglected the differences in educational level in this relationship (Abdelrahman, 2020; Bruine de Bruin & Bennett, 2020; Geldsetzer, 2020; Samadipour et al., 2020). In terms of perceived knowledge, several studies among these indicated positive influences on educational level (Al-Hanawi et al., 2020; Hager et al., 2020; Röösli & Cissé, 2020). Therefore, in line with previous studies, hypothesis 7 was developed as the following.

H7: Educational level (EDU) moderates the relationship between perceived knowledge of COVID-19 (PKC) and willingness to pay more for safety measures (WTP).

Other studies indicated inconsistent results about the relationship between perceived risk and educational level (Harapan et al., 2020; Scott et al., 2020). According to Pititto and Ferreira (2020) and Ciancio et al. (2020), people with no little or no education may have higher perceived risks. In contrast, other studies found that people with lower educational levels often believe they are at lower risk of being infected with Covid 19. Thus, the following hypothesis was formulated.

H8: Educational level (EDU) moderates the relationship between perceived psychological risk (PSR) and willingness to pay more for safety measures (WTP).

3. Methodology

3.1 Research Model

In the present study, we chose willingness to pay more for safety measures (WTP) as the utmost dependent variable. Two factors of the TPB were chosen which are attitude towards the behaviour (ATT) and subjective norms (SUN). Unlike previous studies, we consider perceived psychological risk (PSR) and intention to choose a safe destination (INT) as two mediators. Meanwhile, educational level is also considered as a moderator the in our proposed research model. Details are presented in Fig. 9.1 below.

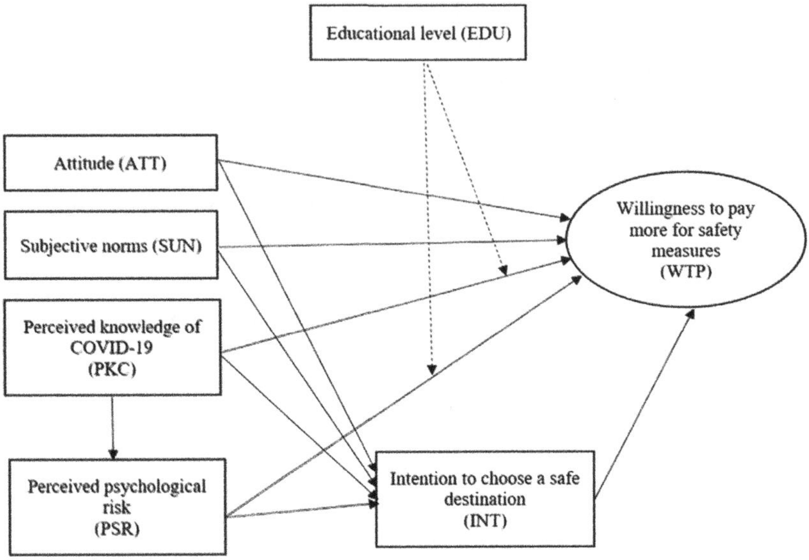

Fig. 9.1. Proposed Research Model. *Source:* Produced by authors.

3.2 Measurements

To ensure the validity of the content, measurement of variables in our study were calibrated from the literature review and through expert interviews with six people working in tourism industry. Based on these interview results, the measurement items were adjusted and revised to fit the context of the current study.

Three TPB variables that are attitude towards the behaviour, subjective norms and intention to choose a safe destination were adopted and adapted from the studies of Ajzen and Driver (1992) and Matiza (2020). The measurement scale of perceived psychological risk was adapted from the studies of Quintal et al. (2010), Reisinger and Mavondo (2005), and Samadipour et al. (2020). Perceived knowledge of COVID-19 measurement was taken from some previous studies including Chan et al. (2014), Reynoso (2010). Lastly, WTP was adjusted from the studies of Wei et al. (2018).

3.3 Sampling Method and Data Collection

The online questionnaire survey method is considered the most appropriate method during the COVID-19 outbreaks with the advantages of low cost, better accessibility to respondents, convenience for respondents and less time-consuming.

A questionnaire designed in Google Form was sent through several popular social networking sites, such as Facebook, Instagram, and LinkedIn, for the best accessibility. In addition, online travel communities and social networks of tourism enterprises have been approached to spread the survey worldwide. The survey opening time is 10 weeks. Upon closing the survey time, we obtained 365 valid responses. SmartPLS 3.3.3 software was used for data analysis.

4. Findings

4.1 Sample Demographics

Table 9.1 shows the comprehensive sociodemographic profile of respondents, including gender, age, educational level, residential area, occupation, and travel frequency. Females made up the majority of respondents (53.7%), and the leading age group was 2,228 and 29–40 years old. A large number of participants had educational levels of Undergraduate (44.7%) and High school (44.4%); In terms of living area, the majority of respondents are from Europe (25.8%); the remainders have similar proportions. Most of the participants' frequency of travel is '2–3 times per year'.

Table 9.1. Demographic Profile of Respondents.

Characteristic	Category	*N*	%
Gender	Male	144	39.5
	Female	196	53.7
	Others	25	6.8
Feature	*Category*	*N*	%
Age	18–21 years	40	11.0
	22–28 years	121	33.2
	29–40 years	97	26.6
	41–50 years	67	18.4
	<50 years	40	11.0
Educational level	High school	162	44.4
	Undergraduate	163	44.7
	Postgraduate	40	11.0
Occupation	Student	94	25.8
	Public sector employees	44	12.1
	Private sector employees	44	12.1
	Public sector managers	35	9.6
	Private-sector managers	43	11.8
	Private business owners	79	21.6

(Continued)

Table 9.1. *(Continued)*

Characteristic	Category	*N*	%
	Others	26	7.1
Living area	Europe	94	25.8
	Asia	79	21.6
	North America	47	12.9
	Latin America	42	11.5
	Africa	35	9.6
	Australia	40	11.0
	Antarctica	28	7.7
Frequency of travel	Once time per year	78	21.4
	2–3 times per year	185	50.7
	4–5 times per year	70	19.2
	Six times and over	32	8.8

Source: Produced by authors.

4.2 Measurement Model Test Results

SmartPLS 3.3.3 software was used to process and evaluate the data. The research model was tested using partial least squares structural equation modelling (PLS-SEM). Factor loadings, composite reliability (CR), and average variance extracted (AVE) were used to examine construct reliability and validity, as well as the correlation between constructs (see Tables 9.2 and 9.3).

Table 9.2. Reliability and Validity of Measurements.

Construct (Items)	Loading Factor (>0.7)	Cronbach's Alpha (>0.7)	Composite Reliability (>0.6)	Average Variance Extracted (>0.5)	Variance Inflation Factor (<5)
Attitude towards the behaviour		0.918	0.948	0.859	
ATB1	0.924				3.150
ATB2	0.928				3.543
ATB3	0.928				3.218
Subjective norm		0.882	0.927	0.809	
SUN1	0.879				2.195

Table 9.2. *(Continued)*

Construct (Items)	Loading Factor (>0.7)	Cronbach's Alpha (>0.7)	Composite Reliability (>0.6)	Average Variance Extracted (>0.5)	Variance Inflation Factor (<5)
SUN2	0.944				3.785
SUN3	0.873				2.739
Perceived knowledge of COVID- 19		0.851	0.909	0.770	
PKC1	0.849				1.892
PKC2	0.899				2.196
PKC3	0.883				2.248
Perceived psychological risk		0.908	0.942	0.845	
PSR1	0.927				3.200
PSR2	0.928				3.153
PSR3	0.902				2.744
Intention to choose a safe destination		0.897	0.936	0.829	
INT1	0.924				3.051
INT2	0.892				2.489
INT3	0.916				2.832
Willingness to pay more for safety measures		0.920	0.949	0.861	
WTP1	0.930				3.559
WTP2	0.923				3.058
WTP3	0.931				3.500

Source: Produced by authors.

The results validate the measurement properties of the indicators used. All item loadings exceed 0.7, confirming robust measures. Reliability is affirmed with CR above 0.6 for all factors. AVE also surpasses 0.5 across the board. Minimal multicollinearity is evidenced by variance inflation factors remaining under 5. Two tests establish discriminant validity: AVE's square root greater than inter-construct correlations and heterotrait-monotrait ratios under 0.85 with 95% confidence [44]. Thus, precise discriminant validity is demonstrated for the research variables. In short, the measurement model and indicators show strong performance without redundancy.

Table 9.3. Discriminant Validity Test.

Fornell and Larcker Criterion						
	ATB	**INT**	**PKC**	**PSR**	**SUN**	**WTP**
ATB	**0.927**					
INT	0.593	**0.911**				
PKC	0.456	0.512	**0.877**			
PSR	−0.284	−0.366	−0.483	**0.919**		
SUN	0.695	0.658	0.384	−0.284	**0.899**	
WTP	0.428	0.564	0.511	−0.525	0.514	**0.928**
HTMT Value						
ATB						
INT	0.652					
PKC	0.513	0.580				
PSR	0.306	0.402	0.545			
SUN	0.773	0.733	0.436	0.312		
WTP	0.464	0.618	0.574	0.572	0.563	

Source: Produced by authors.

4.3 Hypothesis Testing

SEM analysis was run to test the proposed hypotheses. Details of SEM results are presented in Fig. 9.2 and Table 9.4 below.

Attitude towards the behaviour (ATB) has no direct effect (*p*-value = 0.486 > 0.05; t-value = 0.697 < 1.96) on willingness to pay more for safety measures (WTP) but only an indirect effect (*p*-value = 0.036 < 0.05; t-value = 2.100 > 1.96) through the intention to choose a safe destination (INT). Additionally, the confidence interval, denoted as [−0.145; 0.067], implies a 95% likelihood that the actual population value of the direct effect falls within the range of −0.145 to 0.067 (see Table 9.4). Since this interval includes the zero value, it can be inferred that the direct effect lacks statistical significance (Hair Jr et al., 2016). Therefore, hypothesis H1 is accepted, and the indirect relationship between ATB and WTP is indirect-only mediation.

The test results for hypotheses H2, H3, and H6 are similar, showing that all direct or indirect relationships are accepted (*p*-value <0.05, t-value >1.96). Besides, the confidence interval of all relationships with a 95% probability does not include zero value (see Table 9.4). Therefore, it can be concluded that INT has the role of complimentary mediation in the relationship between SUN and WTP.

Similarly, intention to choose a safe destination (INT) also plays a role in complimentary mediation in the relationship between perceived knowledge of COVID-19 (PKC), WTP more for safety measures (WTP), and perceived

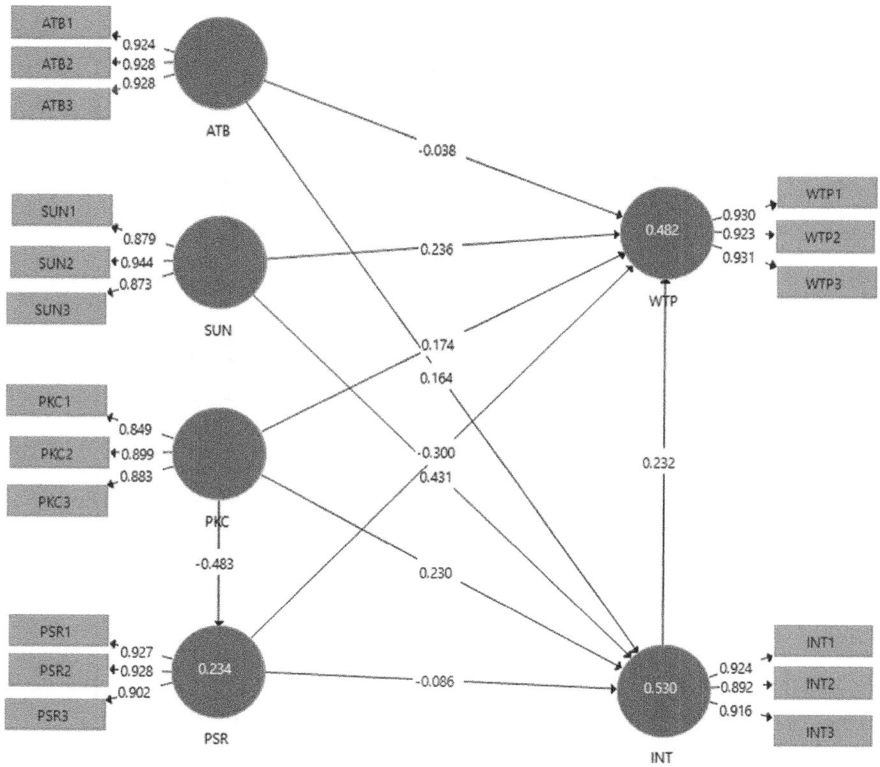

Fig. 9.2. Validated Research Model. *Source:* Produced by authors.

psychological risk (PSR) and WTP more for safety measures (WTP). Hypothesis H6 is accepted because all values are satisfactory ($\beta = -0.483$, p-value = 0.000, $t = 10.583$). This result suggests that the greater one's perception of COVID-19 knowledge, the lower one's perceived psychological risk. Furthermore, the R^2 coefficients are high (see Fig. 9.2), indicating that ATB, SUN, PKC and PSR can explain 48.2% of change in WTP ($R^2 = 0.482$).

In moderation analysis, the changes in the R2 value are important. The previous R2 for the main effect model is 0.482, whereas it is 0.492 in the interaction effect model. Thus, the 1% change indicates the moderating effect of PKC*Edu*WTP and PSR*Edu*WTP. Only the interaction term can be interpreted from the bootstrap result (Table 9.5). The cut off value for this particular test will be 1.96 ($\alpha = 0.05$) and 2.54 ($\alpha = 0.01$). Thus, only moderation of PSR*Education*WTP is significant (t-value = 2.020) whereas PKC*Edu is insignificant (t-value = 1.194). Consequently, hypothesis H7 is rejected but H8 is accepted.

Table 9.4. Results of the Structural Path Model.

Hypo.	Type of Effect	Path	Estimate	Percentile 95% Confidence Interval	*t*-value	*p* value	Conclusion	Results
H1	Direct	ATB → WTP	−0.038	[−0.145; 0.067]	0.697	0.486	No	Indirect-only mediation
	Indirect	ATB → INT → WTP	0.038	[0.009; 0.083]	2.100	0.036	Yes	
H2	Direct	SUN → WTP	0.236	[0.121; 0.371]	3.668	0.000	Yes	Complimentary mediation
	Indirect	SUN → INT → WTP	0.100	[0.047; 0.170]	3.246	0.001	Yes	
H3	Direct	PKC → WTP	0.174	[0.066; 0.283]	3.164	0.002	Yes	Complimentary mediation
	Indirect	PKC → INT → WTP	0.053	[0.020; 0.105]	2.487	0.013	Yes	
H4	Direct	PSR → WTP	−0.300	[−0.404; −0.192]	5.530	0.000	Yes	Direct only
	Indirect	PSR → INT → WTP	−0.020	[−0.049; −0.001]	1.694	0.090	No	
H5	Direct	PKC → PSR	−0.483	[−0.569; 0.393]	10.58 3	0.000	Yes	Supported
H6	Direct	PKC → WTP	0.174	[0.066; 0.283]	3.164	0.002	Yes	Complimentary mediation
	Indirect	PKC → PSR → WTP	0.145	[0.090; 0.208]	4.768	0.000	Yes	

Source: Produced by authors.

Table 9.5. Results of the Moderator Analysis.

Hypo.	Path	Estimate	*t*-value	*p* Values	Results
H7	PKC*Education → WTP	−0.079	1.194	0.116	Insignificant
H8	PSR*Education → WTP	−0.12	2.020	0.022	Significant

Source: Produced by authors.

5. Discussions

This study aims to broaden the TPB model to provide a comprehensive understanding of tourists' behavioural intention to travel post-COVID-19 pandemic by emphasising the importance of tourists' perceived knowledge of COVID-19, perceived psychological risk and WTP more for safety measures.

First, the TPB model in this study confirmed tourists' intention for safer destinations after COVID-19. Tourists believe that they can control the situations of the trip, especially after the COVID-19 pandemic has been brought under control in certain parts of the world; they intend to go to other places after a period of quarantine and lockdown due to Covid outbreaks. Our findings are consistent with the prior studies such as Sánchez-Cañizares et al. (2021) but highlight how tourists' perceived knowledge and risk affect this relationship.

Second, this study analysed tourists' WTP more for safety measures in travel-related activities. Our findings indicate that tourists who intend to travel to a safer destination are also concerned about health-related issues, especially COVID-19. Thus, they are willing to pay extra for tourism products and services that offer them better quality safety measures.

Another significant novel feature of this study is measuring the relationship between tourists' perceived knowledge of COVID-19, perceived psychological risk and WTP more for safety measures. Our findings indicate that tourists' perceived risk mediates the relationship between the perceived knowledge of COVID-19 and WTP more. Since COVID-19's initial global breakout in December 2019, tourists have learnt more about the disease's harmful consequences and wide-ranging dissemination. The higher perceived knowledge of COVID-19, the less perceived risk they confront. Because their experience with uncertainty influences their perception of danger, they are prepared to pay more for higher quality and more safety measures.

This research is among the early studies to examine the moderating factor of educational level towards perceived knowledge, perceived risk, and the WTP more for safety measures. Tourists with high educational levels tend to grasp more knowledge and up-to-date information about COVID-19. As a result, they sense perceived risks and are prepared to pay more for high-quality services to ensure their safety.

6. Conclusion – Implications, Limitations and Suggestions for Future Research

Our extended research framework confirmed the TPB's competence in predicting consumer behaviours, and it can be used in a variety of tourism contexts, especially to explain tourists' behaviour intention for a safe destination.

Our study also contributed to the extant knowledge about disease outbreaks and travel behaviours by adding other factors such as perceived knowledge about COVID19, perceived psychological risk, WTP more and educational level of tourists are analysed in this study to provide a more comprehensive understanding about this decision for destination choice. Thus, our framework provides a broader and better understanding of tourists' post-pandemic decision-making behaviour for safer destination choices. From our findings, tourism practitioners can understand the needs and demands of tourists for travelling to safe destinations and safety measures that are critical in developing marketing strategies and tourism management operations.

This study contains some limitations that can call for further research possibilities. First, our study applies convenience sampling collected through social networking sites. Due to the difficult time affected by COVID-19 outbreaks, the sample size used for data processing is quite small (n = 365). Thus, future research may extend the time for collecting data for a bigger sample size. Second, like other studies in social psychology to measure customer behaviour, the cognitive-centred approach was central to this study.

Some prior studies have found that emotional factors affect tourists' behaviour [46]. Thus, future research may extend the research model with other emotional factors. Finally, in this study, respondents' educational level is measured as a moderating variable between tourists' perceived psychological risk, perceived knowledge about COVID-19, and WTP more. Future studies can measure other factors such as travel frequency or formal and informal information sources about the COVID-19 pandemic.

References

Abdelrahman, M. (2020). Personality traits, risk perception, and protective behaviors of Arab residents of Qatar during the COVID-19 pandemic. *International Journal of Mental Health and Addiction*. https://doi.org/10.1007/s11469-020-00352-7

Ajzen, I. (1985). From intentions to actions: A theory of planned behavior. In *Action control* (pp. 11–39). Springer. https://doi.org/10.1007/978-3-642-69746-3_2

Ajzen, I., & Driver, B. L. (1992). Contingent value measurement: On the nature and meaning of willingness to pay. *Journal of Consumer Psychology*, *1*(4), 297–316. https://doi.org/10.1016/S1057-7408(08)80057-5

Ajzen, I., & Kruglanski, A. W. (2019). Reasoned action in the service of goal pursuit. *Psychological Review*, *126*(5), 774. https://doi.org/10.1037/rev0000155

Ajzen, I., & Peterson, G. L. (1988). Contingent value measurement: The price of everything and the value of nothing. In *Amenity resource valuation: Integrating economics with other disciplines* (pp. 65–75).

Al-Hanawi, M. K., Angawi, K., Alshareef, N., Qattan, A. M. N., Helmy, H. Z., Abudawood, Y., Alqurashi, M., Kattan, W. M., Kadash, N. A., Chirwa, G. C., & Alsharqi, O. (2020). Knowledge, attitude and practice toward COVID-19 among the public in the Kingdom of Saudi Arabia: A cross-sectional study. *Frontiers in Public Health, 8*, 217. https://doi.org/10.3389/fpubh.2020.00217

Arora, S., & Sharma, A. (2021). Covid-19 impact on overtourism: Diversion from mass tourism to alternative tourism. In A. Sharma & A. Hassan (Eds.), *Overtourism as destination risk* (pp. 275–283). Emerald Publishing Limited. https://doi.org/10.1108/978-1-83909-706-520211018

Bruine de Bruin, W., & Bennett, D. (2020). Relationships between initial COVID-19 risk perceptions and protective health behaviors: A national survey. *American Journal of Preventive Medicine, 59*(2), 157–167. https://doi.org/10.1016/j.amepre.2020.05.001

Bruwer, J., & Cohen, J. (2019). Restaurants and wine by-the-glass consumption: Motivational process model of risk perception, involvement and information-related behaviour. *International Journal of Hospitality Management, 77*, 270–280. https://doi.org/10.1016/j.ijhm.2018.07.006

Chan, E. S. W., Hon, A. H. Y., Chan, W., & Okumus, F. (2014). What drives employees' intentions to implement green practices in hotels? The role of knowledge, awareness, concern and ecological behaviour. *International Journal of Hospitality Management, 40*, 20–28. https://doi.org/10.1016/j.ijhm.2014.03.001

Chaulagain, S., Pizam, A., & Wang, Y. (2021). An integrated behavioral model for medical tourism: An American perspective. *Journal of Travel Research, 60*(4), 761–778. https://doi.org/10.1177/0047287520907681

Ciancio, A., Kämpfen, F., Kohler, I. V., Bennett, D., Bruine de Bruin, W., Darling, J., Kapteyn, A., Maurer, J., & Kohler, H. P. (2020). Know your epidemic, know your response: Early perceptions of COVID-19 and self-reported social distancing in the United States. *PLoS One, 15*(9), e0238341. https://doi.org/10.1371/journal.pone.0238341

Cró, S., & Martins, A. M. (2017). Structural breaks in international tourism demand: Are they caused by crises or disasters? *Tourism Management, 63*, 3–9. https://doi.org/10.1016/j.tourman.2017.05.009

del Rio-Chanona, R. M., Mealy, P., Pichler, A., Lafond, F., & Farmer, J. D. (2020). Supply and demand shocks in the COVID-19 pandemic: An industry and occupation perspective. *Oxford Review of Economic Policy, 36*(Suppl. 1), S94–S137. https://doi.org/10.1093/oxrep/graa033

Estevão, C., & Costa, C. (2020). Natural disaster management in tourist destinations: A systematic literature review. *European Journal of Tourism Research, 25*, 2502. https://doi.org/10.54055/ejtr.v25i.417

Frost, I., Van Boeckel, T. P., Pires, J., Craig, J., & Laxminarayan, R. (2019). Global geographic trends in antimicrobial resistance: The role of international travel. *Journal of Travel Medicine, 26*(8). https://doi.org/10.1093/jtm/taz036

Fuchs, G., & Reichel, A. (2006). Tourist destination risk perception: The case of Israel. *Journal of Hospitality & Leisure Marketing, 14*(2), 83–108. https://doi.org/10.1300/J150v14n02_06

Geldsetzer, P. (2020). Knowledge and perceptions of COVID-19 among the general public in the United States and the United Kingdom: A cross-sectional online survey. *Annals of Internal Medicine, 173*(2), 157–160. https://doi.org/10.7326/M20-0912

Global Rescue, World Travel and Tourism Council. (2019). *Crisis readiness: Are you prepared and resilient to safeguard your people and destinations?* World Travel and Tourism Council.

Gössling, S., Scott, D., & Hall, C. M. (2020). Pandemics, tourism and global change: A rapid assessment of COVID-19. *Journal of Sustainable Tourism*, *29*(1), 1–20. https://doi.org/10.1080/09669582.2020.1758708

Guggenheim, N., Taubman-Ben-Ari, O., & Ben-Artzi, E. (2020). The contribution of driving with friends to young drivers' intention to take risks: An expansion of the theory of planned behavior. *Accident Analysis & Prevention*, *139*, 105489. https://doi.org/10.1016/j.aap.2020.105489

Hager, E., Odetokun, I. A., Bolarinwa, O., Zainab, A., Okechukwu, O., & Al-Mustapha, A. I. (2020). Knowledge, attitude, and perceptions towards the 2019 Coronavirus Pandemic: A bi-national survey in Africa. *PLoS One*, *15*(7), e0236918. https://doi.org/10.1371/journal.pone.0236918

Hair Jr, J. F., Hult, G. T. M., Ringle, C., & Sarstedt, M. (2016). *A primer on partial least squares structural equation modeling (PLS-SEM)*. Sage Publications.

Han, H., Al-Ansi, A., Chua, B.-L., Tariq, B., Radic, A., & Park, S.-H. (2020). The PostCoronavirus world in the international tourism industry: Application of the theory of planned behavior to safer destination choices in the case of US outbound tourism. *International Journal of Environmental Research and Public Health*, *17*(18). https://doi.org/10.3390/ijerph17186485

Harapan, H., Anwar, S., Nainu, F., Setiawan, A. M., Yufika, A., Winardi, W., . . ., & Wagner, A. L. (2020). Perceived risk of being infected with SARS-CoV-2: A perspective from Indonesia. *Disaster Medicine and Public Health Preparedness*, 1–5. https://doi.org/10.1017/dmp.2020.351

Kim, J. J., & Hwang, J. (2020). Merging the norm activation model and the theory of planned behavior in the context of drone food delivery services: Does the level of product knowledge really matter? *Journal of Hospitality and Tourism Management*, *42*, 1–11. https://doi.org/10.1016/j.jhtm.2019.11.002

Lam, T., & Hsu, C. H. C. (2006). Predicting behavioral intention of choosing a travel destination. *Tourism Management*, *27*(4), 589–599. https://doi.org/10.1016/j.tourman.2005.02.003

Li, J., Nguyen, T. H. H., & Coca-Stefaniak, J. A. (2020). Coronavirus impacts on post-pandemic planned travel behaviours. *Annals of Tourism Research*. https://doi.org/10.1016/j.annals.2020.102964

Liu, Z. R., Zhang, G. Y., & Zhang, X. M. (2014). Urban street foods in Shijiazhuang city, China: Current status, safety practices and risk mitigating strategies. *Food Control*, *41*, 212–218. https://doi.org/10.1016/j.foodcont.2014.01.027

Matiza, T. (2020). Post-COVID-19 crisis travel behaviour: Towards mitigating the effects of perceived risk. *Journal of Tourism Futures*. https://doi.org/10.1108/JTF-04-2020-0063

Nowacki, M. (2013). The determinants of satisfaction of tourist attractions' visitors. *ACTIVE, Zdzisławy*, *11a*. 61-054 Poznań. https://depot.ceon.pl/handle/123456789/8794

Pititto, B. A., & Ferreira, S. R. G. (2020). Diabetes and covid-19: More than the sum of two morbidities. *Revista de Saúde Pública*, *54*, 54. https://doi.org/10.11606/s1518-8787.2020054002577

Prelims. (2022). In P., Mohanty, A., Sharma, J., Kennell, & A., Hassan (Eds.), *The Emerald handbook of destination recovery in tourism and hospitality* (pp. i–xxxvii). Emerald Publishing Limited. https://doi.org/10.1108/978-1-80262-073-320221030

Quintal, V. A., Lee, J. A., & Soutar, G. N. (2010). Risk, uncertainty and the theory of planned behavior: A tourism example. *Tourism Management*, *31*(6), 797–805. https://doi.org/10.1016/j.tourman.2009.08.006

Ramdas, M., & Mohamed, B. (2014). Impacts of tourism on environmental attributes. In *Environmental literacy and willingness to pay: A conceptual and theoretical review*. https://doi.org/10.1016/j.sbspro.2014.07.307

Reisinger, Y., & Mavondo, F. (2005). Travel anxiety and intentions to travel internationally: Implications of travel risk perception. *Journal of Travel Research*, *43*(3), 212–225. https://doi.org/10.1177/0047287504272017

Reynoso, J. (2010). Satisfaction: A behavioral perspective on the consumer. *Journal of Service Management*, *21*(4), 549–551. https://doi.org/10.1108/09564231011066132

Röösli, M., & Cissé, G. (2020). Towards health for future. *International Journal of Public Health*, *65*(1), 1–2. https://doi.org/10.1007/s00038-019-01314-0

Samadipour, E., Ghardashi, F., & Aghaei, N. (2020). Evaluation of risk perception of COVID-19 disease: A community-based participatory study. *Disaster Medicine and Public Health Preparedness*, 1–8. https://doi.org/10.1017/dmp.2020.311

Sánchez-Cañizares, S. M., Cabeza-Ramírez, L. J., Muñoz-Fernández, G., & Fuentes-García, F. J. (2021). Impact of the perceived risk from Covid-19 on intention to travel. *Current Issues in Tourism*, *24*(7), 970–984. https://doi.org/10.1080/13683500.2020.1829571

Scott, J., Dardas, L., Sloane, R., Wigington, T., Noonan, D., & Simmons, L. A. (2020). Understanding social determinants of cardiometabolic disease risk in rural women. *Journal of Community Health*, *45*(1), 1–9. https://doi.org/10.1007/s10900-019-00710-0

Wang, C., Zhang, J., Sun, J., Chen, M., & Yang, J. (2020). Public environmental facilities: Hygiene factors for tourists' environmental behaviour. *Environmental Science & Policy*, *106*, 40–47. https://doi.org/10.1016/j.envsci.2020.01.009

Wei, S. M., Ang, T., & Jancenelle, V. (2018). Willingness to pay more for green products: The interplay of consumer characteristics and customer participation. *Journal of Retailing and Consumer Services*, *45*, 230–238. https://doi.org/10.1016/j.jretconser.2018.08.015

Wen, J., Kozak, M., Yang, S., & Liu, F. (2020). COVID-19: Potential effects on Chinese citizens' lifestyle and travel. *Tourism Review*. https://doi.org/10.1108/TR-03-2020-0110

Wen, J., Kozak, M., Yang, S., & Liu, F. (2021). COVID-19: Potential effects on Chinese citizens' lifestyle and travel. *Tourism Review*, *76*(1), 74–87. https://doi.org/10.1108/TR-03-2020-0110

Williams, A. M., & Baláž, V. (2013). Tourism, risk tolerance and competences: Travel organization and tourism hazards. *Tourism Management*, *35*, 209–221. https://doi.org/10.1016/j.tourman.2012.07.006

World Travel and Tourism Council. (2020). *Coronavirus brief: The 15th of April 2020*. https://wttc.org/Portals/0/Documents/WTTC%20Coronavirus%20Brief%20External%2015_04.pdf?ver=2020-04-15120258-850. Accessed on 21 April 2020.

Yang, Y., Altschuler, B., Liang, Z., & Li, X. R. (2021). Monitoring the global COVID-19 impact on tourism: The COVID19 tourism index. *Annals of Tourism Research*, *90*, 103–120. https://doi.org/10.1016%2Fj.annals.2020.103120

Chapter 10

Branding India as a Sustainable Blue Economy Destination in the Indian Ocean Region

Aishwarya Singh Raikwar and David T. Easow

Hindustan Institute of Technology and Science, India

Abstract

Indian Ocean region (IOR) is a coveted maritime space in the international arena because of its unique positioning and importance. The third largest water body on the planet Earth holds great economic significance and sustains the livelihood of its people. While the blue economy (BE) approach gains momentum worldwide, following this for the responsible consumption of ocean resources, India too pursues and has developed a draft policy on the BE. Maritime and coastal tourism is one of the prime areas contributing to this framework of BE. This study will explore the advances in BE prospects of Indian Tourism and highlight the elements of branding a BE Destination in the region. India being a key player and an emerging economy, situated at a transitional maritime position is strategic in nature. India as a country is a strong contestant for the title of tourism hotspot but with its highly regarded destinations, it surprisingly attracts fewer tourists. The paper examines secondary sources of data and attempts to review the untapped potential of the brand India in this vast maritime space. There is enough literature available on destination branding but this paper stands out with its unique and innovative approach that combines economic aspects with maritime sustainability and surely add value to this field of knowledge. This chapter presents the recommendations for BE-led repositioning of India tourism in the IOR. To achieve higher productivity and sustainability, a socioeconomic transformational shift is required in the segments of the tourism sector.

Keywords: Blue economy; destination branding; reimagining; Indian Ocean region; maritime tourism; sustainability

The Emerald Handbook of Tourism Economics and Sustainable Development, 189–210

doi:10.1108/978-1-83753-708-220241010

1. Introduction and Background

Probably more than in earlier times, the Indian Ocean matters today. It is an essential link for international trade, especially energy. Its littoral is large, heavily populated and home to some of the most rapidly developing regions in the world (Jaishankar, 2016). For India, the most populous state in the region and a prominent player in international diplomacy, the Indian Ocean region (IOR) is significant. India has always been engaged in regional issues and recently has started to reassess its priorities. The IOR's strategy, which is currently developing in its early phases, is closely aligned with international goals for protecting the ocean as a common resource: a vital trade route, a sustainable resource base, and a region free from increased military competition, non-state actors and devastating natural disasters. Further capacity building, increased transparency and confidence-building measures and improved institutional cooperation are all necessary for achieving these goals. The term 'blue economy (BE)' is increasingly being used to describe economic activity that depends on marine ecosystems or the seabed (Llewellyn et al., 2016).

Although a number of stakeholders (both state and non-state) have begun to use the phrase 'BE', a working definition of the idea at the regional or global level has not yet been established (Gamage, 2016). 'The Blue Economy' philosophy was coined by Prof. Gunter Pauli in 1994 when asked by the United Nations to reflect on the business models of the future in preparation for COP3 in Japan (blueeconomy.org). Later in 2010, he introduced the BE business model in his book, 'The Blue Economy: 10 years – 100 innovations – 100 million jobs' (Gunter Pauli, 2011). The advancement of the 'Green Economy' idea and the continued development of the Institutional Framework for Sustainable Development were the two main focuses of the 'Rio +20' United Nations Conference on Sustainable Development (UNCSD), which took place in Rio de Janeiro from June 20–22, 2012 (UNCSD, 2012). There has been a growing understanding that the world's oceans and seas need more in-depth attention and concerted action in the aftermath of the Rio +20 process.

Many coastal nations questioned the focus of the Green Economy and its relevance to them during the Rio +20 preparation phase. Strong arguments were made for a 'BE' strategy to be given more prominence in the Rio+20 planning process (UNCSD, 2012). So, the concept of BE in practice is, just a decade old but due to its relevance in the changing dynamics of world oceans, it is gaining popularity. Maritime and coastal tourism constitutes an important component of BE along with sectors such as renewable energy, fisheries, maritime transport, climate change, and waste management. It is the second largest industry in BE after oil and natural gas (OECD report agenda 2030). Jobs and economic prosperity can result from ocean and coastal tourism.

More than 41 million tourists visited Small Island Developing States (SIDSs) and Coastal Least Developed Countries (CLDCs) annually (World Bank, 2018). In contrast, despite its significance in the BE, maritime tourism receives modest attention in the literature (Kabil et al., 2021). Many developing nations put a high priority on marine and coastal tourism. The world economic crisis hasn't prevented international tourism from flourishing until the global shutdown of recent times due to COVID-19 pandemic. The tourism sector was the first one to get affected by the pandemic but it is also expected to grow significantly in the post-pandemic world. The World Travel and Tourism Council represents the Travel & Tourism sector globally over year growth of 20.7%, the travel and tourism sector's contribution to the Indian economy may surpass pre-pandemic levels this year (WTTC, 2022).

This presents a unique opportunity for India tourism to brand itself as a promising maritime destination in the global travel market. The Government of India has rolled out draft policy 2021 for BE. This research aims to explore the dimensions of reaching out to the world in a new era of sustainable tourism development showcasing blue growth prospective of brand India. As dealing with fragile ocean systems comes with great responsibility whereas the growing threats of climate change even complicate it more. Together with issues and difficulties come challenges and possibilities, and the BE presents a number of them for equitable, clean, and sustainable blue growth in both established and growing industries (UNCSD, 2012).

There is no clear-cut definition of a brand, despite several specialized research, and each tourist destination brand has unique characteristics (Naydenov, 2019). 'Incredible India' has so far not reached the masses it desired to reach. Establishing a brand requires the owner to communicate consistently, intentionally, and sustainably. Every marketing action must include the brand (Naydenov, 2019). The BE in the coming times will be defining feature for the states aspiring to establish themselves as sustainable destinations that are future-ready.

Trends like globalization, regionalization and the knowledge-based economy are all related to modern economic development. To identify the source, the creator, or the owner of a specific object, the Old Norse word 'brand' was introduced into the English language in the 10th century. Today, it is employed to distinguish a product or service from others of a similar nature while also identifying it with the producer, the intermediary, and the dealer. This is done mostly in the minds of the customers.

By launching a new 'product' under the same name, a destination brand can be enhanced (Naydenov, 2019). According to Ritchie and Ritchie (1998), a destination brand is any name, symbol, logo, word, or other graphic sign that serves to identify and set the destination apart; it also promises a memorable experience that is exclusively connected to the destination; and it strengthens and authenticates the assortment of good memories amassed by the destination's experiences in order to highlight the significance of branding through symbols. Under the Incredible India Tagline, India needs to create a BE brand.

The Ministry of Tourism compiles monthly Foreign Tourist Arrivals (FTAs) based on information received from Bureau of Immigration (BOI) (refer Fig. 10.1). It suggests the India receives very few tourists through sea whereas air being the predominant one as mode of travel among all.

Tourism, when managed properly, may play a positive role in the region's social, cultural, economic and environmental development, and as such represents a substantial development opportunity for countries in the IOR. Unchecked tourism development, on the other hand, can harm natural resources, consumer

Year	Arrivals	% distribution by mode of travel		
		Air	Sea	Land
2001	2537282	87.1	0.9	12.0
2002	2384364	81.9	0.6	17.5
2003	2726214	83.1	0.5	16.4
2004	3457477	85.6	0.5	13.9
2005	3918610	86.5	0.4	13.1
2006	4447167	87.1	0.6	12.3
2007	5081504	88.4	0.6	11.0
2008	5282603	89.1	0.7	10.2
2009	5167699	89.8	1.0	9.2
2010	5775692	91.8	0.7	7.5
2011	6309222	92.0	0.8	7.2
2012	6577745	91.7	0.7	7.6
2013	6967601	91.0	0.5	8.5
2014	7679099	86.1	0.4	13.5
2015	8027133	84.5	0.7	14.8
2016	8804411	84.1	0.9	15.0
2017	10035803	79.6	0.7	19.7
2018	10557976	79.6	0.8	19.6
2019	10930355	77.4	0.9	21.7
2020	2744766	79.2	1.5	19.3

Fig. 10.1. FTAs in India According to Mode of Travel, 2001–2020.
Source: India Tourism Statistics (2021).

patterns, pollution, and social systems. Sustainable planning and management are thus required for countries in the region to reap the full benefits of a positive and sustainable tourism economy (Indian Ocean Rim Association [IORA]).

The purpose of this study is to develop and explore the potential of Indian tourism as a preferred destination brand. The study has concluded and recommend a theoretical model of destination branding through BE approach.

2. Literature Review

The late 1990s saw the beginning of the rise in popularity of destination branding (Oppermann, 2000). Some of its prominence was spurred by its key theme for the Travel & Tourism Research Association Annual Conference in 1998 (Ritchie & Ritchie, 1998). According to Gartner (2014), destination branding is a relatively recent field of study in tourist studies. However, for a number of reasons, brand positioning is a difficult and complex task. The first – and possibly most important – point is that destinations have many more dimensions than other kinds of services or consumer goods. But for most places, a slogan serves as an essential means of communicating a destination's brand strategy to the public (Pike, 2005). The key to selling in the new economy is branding, which unites quality and destination branding viewpoints and goes beyond simple marketing speak (Manhas & Dogra, 2013).

A destination is a geographical place as well as a metaphysical space that is defined by the network of values and meanings associated with it. Because tourism has an impact on the economy, society and environment, locations need to be carefully considered, and help with developing and managing the brands that represent them, is needed (Compelo et al., 2013).

Many anticipate that the Indian Ocean will become a significant geopolitical and economic force in the 21st century. Currently, just three Indian Ocean states (Australia, Singapore, and the United Arab Emirates) are among the top 20 countries in terms of per capita gross national income (World Bank, 2015), and this must improve if the Indian Ocean is to realize its potential. Coastal and island nations in the Indian Ocean rely on the oceans for economic opportunities, with some nearly entirely dependent on marine resources due to a limited land-based resource base. Although the 'BE' term seems to be new, the water resources and the awareness of their importance have been around since prehistoric times (Kabil et al., 2021). It is also important to recognize that developing nations face systemic issues including inadequate administrative infrastructure and poor governance, which impede their progress towards achieving the Sustainable Development Goals (Hunjra et al., 2023) and BE is considered a prominent tool towards advancement of SDGs in developing countries.

There were roughly 5 million new employment and EUR 218 billion in GVA generated by BE sectors in 2018, As per the European Union BE Report 2022. Maritime and coastal tourism is at the forefront of BE's economic sectors, providing job opportunities and sharing in the gross value added (GVA) (European Commission, 2018).

Ritchie and Ritchie (1998) defined destination branding as '. . .a name, symbol, logo, word mark or other graphics that identifies and differentiates the destination; conveys the promise of a memorable travel experience that is uniquely associated with the destination, and consolidates and reinforces the recollection of pleasant memories of the destination experience'. It is essential to comprehend the characteristics that distinguish a destination brand from the branding of traditional products or services for it to conform to the definition's theme. 'The place product is a distinctive set of structures, amenities, and locations that embody a variety of independent service enterprises, both public and private' (Hankinson, 2009).

Oceans, seas, islands, and coastal areas are vital to the health of many national economies, especially those in developing nations, as well as to global food security and sustainable economic prosperity, according to the Rio+20 outcome document (DESA, 2012). The growth of tourism has been proposed as a substitute revenue stream to assist tiny island economies in reducing their dependency on diminishing marine resources (Praptiwi et al., 2021).

Since the oceans make up over three-fourths of the planet's surface, they are the largest ecosystem on the planet and offer a vast platform for the discussion of increasingly complex and interrelated development concerns like security, trade, livelihoods, and climate change (Larik et al., 2017). Since many of the littoral regions are in poor nations, the problems with sustainable development and the economy along the Indian Ocean rim are especially difficult (Roy, 2018). Furthermore, according to Hunjra et al. (2022) sustainable development is a macro-, systemic, multi-level, and trans-facet concept. There is daunting pressure on ocean resources as the IOR is abode to 60% of the world's total population. The necessity to incorporate sustainability and conservation into the management of the maritime domain gives rise to the BE concept, which is now in line with commerce and economic activity (Smith-Godfrey, 2016).

3. Research Objectives and Questions

- To study the role of destination branding and reimagining the Indian Tourism.
- To explore the opportunities provided by the BE specifically in the Indian Ocean region.
- To synergize the BE benefits and Indian tourism towards destination branding of India.

4. Methodology

The study was conducted through a combination of desk research and content analysis of various secondary sources of data as magazines, annual reports, articles, opinion pieces followed by a detailed literature review. The BE is a concept which advocates the sustainable stewardship of the ocean resources and promises sustained economic growth. It has been a decade now since this term started gaining traction in maritime domain. Considering this, the article provides a conceptual paradigm for the strategic positioning of India as a vibrant BE destination in the IOR.

Naturalistic inquiry is used in qualitative research, which inductively examines real-world contexts to produce detailed narrative descriptions and develop research ideas (Patton, 2005). Destinations must strategically position themselves to build a well-known brand image before they can present themselves to the world. Exploring the possibilities of BE as a sustainable development tool to brand India as a new destination. Developing it as a BE destination will require holistic planning and concerted efforts and most important strategic partnership between the public and the private sectors. India's image as a tourist destination has always been backed up by its cultural and historical background. But the tourists have always had fewer glimpses of what the real India is. Destination branding is a significant field of study and an effective tool for marketing travel destinations (Ruiz-Real et al., 2020). As rationale for this we propose rebranding of India as a BE Destination.

5. The Beginnings

Coastal and island developing nations have continued to lead this BE movement because they understand the importance of the oceans to humankind's future and that the BE provides a sustainable development strategy that is more appropriate for their unique set of conditions, limitations, and difficulties (UNCSD, 2012). Establishing inclusive processes and measurable outcomes for people who may be seriously impacted by measures but have little means to participate in participatory processes is essential to the advancement of the BE (UNEP, 2015).

A mutually beneficial relationship between destinations and travelers is built through meeting their basic and emotional needs (such as 'relaxing' and 'pretty') through effective destination branding (e.g. eating) (Ekinci, 2003). Branding and developing India as a BE destination brand will necessitate 'Blue Thinking', which will demand targeted actions on a variety of BE aspects. India has a fair share of natural resources in the Indian Ocean Rim; all it needs to do is manage them wisely (OECD, 2016).

In response to the increasing market competition for building a distinct brand identity, corporations have increased their efforts to strengthen their brands by implementing experiential strategies that provide clients with a distinctive and personalized experience. As the BE transforms from an idea to a time-testing reality, it is fast evolving and diversifying. BE developments are often both highly opportunistic, and highly strategic in initiation, approach and execution (UNEP, 2015). Branding has become an essential tool for tourism destinations, as it allows them to communicate more clearly the richness of experiences that tourists visiting a region might expect.

The BE is an economy that 'comprises a range of economic sectors and related policies that together determine whether the use of ocean resources is sustainable', according to a recent definition provided by a UN representative. Understanding and properly managing the many facets of oceanic sustainability – from pollution prevention to healthy ecosystems to sustainable fisheries – is a key task for the BE. Second, the BE compels us to acknowledge that unprecedented levels of cross-sector and cross-border cooperation through a range of partnerships that will be necessary for the sustainable management of ocean resources. Small Island Developing States (SIDS) are a group of small island nations that face unique sustainable development challenges due to their small size, limited resources, vulnerability to natural disasters, and exposure to the impacts of climate change. These states are typically characterized by their low-lying coastal areas, making them particularly susceptible to rising sea levels and extreme weather events. The UN observes that the BE will aid in achieving the UN Sustainable Development Goals, of which one goal, 14, is 'Life Below Water'. Fig. 10.2 shows BE definitions by several agencies.

Adoption of a clear definition of BE is the first step towards attaining an insight into the potential of BE in the IOR and taking collective action to explore it (Sakhuja, 2015).

6. Indian Ocean Rim Association

It is an intergovernmental organization comprising 23 member states and 10 dialogue partners working to strengthen regional cooperation and drive

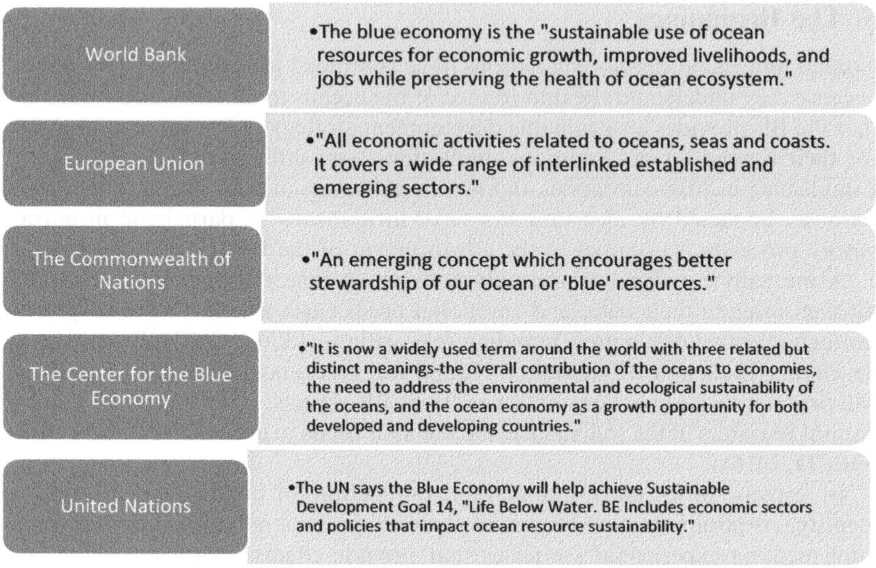

Fig. 10.2. Blue economy Definitions by Several Agencies. *Source:* Compiled by the author from the above-mentioned agencies.

sustainable development in the IOR. The organization at the 14th IORA Ministerial Meeting in Perth, Australia, on 9 October 2014, recognized this special priority area. Increasing worldwide interest and potential in the BE drew the attention of all IORA Member States, who now see it as a major priority for creating jobs, reducing poverty, and assuring the sustainability of business and economic models in the IOR. The many IORA Member States are becoming increasingly interested in developing a unified vision for the BE, which would make this sector an engine for balanced economic growth throughout the Indian Ocean Rim. The BE has a wide range of rich resources. A variety of capacity-building programs have been implemented since 2014 in a variety of fields. However, tourism is still waiting its turn; there is an urgent need to revitalize maritime and coastal tourism, given its contribution to the GDPs of the IORA states (IORA website).

The Indian Ocean is more than just a commercial waterway. The large drainage basin of the ocean is significant in and of itself, as it is home to around two 60% of the total world population. This opens up new prospects, especially considering the strong rates of economic growth throughout the Indian Ocean's rim, which includes India, Bangladesh, Southeast Asia, and Eastern and Southern Africa. The heavily populated littoral, on the other hand, is subject to natural or environmental disasters (Jaishankar, 2016).

7. India's Initiative to BE

The draft BE strategy was recently released by the Ministry of Earth Sciences (MoES), who is seeking feedback and suggestions from a range of stakeholders. It is consistent with the Indian government's 2030 vision for a 'New India'. The strategy plan identified the BE as one of the 10 critical components for the country's growth. It stresses policy in numerous critical sectors to achieve overall economic growth in India. It identifies the seven thematic areas listed below.

(1) BE and ocean governance accounting framework at the national level.
(2) Tourism and coastal marine spatial planning.
(3) Marine fisheries, aquaculture, and fish processing are all examples of industries.
(4) Manufacturing, emergent industries, trade, technology, services and skill development are all areas of focus.
(5) Logistics, infrastructure and transportation, especially transshipments.
(6) Offshore energy and coastal and deep-sea mining.
(7) Security, strategic dimensions, and international engagement are all important considerations.

8. Rising Diplomatic Profile in Indian Ocean

India's diplomatic outreach in the IOR has increased in the 21st century but it lacks the economic and military means to position itself. The BE will be an addition to the economy prospect and for the military might Indian Navy is much capable of securing our interests. The Indian Navy has re-oriented its mission-based deployments and have increased its naval presence in IOR and they have got the 'Indian Ocean covered' as stated by the officials (Ministry of Defence, 2024).

Being the most populated nation along the coast, India places a high value on the Indian Ocean. India's leadership position will be crucial in shaping the strategic future for the other states around the Indian Ocean (see Fig. 10.3), as well as for those outside of it (Jaishankar, 2019).

The IOR has been an important trade arena for centuries (Baruah et al., 2023). India's strategic location in the IOR, positioned along the primary shipping routes, presents significant advantages. It strategically positions the outer edges of the IOR and key narrow passages at approximately equal distances from India. This arrangement enables India's naval forces to easily access, maintain, and move around the region. India is strategically situated to have significant influence over the maritime domain and effectively advance and protect its national maritime interests throughout the IOR (IOR). Simultaneously, India's extensive coastline and maritime zones necessitate substantial resources and investments to assure their security (India Maritime Security Strategy, 2016).

9. What Is the Need to Go Blue?

Because the future belongs to this! Considering the impact of damage, we have already caused to the oceans worldwide, if we don't come up with a solution to this, the future is uncertain. But we cannot also afford to ground our all-maritime operations we heavily depend on oceans for our economic growth. Here comes

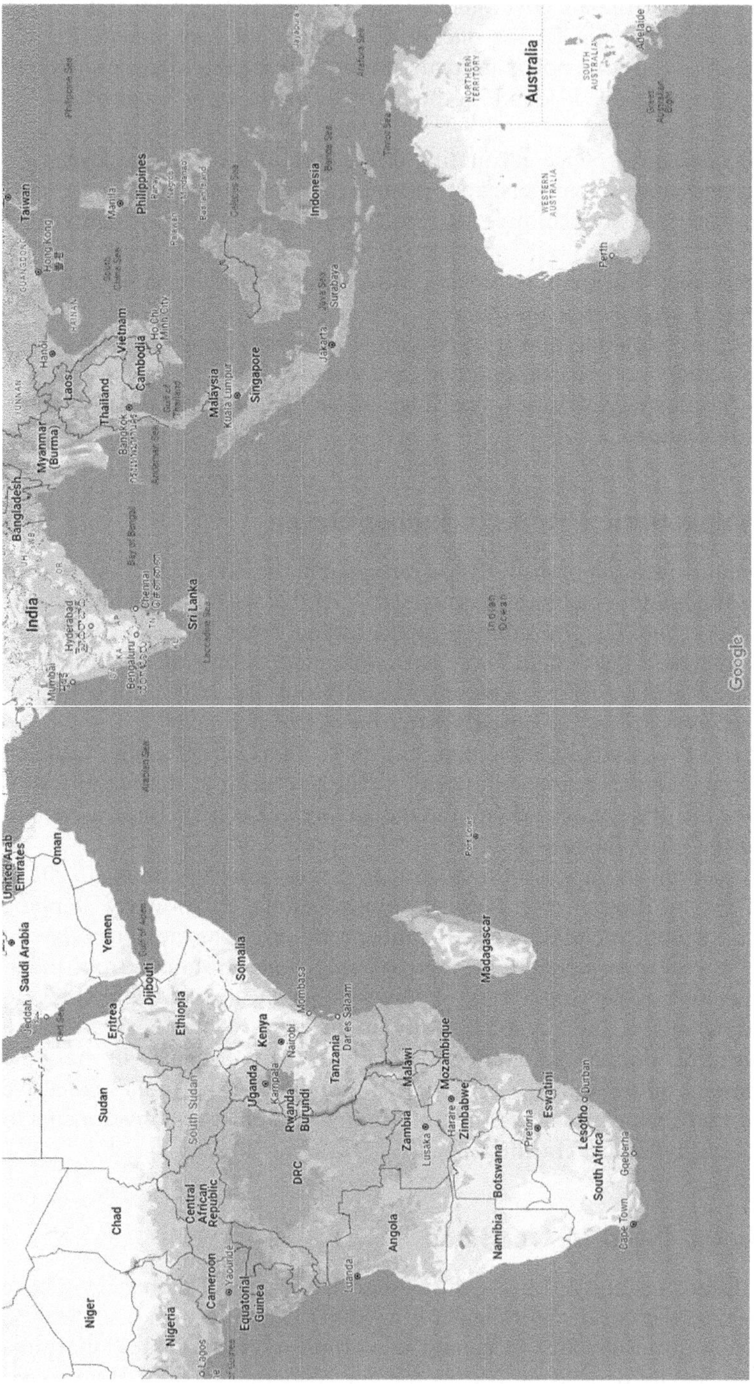

Fig. 10.3. Geography of the Indian Ocean Region. *Source:* Google Maps (2024).

the BE as a saviour. BE is the only solution to this, creating meaningful development spaces without compromising ocean health and the ecosystem.

China, which emerged as the world's workshop at the beginning of the 21st century and declared its maritime destiny, to develop a blue water navy and assert what it considered its maritime rights, is another country whose current maritime imperative can be compared to that of the United States at the dawn of the 20th century when the latter became an industrial power and turned to the sea for global markets and exercising its maritime capabilities (Basu, 2021).

Developing its BE potential will also help India to contribute towards the United Nations Sustainable Development Goals- 13 & 14 which addresses the 'Climate Action' & 'Life below Water' respectively (Upadhyay & Mishra, 2020). India is a critical player in IOR and can influence and direct joint actions on some of the biggest challenges of our century. The world has very fewer glimpses of our country to change that image India needs to reposition itself in the global market. BE can be a fantastic tool to progress in this direction, an economic approach which is viable and has a promising future. The international community widely acknowledged the significance of oceans for sustainable development, which was reflected in Agenda 21, the Johannesburg Plan of Implementation, and a number of Commission on Sustainable Development decisions, among other documents (DESA, 2012).

Oceans account for more than 15% of the annual animal protein consumption for about 4.3 billion people worldwide and are critical to global food security. Fisheries have been the fastest expanding animal-food producing sector since 2011. Governments are increasingly cognizant of the significance of sustainable development of coastal and marine resources to safeguard long-term national interests in the IOR, a region that is so reliant on its resources.

10. Blue Economy: The Future Economy

Given that India is a peninsula, trade, religion, and culture have always been linked to it through the sea; these early ties, however, were broken over time (Basu, 2021). While there is not a single accepted definition of a BE India has to define its notion. As land resources reach their limit human activities are shifting to oceans. But oceans cannot be treated as they are perceived as 'limitless' or 'Dumping Grounds' for many decades. Oceans can serve as a great platform for Maritime security, responsible tourism, climate change mitigation, sustainable fisheries and renewable energy. India has to develop the necessary infrastructure and manpower resources to mobilize this potential.

Fig. 10.4 shows the two major facts of BE development approach. The increasing threat of climate change to world oceans is evident in the sea level rise resulting in the loss of marine life and the sinking of several island destinations worldwide. This problem asks for an innovative solution and immediate action and the answer to this is the BE framework. BE will also provide a boost to Brand India Tourism in marketing it as a destination of the future. A BE destination in IOR.

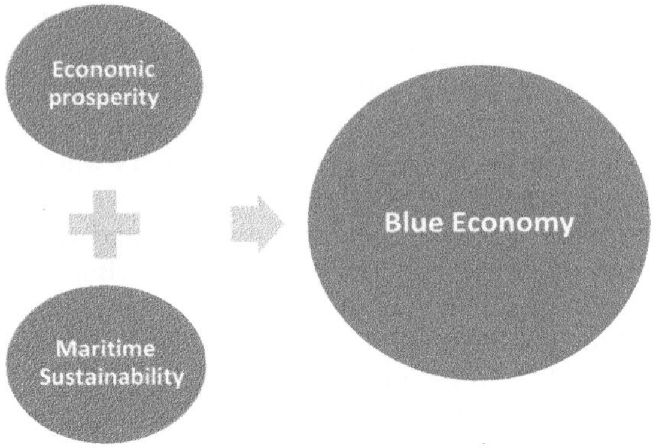

Fig. 10.4. The Two Major Facts of Blue Economy Development Approach. *Source:* Authors own elaboration.

A very good example is Bangladesh a major conference on the advancement of the BE was held in Bangladesh in 2014, making it the first country to convene such a gathering in a regional context. With nearly half of the world's population expected to live in Indian Ocean Rim (IOR) countries by 2050, the region is transitioning from its identity as the 'Ocean of the South' to the 'Ocean of the Centre', and then to the 'Ocean of the Future', as its core position in international trade, industry, labour, environment, and security is likely to shape the world in the 21st century (Doyle & Seal, 2015).

11. Findings and Discussions

India today realizes its vast coastline. Although the major focus will be on maritime transport as it supports 90% of the trade that the country does with the rest of the world. The country is bestowed with rich Maritime and Coastal Tourism advantages. Now, the question lies in its effective and responsible actions towards utilising them.

Incredible India began as a marketing campaign by the Government of India in 2002 to encourage tourism and project India as a viable tourist destination. Though much work remains to be done on the ground to make India more tourist-friendly, the government is taking the correct steps by recognising the vast potential of tourism. With a budget of $46 million, the Ministry of Tourism launched the 'Incredible 2.0 Campaign' in 2017–18 to promote diverse places and areas such as spiritual, medical and wellness tourism. Another concern that the government must address is the development of infrastructure to accommodate the needs of developing tourism. It is important to note that rebranding can be a risky and challenging strategy but if managed well, it can change a destination's future.

It is challenging to distinguish between destinations based on their 'hard aspects', which are infrastructure, according to recent studies (Yuksel, 2003). Numerous studies have demonstrated that a number of locations are scaling and functioning well in the 'hard aspects', which include the economy, the infrastructure for tourists, and other related services. The 'soft aspect' of a location, which encompasses its surroundings, the friendliness of the locals, art and cultural customs and recreational opportunities, is becoming increasingly significant and distinctive (Morgan et al., 2011). India must increase its efforts in both areas. Despite our history of using soft power, we must quickly capitalize on our hard power potential in this sector if we want to compete with other top travel destinations. It involves committing to goals and ensuring they are met. It is about making promises and ensuring that they are delivered. Rebranding is about adding value to an existing brand, in other words (Jesca et al., 2014).

The way the brand makes an impact on consumers. People's views of a branded area are said to be influenced by the symbolic features of the brand, such as the name, logo, and slogan. These markers give people a foundation for recalling the location and forming favourable perceptions of it (Kladou et al., 2016). India must overcome its existing image as a chaotic overpopulated Asian country (Gartner, 1989). Since it got independence, it has come a long way, developed, and became self-reliant.

11.1 Conceptual Paradigm for Blue Economy Brand Building

Fig. 10.5 shows conceptual paradigm for BE brand building.

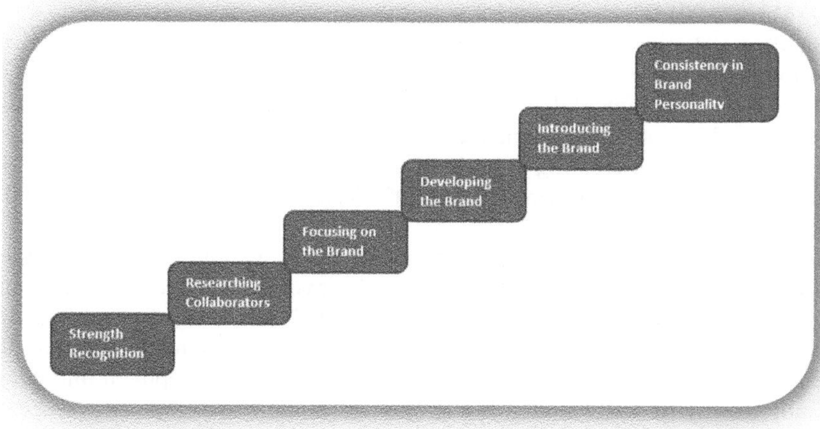

Fig. 10.5. Conceptual Paradigm for Blue Economy Brand Building. *Source:* Authors own elaboration and https://seosandwitch.com/branding-cycle/.

Strength Recognition: India is essentially a maritime nation, its 7,500 kms long coastline holds potential for maritime activities and development initiatives. With a GDP growth rate of 6.1% in 2018–2019, India is one of the world's fastest-growing and largest economies. Approximately, 95% of India's merchandize trade takes place via seas. It is important to boost country's strategic capacity in the IOR.

Researching Collaborators: Developing maritime infrastructure will require regional integration and collaborative efforts with the partners having common stakes in the IOR. Australia is one such partner, Bangladesh is also doing well in this front.

Focussing on the Brand: It will take time to shift the focus from land-based activities to the ocean-based one but several countries are turning 'Blue' from being a Brown economy.

Developing the Brand: A BE destination brand building will require long term investments, 'Blueing' of the economic activities, creating ocean-based livelihood opportunities and organising and implementing the policies. India lacks in infrastructural capabilities as well as trained manpower dedicated to this service sector.

11.1.1 Introducing the Brand

Challenges of Branding a Maritime Destination – Despite the challenges to collective action presented by the 'ASEAN Way', there is still hope for a BE in Southeast Asia (Gamage, 2016). The intricacy of tourists' decision-making is a distinguishing feature of destination branding. Travelers are purchasing an assortment of products and services, most of which have a high price tag and inherent unpredictability (Cai, 2002). Additionally, before purchasing a vacation package, tourists are unable to preview the destination (Cai, 2002; Eby et al., 1999; Martin, 2016).

Building a destination brand is challenging since destinations involve various stakeholders in the brand-building process (Almeyda-Ibáñez & George, 2017). The external environmental factors of a country such as political, economic, social and cultural should be considered well in advance.

The complexity of creating a destination brand is correlated with creating the experiential aspect and knowing the tourist's decision-making process (Almeyda-Ibáñez & George, 2017).

The BE model generates economic opportunities without compromising on ocean health, and those economic gains should involve and reach local communities then only the purpose of BE will be served. It has been reported that the maritime and coastal communities do not receive the adequate share of the benefits instead they are affected by the ill effects of these development initiatives. It is noteworthy that BE model dwells upon sustainable development of oceans resources and as stated by Priyakrushna Mohanty et al. (2020) Sustainable marketing of the destination is futuristic thinking which will help to induce the values of sustainability in the minds of both tourists and hosts.

Olins and Hildreth (2011) state that a further difficulty may be the ongoing misinterpretation of country branding by experts and government officials as a result of the latter's ignorance of the former. Government representatives are drawn to nation branding because it promotes internal unity and foreign political and economic advancements, but they are unaware of the process involved in 'nation branding' (p. 57). Tourism is a highly competitive environment with most destinations offering the same product; tourists are facing a wide range of variety (Hosany et al., 2006). There has also been damage and degradation of ocean resources by the BE-led development model around the world. The extinction of the fishing community in South Africa is one such example.

National Security – Maritime security is essential to supporting the BE (Voyer et al., 2018) and it is also an enabler to BE. The maritime resources of a nation need to be protected from the various threats of piracy, overexploitation and terrorism also sea becomes a medium in illegal fishing, human trafficking and smuggling.

Territorial Claims – Defining national boundaries in waters is difficult and a major cause of dispute among nations to exploit the resources it has underneath. Access to and management of ocean resources are frequently made more difficult when there are conflicting maritime claims because of the ensuing confusion concerning jurisdiction. When maritime borders are unclear and there may be large disputed water areas, the sustainable management of marine life resources may be seriously impeded (Voyer et al., 2018).

A Lack of Regional Cooperation – It has also impacted Indian Ocean tourism. While the elusive mechanics of effective regional integration remain a focus of economic and policy specialists in targeted regions around the world, the question of whether and how regional integration efforts, particularly in the tourist sector, can affect substantial or transformative change continues. With limited success, the Caribbean and South Pacific have worked for decades to develop a regional brand and integrate their various products. On a comparable basis, Indian Ocean destinations have made no real regional tourism integration attempts (World Bank Report, 2013).

Ecological Threats – The increasing ocean acidification leading to dead zones and incidents of coral reef bleaching is becoming common. The loss of biodiversity and extinction of several species makes it even worse.

Climate Change – The most challenging issue of 21st Century is threatening and its impacts are quite visible in Ozone depletion, melting of glaciers, rising sea levels causing existential threats to the low-lying countries and SIDS. The increase in extreme weather events such as tsunami, typhoons, floods, droughts, landslides and change are seasonal patterns are being experienced at almost every corner of the planet.

Marine Litter and Plastic Pollution – Oceans cover 71% of the surface of the planet and are an important carbon sink. Today, they are under threat because of human activities, but oceans can recover if we allow them to. Tourism is the major activity which intensifies this issue. Despite their economic, social, and environmental advantages, oceans, seas, and their associated ecosystems face a slew of local and global dangers from anthropogenic activities and climate change. Assuring healthy and productive oceans is thus critical to attaining sustainable development.

Financing – To compete with the established tourism brands with a limited budget is another challenge to overcome these destinations have to be smart enough to market themselves. Creating a Unique Selling Proposition (USP) and targeting the potential market will help destinations in the longer run. It is critical to allocate expenditures for tourism promotion in order to establish a successful destination brand.

Consistency in Brand Personality – Once the brand is established maintaining consistency in brand personality is crucial for building trust, loyalty, and recognition among the tourists. Here are some key strategies for achieving consistency.

- *Defining the Brand Personality:* Starting by clearly defining the personality traits that will differentiate the destination brand India from the rest of the world. Is it strategic, sustainable, innovative, adventurous, or something else? Ensuring that these traits align with the brand values and resonate with the target audience will be important.
- *Consistent Visual Identity:* Maintaining consistency in the visual identity, including logos, colours, fonts, and imagery. Using these elements consistently across all marketing materials, website, social media profiles, and other brand assets will create a strong appeal for travellers.
- *Coordinated Marketing Campaigns:* Aligning the marketing campaigns with the brand personality to ensure coherence in messaging and imagery. Whether it is a social media campaign, email marketing, or advertising, all content should reflect the established brand personality.
- *Monitor and Adjust:* Regularly monitoring the brand's communication efforts to ensure that they align with the defined brand personality. Collecting feedbacks from customers and stakeholders, and adjusting the approach if necessary to maintain consistency.
- *Adaptability:* While consistency is essential, keeping in mind of the need to adapt the brand personality to different contexts and platforms is essential. For example, the tone on social media might be more informal than in corporate communications, but it should still reflect the core attributes of your brand.
- *Staying True to the Core Values:* Ultimately, consistency in brand personality should be rooted in the brand's values and mission. Ensuring that every communication reflects these core principles to maintain authenticity and trustworthiness.

By implementing these strategies, we can effectively maintain consistency in BE destination India brand personality, which in turn will help to strengthen the brand identity and will help to establish India as a sustainable maritime destination is the IOR.

11.2 Imperatives to India's Blue Growth

Geoeconomic Opportunity – IOR boasts some of the most vital 'chock points' of the world trade, this is reason behind contestations of its waters among major superpowers. India's location at an important maritime stretch is the biggest advantage and it must tap the potential it presents. As land resources reach their limit, we are eying the oceans and world is now opening up to harnessing the ocean opportunities as the nations are investing in BE.

A Regional Branding Strategy – Along with the littorals India must work on developing a regional branding strategy. The increasing competition among IOR nations over marine resources for their development stresses the need for improved regional collaboration that enables more sustainable exploitation of these resources. Spite of significant accolades for the region's tourist hotspots, the countries surrounding the Indian Ocean receive comparatively few visits. These countries have a lot to offer, from some of the best beaches in the world to a fascinating cultural melting pot and a position as one of the planet's most important biodiversity hotspots. Because of the destination's varying degrees of development, this region has not yet gained widespread attention. Joint marketing can assist in raising the region's profile, particularly in emerging markets. This can be accomplished through collaboration on advertising materials and trade show representation.

Cruise Tourism – The growth of cruise tourism can be another step towards regional collaboration and could eventually serve as a major hub for integration initiatives. By definition, cruise tourism necessitates regional cooperation and unity. The nations in the area must work together to create a cruise tourism plan in order to draw in more cruise ships (World Bank Report, 2013). The countries in the region must realize the huge potential in this area, and collaborate in capacity building for cruise terminals. An excellent example is Europe and its world-class development of cruise liners. The majority of the cruise liners surpass this region and harbour in and around Europe. To develop this industry countries in the region, need effective collaboration and joint efforts to place themselves in the market and compete with modernized Europe which currently has world class infrastructure and a strong brand image. Fig. 10.6 shows BE management framework for India.

12. Limitations

The study has its own set of limitations the authors recommend a strong case for BE as a branding tool but how it will turn out in a complex Indian environment remains a key question. Because the Government of India is developing a BE policy draft now only, its approval and effective implementation will decide the future course of action. Creating and developing a brand is not an easy task, so even developed countries do not find brand management an easy mission. Very few countries have successfully launched a national brand (Petrevska & Cingoski, 2017).

13. Conclusion

We may infer that India as a destination has a low degree of environmental quality, resulting in a poor and insufficient foundation for launching a national

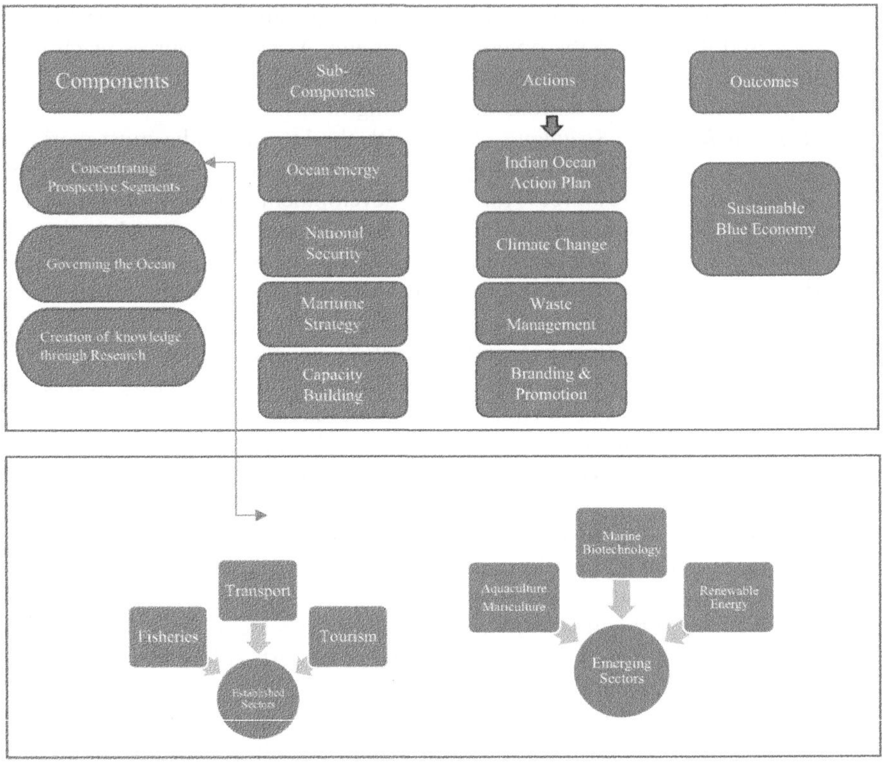

Fig. 10.6. Blue Economy Management Framework for India. *Source:* Authors own. https://doi.org/10.1016/j.ocecoaman.2018.03.001.

BE brand. Despite all its glowing reviews and accolades, the country has not performed well in the global tourism marketplace. The modest number of people who visit each year is indicative of this. The country has not yet achieved a 1% share of the International Tourist Arrivals. Additionally, COVID-19 pandemic and the closing of international borders have caused severe damage to this service industry. As now world reopens for tourists, there is a need to develop a comprehensive development framework which is future ready to be ahead in the blue race. This paper contributes to the development of BE and presents a unique dimension to branding India as a destination of the future in IOR. Building a future destination and that too a sustainable one which will effectively use its Blue Resources will take time because it is a long-term and continued growth process. It also involves multiple stakeholders which adds to its complexity. Despite the obstacles, this objective is achievable if the ocean decade opportunities are realized developed and managed well the expected outcomes will follow. To achieve these objectives concerted actions are required. India requires a strategic repositioning in the IOR to be ahead in the 'Blue Race'. The BE brings the momentum and accentuates that now is the time to embrace regional cooperation initiatives in tourism, which can result in a sustainable platform for leveraging future regional integration efforts. Before meeting these preconditions for regional integration, several gaps must be filled through enhanced cooperation. It will need concerted action from countries in the Indian Ocean to promote worldwide interest in visiting the region's high-quality ecotourism destinations and draw a reaction from the aviation sector. Port facilities and an adequate supply of interesting things to do are necessities if the cruise industry is to see a growth in its presence in the Indian Ocean. Again, regional coordination is needed to determine which cultural and ecotourism offerings will appeal to the cruise ship industry and which ports could serve as resupply stops for these vessels.

Abbreviation

BE—Blue Economy
IOR—Indian Ocean Region
IORA—Indian Ocean Rim Association
LDC—Least Developed Countries
OECD—Organization for Economic Co-operation and Development
SIDS—Small Island Developing States
UN—United Nations
USP—Unique Selling Proposition

References

Almeyda-Ibáñez, M., & George, B. P. (2017, March 27). The evolution of destination branding: A review of branding literature in tourism. *Journal of Tourism, Heritage & Services Marketing*. http://doi.org/10.5281/zenodo.401370

Baruah, D., et al. (2023, July). *Mapping the Indian Ocean Region*. Carnegie Endowment for International Peace. https://carnegieendowment.org/2023/06/15/mapping-indian-ocean-region-pub-89971

Basu, P. (2021, November). *Maritime India: The quest for a steadfast identity*. ORF Occasional Paper No. 339. Observer Research Foundation. https://policycommons.net/artifacts/1900129/maritime-india/2651292/

Cai, L. A. (2002). Cooperative branding for rural destinations. *Annals of Tourism Research*, *29*(3), 720–742. https://doi.org/10.1016/S0160-7383(01)00080-9

Compelo, A., Robert, A., Maree, T., & Juergen, G. (2013). Sense of place: The importance for destination branding. *Journal of Travel Research*, *53*, 154. http://jtr.sagepub.com/content/53/2/154; https://doi.org/10.1177/0047287513496474

DESA. (2012). Sustainable Development in the 21st century. Review of implementation of agenda 21. https://sustainabledevelopment.un.org/content/documents/1126SD21%20Agenda21_new.pdf

Doyle, T., & Seal, G. (2015). Indian Ocean futures: New partnerships, new alliances and academic diplomacy. *Journal of the Indian Ocean Region*, *11*(1), 2–7. https://doi.org/10.1080/19480881.2015.1019994

Eby, D. W., Molnar, L. J., & Cai, L. A. (1999). Content preferences for in-vehicle tourist information systems: An emerging travel information source. *Journal of Hospitality & Leisure Marketing*, *6*(3), 41–58. https://doi.org/10.1300/J150v06n03_04

Ekinci, Y. (2003). An investigation of the determinants of customer satisfaction. *Tourism Analysis*, *8*(2–4), 193.

European Commission Report. (2018). https://op.europa.eu/webpub/com/general-report-2018/en/

European Union Blue Economy Report. (2022). https://op.europa.eu/en/publication-detail/-/publication/156eecbd-d7eb-11ec-a95f-01aa75ed71a1

Gamage, R. N. (2016). Blue economy in Southeast Asia: Oceans as the new frontier of economic development, *Maritime Affairs: Journal of the National Maritime Foundation of India*. https://doi.org/10.1080/09733159.2016.1244361

Gartner, W. C., (1989). Tourism image: Attribute measurement of state tourism products using multidimensional scaling techniques. *Journal of Travel*, *28*(2), 16–20. https://doi.org/10.1177/004728758902800205

Gartner, W. C. (2014). Brand equity in a tourism destination. *Place Branding and Public Diplomacy*, *10*, 108–116. https://doi.org/10.1057/pb.2014.6

Google Maps. (2024). *Map of 7.1299147,71.7200368*. https://www.google.com/maps/-7.1299147,71.7200368,4z?entry=ttu. Accessed on 25 March 2024.

Gunter Pauli. (2011). The blue economy. *Japan Spotlight*. Cover story.

Hankinson, G. (2009). Managing destination brands: Establishing a theoretical foundation. *Journal of Marketing Management*, *25*(1–2), 97–115. https://www.tandfonline.com/doi/abs/10.1362/026725709X410052

Hosany, S., Ekinci, Y., & Uysal, M. (2006). Destination image and destination personality: An application of branding theories to tourism places. *Journal of Business Research*, *59*(5), 638–642. https://doi.org/10.1016/j.jbusres.2006.01.001

Hunjra, A. I., Azam, M., Bruna, M. G., & Bouri, E. (2023). A cross-regional investigation of institutional quality and sustainable development. *Journal of International Financial Markets, Institutions and Money*, *84*, 101758. https://doi.org/10.1016/j.intfin.2023.101758

Hunjra, A. I., Azam, M., Bruna, M. G., Verhoeven, P., & Al-Faryan, M. A. S. (2022). Sustainable development: The impact of political risk, macroeconomic policy uncertainty and ethnic conflict. *International Review of Financial Analysis*, *84*, 102370. https://doi.org/10.1016/j.irfa.2022.102370

India Maritime Security Strategy Document. (2016). https://www.indiannavy.nic.in/sites/default/files/Indian_Maritime_Security_Strategy_Document_25Jan16.pdf

India Tourism Statistics, Ministry of Tourism, Government of India report. (2021). https://tourism.gov.in/sites/default/files/2022-04/India%20Tourism%20Statistics%202021.pdf

Jaishankar, D. (2016, September 12). Indian Ocean region: A pivot for India's growth. *Brookings*. www.brookings.edu. https://www.brookings.edu/opinions/indian-ocean-region-a-pivot-for-indias-growth/#_ftn3

Jaishankar, D. (2019). *Acting East: India in the Indo-Pacific*. Brookings India, Impact Series 102019-02, October 2019. Brookings Institution India Center. https://policycommons.net/artifacts/4140870/acting-east/4949630/

Jesca, C., Kumbirai, M., & Brighton, H. (2014). Destination rebranding paradigm in Zimbabwe: A stakeholder approach. *International Journal of Advanced Research in Management and Social Sciences*, *3*(1), 30–41. www.garph.co.uk

Kabil, M., Priatmoko, S., Magda, R., & Dávid, L. D. (2021). Blue economy and coastal tourism: A comprehensive visualization bibliometric analysis. *Sustainability, 13*(7), 3650.

Kladou, S., et al. (2016). The role of brand elements in destination branding. *Journal of Destination Marketing & Management.* http://doi.org/10.1016/j.jdmm.2016.06.011i

Larik, J., et al. (2017). Blue growth and sustainable development in Indian Ocean governance. *Policy Brief, 18.*

Llewellyn, L. E., English, S., & Barnwell, S. (2016). A roadmap to a sustainable Indian Ocean blue economy. *Journal of the Indian Ocean Region, 12*(1), 52–66. http://doi.org/10.1080/19480881.2016.1138713

Manhas, P. S., & Dogra, J. (2013). Quality management practices and tourism destination branding: Inter-relationship and preferential study of the components. *Journal of Services Research, 13*(1), 75.

Martin, R. (2016, December). The future of branding. *Vision, 20*(4). https://martinroll.com/books/the-future-of-branding/

Ministry of Defence. (2024). PIB. https://pib.gov.in/PressReleasePage.aspx?PRID=2001842#:~:text=by%20PIB%20Delhi-,Indian%20Navy%20units%20are%20regularly%20deployed%20on%20Mission%20Based%20Deployments,maritime%20security%20in%20the%20IOR

Morgan, N., Pritchard, A., & Pride, R. (2011). Tourism places, brands, and reputation management, *Destination brands: Managing place reputation* (3rd ed., pp. 3–19). Routledge-Taylor & Francis Group.

Naydenov, K. (2019, March). Theoretical bases of tourism branding. *Knowledge – International Journal, 30*(6).

OECD Report. (2016). *The Ocean Economy in 2030.* Ocean Economy Group/OECD Space Forum.

Olins, W., & Hildreth, J. (2011). *The evolution of destination branding: A review of branding literature in tourism.*

Oppermann, M. (2000). Tourism destination loyalty. *Journal of Travel Research, 39,* 78. https://doi.org/10.1177/004728750003900110; http://jtr.sagepub.com/content/39/1/78

Patton, M. Q. (2005). *Qualitative research.* John Wiley & Sons, Ltd.

Petrevska, B., & Cingoski, V. (2017). Branding the green tourism in Macedonia. *Sociologija i prostor: časopis za istraživanje prostornoga i sociokulturnog razvoja, 55*(1(207)), 101–116. https://hrcak.srce.hr/file/268703

Pike, S. (2005). Tourism destination branding complexity. *The Journal of Product and Brand Management, 14*(4), 258–259. http://eprints.qut.edu.au

Praptiwi, R. A., Maharja, C., Fortnam, M., Chaigneau, T., Evans, L., Garniati, L., & Sugardjito, J. (2021). Tourism-based alternative livelihoods for small island communities transitioning towards a blue economy. *Sustainability, 13,* 6655. https://doi.org/10.3390/su13126655

Priyakrushna Mohanty, P. M., Rajamanicam, H., & Anukrati Sharma, A. S. (2020). Exploring the future of sustainable tourism. In *Sustainable destination branding and marketing: Strategies for tourism development* (pp. 216–227). CABI. https://doi.org/10.1079/9781786394286.0216

Ritchie, J. R. B., & Ritchie, J. R. R. (1998, September). The branding of tourism destinations. In *Annual congress of the international association of scientific experts in tourism* (pp. 1–31).

Roy, A. (2018). *Blue economy in the Indian Ocean: Governance perspectives for sustainable development in the region*. Observer Research Foundation. European Commission. Annual Economic Report on the EU Blue Economy. European Commission.

Ruiz-Real, J. L., Uribe-Toril, J., & Gázquez-Abad, J. C. (2020). Destination branding: Opportunities and new challenges. *Journal of Destination Marketing & Management*, *17*, 100453. www.elsevier.com/locate/jdmm

Sakhuja, V. (2015). Harnessing the blue economy. *Indian Foreign Affairs Journal*, *10*(1), 39–49. http://www.jstor.org/stable/45341010

Smith-Godfrey, S. (2016). Defining the blue economy. *Maritime Affairs: Journal of the National Maritime Foundation of India*, *12*(1), 58–64. http://doi.org/10.1080/09733159.2016.117513

UNEP. (2015). *Blue economy: Sharing success stories to inspire change*. www.unep.org/greeneconomy

United Nations Conference on Sustainable Development (UNCSD). (2012, June 20–22). *Future we want – Outcome document*. https://sustainabledevelopment.un.org/index.php?menu=1298

Upadhyay, D. K., & Mishra, M. (2020). Blue economy: Emerging global trends and India's multilateral cooperation. *Maritime Affairs: Journal of the National Maritime Foundation of India*, *16*(1), 30–45. https://doi.org/10.1080/09733159.2020.1785087

Voyer, M., Schofield, C., Azmi, K., Warner, R., McIlgorm, A., & Quirk, G. (2018). Maritime security and the Blue Economy: Intersections and interdependencies in the Indian Ocean. *Journal of the Indian Ocean Region*, 1–21. https://doi.org/10.1080/19480881.2018.1418155

World Bank. (2015). *World development indicators*.

World Bank. (2018). Summary report, 26–28 November 2018. Sustainable Blue Economy Conference. https://enb.iisd.org/events/sustainable-blue-economy-conference/summary-report-26-28-november-2018

World Bank Report. (2013). The way forward for Indian Ocean Island tourism economies: Is there a role for regional integration? *Indian Ocean Tourism: Regional Integration or Cooperation?*

World Travel and Tourism Council. (2022). https://wttc.org/Portals/0/Documents/Reports/2022/EIR2022-Global%20Trends.pdf

Yuksel, E. (2003). From destination image to destination branding: An emerging area of research. *E-Review of Tourism Research (eRTR)*, *1*(2).

Chapter 11

Fostering Resilience Through Travel: Prioritising Well-being, Accessibility and Sustainability

Manpreet Arora

Central University of Himachal Pradesh, India

Abstract

The outbreak of COVID-19 endemic forced people not only to think but also to pause and objectively reflect how to deal with the situation that has arisen and how to develop well-being and resilience strategies for the welfare of humanity. The endemic brought a global economic shock of enormous magnitude in most of the countries. Policymakers across the globe today have been facing alarming situations and unpredictable challenges as they try to find solutions to problems in the areas of tourism, at macro-economic levels and in socio-cultural arena. People in developing countries are concerned with earning livelihood and supporting their families and find some opportunities to survive by increase in tourist arrivals. Conversely, industralized countries struggle to improve the general psychological and physical health of their citizens. Travelling for well-being appears to be a sensible and uplifting resilience tactic in such a situation. This piece discusses how, in such a situation, cultivating resilience and well-being techniques may enable us to overcome a variety of obstacles in the travel and tourist industry. According to the author, policies aimed at enhancing well-being through travel should give priority to factors like affordability, accessibility, sustainability, safety and education. By promoting an environment where travel is affordable, accessible and sustainable, policymakers can guarantee that more people can take advantage of the positive experiences and enrichment that travel can provide to their lives, while also benefiting the greater community and environment.

The Emerald Handbook of Tourism Economics and Sustainable Development, 211–232

Published under exclusive licence by Emerald Publishing Limited
doi:10.1108/978-1-83753-708-220241011

Keywords: Well-being; resilience; COVID-19; strategies of well-being and resilience; wellness

1. Introduction

Different individuals exhibit varied relationships with wealth; some feel content to collect as much as possible, while others believe it robs us of true happiness and inner serenity. However, because word tends to refer to general wellness with numerous aspects, including physical, spiritual, emotional and intellectual, well-being is valued and embraced by the majority of people. In order to realize the objective of health for all, healthy lives and the promotion of well-being are included in the World Health Organization (WHO) Sustainable Development Goals (SDGs) that look forward to 2030. Numerous research on well-being have been conducted, with an emphasis on the psychological aspects of human life as well as the general health and well-being of humanity (Ryff, 1995; Ryff & Singer, 2008; Sirgy, 2012). The idea of well-being encompasses a wide range of elements, including health, human development, education, gender equality and happiness (Daniels, 2000; De Ruyter, 2015; Field, 2009; Rechtschaffen, 2014; Stone & Mackie, 2013). Policymakers are attempting to address the socioeconomic situation that has developed as a result of the absence of adequate infrastructure and corrective actions. Despite greater infrastructure and lower population, even the most developed economies in the world have not been able to contain the damage caused by COVID-19.

The question that follows is: Can we actually consider our well-being in such situations? Given the predicament the globe finds itself in as a result of the COVID-19 epidemic, it may be challenging to provide an answer. Global economy has undoubtedly plunged into poverty during and after endemic, some sectors are worst hit (Bargain & Aminjonov, 2020; Bargain & Ulugbek, 2020; Finch & Hernández Finch, 2020; Tavares & Betti, 2020).

The crisis has given us the opportunity to rethink that how the overall well-being of a person can be enhanced as life is appearing to be more uncertain. Simultaneously life has so much to offer in material as well as non-material terms. Our natural resources, ecosystems have many benefits to offer and the transition phase can be interesting and soul satisfying through life experiences by visiting various destinations around the world. The governments all over the world are taking steps to revive their economies by boosting various sectors. According to World Tourism Organization, (UNWTO) a specialized agency of the United Nations,

> The economic contribution of tourism in 2021 (measured in tourism direct gross domestic product) is estimated at US $ 1.9 trillion, above the US $ 1.6 trillion in 2020, but still well below the pre-endemic value of US $ 3.5 trillion. Export revenues from international tourism could exceed US $700 billion in 2021, a small improvement over 2020 due to higher spending per trip, but less than half the US $1.7 trillion recorded in 2019.

Table 11.1. Unprecedented Impact From COVID-19.

	2019 (Pre-endemic Year)	2020	2021 (Preliminary Results)
International tourist arrivals (overnight visitors)	1.5 billion	400 million	415 million
Export revenues from international tourism (international tourism receipts + passenger transport)	USD 1.7 trillion	USD 638 billion	USD 700–800 billion
Tourism direct gross domestic product	USD 3.5 trillion	USD 1.6 trillion	USD 1.9 trillion

Source: UNWTO, https://www.unwto.org/impact-assessment-of-the-covid-19-outbreak-on-international-tourism

The impact of endemic on the tourism sector around the world is alarming calling for resilience strategies. The UNWTO estimates/impact are shown with the help of following Table 11.1.

2. Review of Literature

When lockdowns were imposed forcibly in many parts of the world, many nations suffered in various ways. Due to the endemic, there were numerous forced modifications that interrupted our daily routines, including house confinement, working from home, social isolation, etc. These changes caused difficulties for all social classes. It involves negative health outcomes, diminished well-being, sleeplessness and disruptions to day-to-day activities. A person's general well-being should be of significant concern due to variables including anxiety, worry, loneliness, increased family and work stress and more screen time (Majumdar et al., 2020). Zhang et al. (2020) are of the view that the adults who were unemployed during the corona virus outbreak were shown to have worse physical and mental health. They were discovered to be in anguish and to be unsatisfied with their lives.

Caffo et al. (2020) analysed that one of the first nations in Europe to cope with COVID-19 was Italy. However, numerous efforts were implemented by the nation to stop the corona virus from spreading, all of which were focused entirely on isolation, social seclusion and quarantining. For the nation, it has had a great deal of negative economic, social and psychological effects. Children's psychological well-being has been negatively impacted by a protracted time of social isolation. It will have a greater impact on children and adolescents who already had medical conditions and those who had mental health issues. Due to the dramatic rise in unemployment, teenagers are particularly vulnerable to this endemic's effects on their mental health. According to authors (Moen et al.,

2020), women in their 50s and 60s also experience significant rises in unemployment.

Due to technology, face-to-face social encounters now last for fewer than 30 minutes (Giuntella et al., 2020). Sleep duration reduced, but daily screen time rose sharply to five hours, posing a serious threat to people's general well-being. In addition, it was discovered that interruptions to physical exercise were the main risk factor for depression. While middle-aged and older persons were discovered to be able to control their negative emotions better than the younger generation, younger adults were found to be more influenced by the endemic (Knepple Carney et al., 2020). It was discovered that the mental health of the families who endured difficulties throughout the crises were majorly the poorer section of society (Gassman-Pines, et al., 2020). Researchers have also identified social alienation as a factor in the current situation that has an impact on the well-being of employees, particularly those who work in service fields (Tuzovic & Kabadayi, 2020). The researchers are deeply concerned about the urgent need to fully know and appreciate the consequences of unemployment on employee well-being. Caruana et al. (2020) also discuss the global disruption of everyday healthcare. They can aid in the formation of policies and have the potential to influence the practices used during the endemic period, concerns and expectations must be understood in order to take prompt action. Krendl and Perry (2020) stated that in the short term, the COVID-19 endemic had a detrimental effect on the social and mental health of older persons. According to the results, older persons were more unhappy and lonely at the beginning of crises. Their physical and mental health were both being badly impacted by loneliness.

3. Research Gap

The psychological well-being takes centre stage when we consider the concept of well-being in all spheres of human life during the COVID-19 pandemic. Given that it is the centre of human life, we cannot afford to disregard it. Our ability to survive the endemic's cycle of shocks could be in jeopardy if we lack the psychological toughness and resilience needed. The literature that is now available also hints that its effects have been disastrous, particularly because people's means of subsistence were negatively impacted (Hunjra, Azam, Bruna, & Taskin, 2022). However, the development of the resilience and well-being initiatives has not received the targeted attention needed for them to act as a guidepost for policymakers. There isn't a lot of literature in this field explaining how to deal with this issue and create a future path for ourselves so that we can live more assuredly and intelligently. Travelling for health reasons could be a way to escape the isolation and abnormal ways of living that can occur during COVID-19. To better negotiate our future, we also need to live in an environmentally friendly way. The adoption of sustainable living habits is urgently required. But still risk and uncertainty affect sustainable development negatively (Hunjra, Azam, Bruna, Verhoeven, & Al-Faryan, 2022). The methods we apply and the resilience techniques we choose must be ethical, workable and fundamentally a part of how we live. They must also not hurt the environment or ourselves.

4. Objectives of the Study

This paper will aim at exploring conceptually: - What are the most effective resilience and well-being tactics to combat the endemic and to cope with the current situation?

5. Research Methodology

This conceptual, qualitative work tries to support the conceptual postulations with data from secondary sources, such as websites and published reports of international reputation. We can develop resilience and well-being methods that can help us deal with the COVID-19 endemic more effectively with the aid of a thorough idea development. A critical conversation on various aspects of well-being would broaden the scope of critical study with reference to resilience strategies.

5.1 Why Well-being?

Well-being has become an important concept, especially after facing the pandemic (Etheridge & Spantig, 2020). We find many evidences which establish that being physically active helps to manage stress and anxiety and they are more prevalent during the uncertain times (Kim & McKenzie, 2014; Popov et al., 2021; Zach et al., 2021. Therefore, cultivating resilience becomes a foremost requirement in such a scenario (Arora et al., 2023). In order to navigate uncertainty, it becomes essential for everyone to build resilience. It may involve developing one's ability to adapt towards the changing circumstances and then bouncing back from the setbacks. We should engage in practices like mindfulness meditation or yoga which can help to manage the stress and build emotional resilience. In order to build resilience, travel and living near the nature can be a great source of help. When one is physically active, it not only helps in controlling stress and worry, but is also important in the development and stability of emotional health especially in unpredictable situations. Resilience development becomes crucial for dealing with the unpredictability. It also entails with learning that how one should adjust to changing conditions and recover from setbacks. Many exercises or practices like yoga, meditation and mindfulness can also assist in reducing stress. We all have witnessed this during COVID-19 endemic that one must improve their overall well-being. It becomes important to focus on not only physical health, but on mental health also. One can develop a comprehensive plan that takes into account the physical, mental, emotional and social aspects of life. One of the important strategies to build resilience in various aspects can be by enhancing human connections (Zautra et al., 2010). Human connections are a fundamental need or necessity of every human being, and during uncertain times, it becomes even more important. In the world of digitalization, when it is not possible to physically visit every loved one, then it becomes important that we may use digital platforms for communication in order to keep and improve our social ties with friends and family. It has also the ability to enhance our well-being if wisely used. When we share our feelings and experiences with our closed ones, it can act as a safety net during the trying times. Further, adopting a growth mentality is also very necessary, as it may help in viewing obstacles as chances for development and education. We should always put an emphasis on self-improvement and adaptation, which can help in succeeding in the challenging circumstances. It is also very important that one should embrace a mindset which focuses on growth to handle the challenges of life. Sometimes while facing the challenges of life, we quite often forget that growth is indispensable to handle challenges of life. But we should not be harsh on ourselves. As self-realization is important, so is self-compassion. Self-compassion exercises can work as wonders for holistic, well-being of a human being. It is very important to treat ourselves nicely and compassionately. We must understand and ponder upon the fact that it is normal to occasionally experience uncertainty and anxiety. If we are treating ourselves with the same compassion and understanding that we often extend to a friend going through similar situations, we are demonstrating self-compassion. The general well-being of a person can be improved by participating in several activities that offer significance to life and provides direction to a free and wavering mind. A feeling of purpose can inspire and satisfy us whether it is through our job or a personal interest or volunteer work or something that makes us happy. The long-term sustainability of every human being relies on overall well-being. It is the vital component of human existence for a number of reasons. The significance of well-being cannot be underemphasized, and it includes various aspects of physical, mental, emotional and social health. The following are some justifications for the need for why well-being is important.

Well-being is essentially very important for enhancing the quality of life (Estes & Sirgy, 2019; Marquez et al., 2020). Fundamentally, it has the potential of making a substantial contribution to life's entire quality. People who are in good health typically feel happier, more content and more satisfied with their lives. People who carry high sense of well-being are better able to indulge in life's joys, follow their hobbies and take their passions to a higher level. Well-being has the ability to lay the groundwork for a fruitful and pleasant life. One of the important elements of holistic well-being is being physically fit i.e. physical well-being. When we engage in regular physical activities and have a healthy dietary habit, then the possibility of developing chronic diseases is reduced. By eating a healthy diet and getting enough sleep we can enhance our life expectancy.

Those people who put their physical health as a priority are more likely to live longer, healthier lives and have fewer health problems. Healthy people are also regarded as productive people (McCunney, 2001). Those people who are in good health, they typically exhibit higher levels of productivity as well as creativity. When people are at their best physical and mental health, they are able to devote full attention to their work and creative endeavours. It is most likely that they will develop novel solutions to be driven to accomplish their objectives and make constructive contributions to their communities and society at large. Another critical component of general well-being is social well-being. Well-being is highly correlated with strong social networks and holds them interactions with family, friends and community (Burke et al., 2010). These relationships offer chances for personal development, emotional and a sense of belongingness. The capacity to establish and keep meaningful relationships boosts one's overall fulfilment and happiness, which has fostered by well-being. COVID-19 has brought the human fragility in front of us and has revealed the unpredictable nature of life to the forefront of everyone's consciousness (Arora & Sharma, 2021b, p. 201). Uncertainty in many respects is now a very recurring subject that effects people in different ways. There might be certain ways by which one can improve their general well-being in the atmosphere of uncertainty. The social, emotional, mental, and physical facets of life are all important for following the holistic approach of well-being (Steele, 2020). But overall, the foundation of comprehensive and holistic well-being is the physical health. It's a well-known fact that a healthy immune system and stress management of a person depends on regular exercise, workout, balanced diet and enough sleep. It is very crucial to invest in the physical well-being where the uncertainty is bound to occur. One of the crucial aspects of well-being is the capacity of a person to overcome hardships. It is possible to learn people are emotions and adjust according to unpredictable changing circumstances. Yogic activities, meditation and practicing mindfulness plays a very important role in the stability of mind. Resilience is a defence against the shocks of unpredictable changing world. That is why in order to become emotionally, spiritually and physically strong one has to focus on the overall well-being.

Various digital transformations can also act as a source of building resilience (Aydın & Baykal, 2023). In the times of isolation and distance, we learnt that technology can play an important role in making social ties with friends and family. It depends upon a person's mindset that how the technology is to be used. A growth mindset can encourage people to see obstacles as learning experiences when faced with uncertainty. People who embrace self-improvement and adaptation not only survive but also prosper through difficult circumstances. At any circumstance, if we realize that mental and emotional health is being jeopardized, then it becomes crucial to take appropriate steps for enhancing mental well-being. Certain proactive steps in managing anxiety and preserving a sound mind will help to maintain our physical and mental well-being. COVID-19 endemic has brought the fragility of human life in front of us and has revealed the unpredictable nature of life to the forefront of everyone's consciousness. Uncertainty in many respects is now a recurring subject that affects people in many different ways. There can be various ways by which one might improve their general well-being in this atmosphere of uncertainty. The social, emotional, mental and physical facets of life are all important for following the holistic approach of well-being. The basic foundation of comprehensive well-being is a sound physical health. Therefore, it becomes very crucial to invest in the physical well-being where the pandemic has highlighted already that health uncertainty is abound. In spite of the unpredictability in the external environment, people can seek fulfiled life and balance the multiple challenges posed by the external environment. The four major aspects of psychological, physical, social and economic well-being are included in the proposed well-being or wellness model shown below in Fig. 11.1 and are further highlighted in Table 11.2.

5.2 How Travel Enhances Well-being?

Travel has the potential to significantly enhance well-being in various ways, offering individuals unique opportunities for personal growth, relaxation and the development of meaningful experiences. Individuals can momentarily detach themselves from their daily routines, duties and anxieties by travelling to exotic and tranquil locations. Taking a break from the routine can encourage relaxation, lower stress levels and revitalize the body and mind. Leisure time activities, like relaxing on a beach, going on a hike, or visiting historical sites, offer a great chance to unwind and refresh. People are exposed to many cultures, languages and traditions through travel. One's perspective can be broadened, one's cultural awareness can be improved, and one's curiosity and open-mindedness can be fostered through this cultural research. Understanding various ideas and learning about various ways of life can help one grow personally and develop a deeper

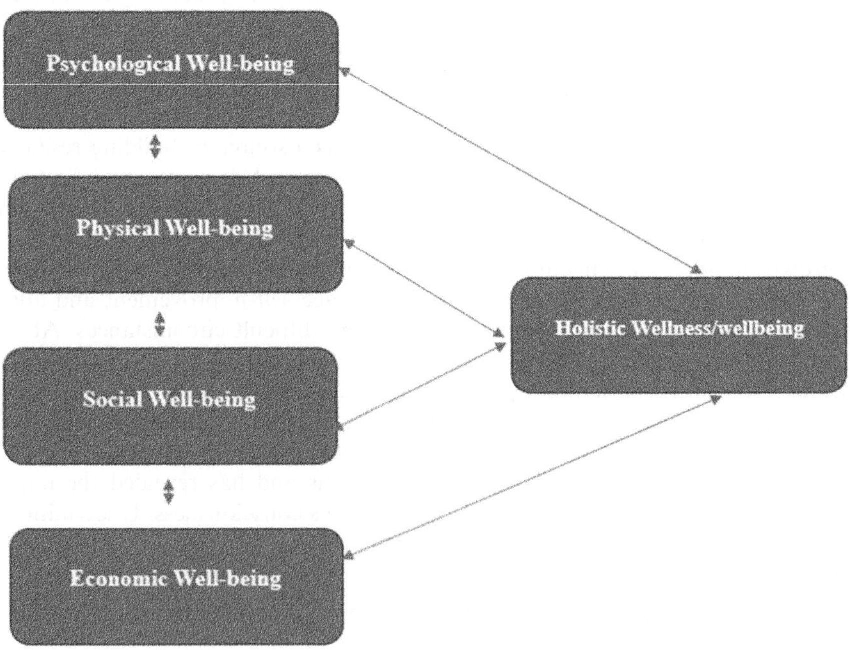

Fig. 11.1. The Four-Dimensional Well-Being Model. *Source:* Created by Author.

Table 11.2. Dimensions of Four-Dimensional Well-Being Model.

	Main Elements	Factors Affecting/Governing
Psychological well-being	Emotional health	Control and regulation of emotions, emotional resilience, effective coping strategies Tactics to improve emotional health – Mindfulness based exercise, meditation, counselling etc.
	Mental wellness	Development of cognitive abilities, mental clarity, ability to analyse and solve problems Tactics to improve mental health – Staying positive, practicing gratitude. Better physical health, meditation, relaxation techniques
	Stress management	Family stress, emotional stress, financial stress and social stress Tactics to improve mental health – Having a daily routine, better sleeping habits, adopting healthy eating practices, practicing consistency, exercising regularly, time management skills, stress-relieving exercises like yoga, breathing exercises, relaxation techniques
Physical well-being	Physical fitness	Unhealthy eating practices, Immunity level, poor sleeping habits Tactics to improve physical fitness – Balanced diet, appropriate intake of water, participating in various physical activities, exercising
	Nutritional health	Diet, spending on junk food, eating techniques, nutritional education Tactics to improve nutritional health – A balanced diet, food planning, learning mindful eating techniques
	Sleep hygiene	Resting habits, sleep time Tactics to improve nutritional health – Getting enough sleep, maintaining regular sleep patterns, eating light at night
Social well-being	Social connections	Demographic factors, social cohesiveness, social connections, communication skills

(Continued)

Table 11.2. *(Continued)*

	Main Elements	Factors Affecting/Governing
		Tactics to improve social connections – Joining clubs, volunteering, social service acts
	Emotional support	Emotional behaviours, social networks, emotional conditioning. Tactics to increase emotional support – Good communication strategies, active listening, practicing gratitude and empathy.
	Cultural engagement	Customs, traditions and cultures, sense of community, social enrichment Tactics to improve cultural ties – Visiting cultural events, travelling, exchanging cultures, practicing gratitude, respecting all cultures
Economic well-being	Financial literacy	Knowledge of personal money, personal finance, budget, saving habits, investment opportunities Tactics to improve financial literacy – Developing habits of financial planning and financial education
	Career satisfaction	Career goals, education level, skills Tactics to improve that is satisfaction – Skill improvement. Work life balance
	Financial stability	Financial objectives, financial health, debt level, financial literacy Tactics to improve financial stability – Monitoring financial expenses, managing debt, preparing financial plans, taking financial advisors

Source: Created by author.

appreciation for the diversity of the world. Long after the journey has finished, travel may generate happy and fulfiling memories and experiences. These recollections provide a source of joy and captivating narratives to share, fostering relationships with others and improving general well-being. Physical activities like swimming, hiking and exploring unfamiliar regions are frequently involved with travel. Participating in these activities can increase energy and vitality while simultaneously improving physical fitness. It promotes leading a better and more active lifestyle. Togetherness and shared emotions can be deepened by taking an excursion with friends or family. The shared moments and recollections can foster greater ties and a sense of belonging, which benefits social well-being. Creative thinking and problem-solving abilities can be stimulated by exposure to a variety of destinations, cultures and landscapes. People are prompted to think outside of their typical perceptions and emotions are fostered. People are frequently forced out of their comfort zones through travel, which motivates them to acclimate to novel circumstances and difficulties. This process of self-discovery can promote personal development, self-awareness and self-confidence.

People can be fully present in the moment while travelling since they are not bothered by distractions or the obligations of daily life. Being observant while travelling can help one appreciate the beauty around them and feel more connected to their surroundings. It promotes wellness and overall well-being of an individual. Regular travel breaks can improve work-life balance and help prevent burnout. It provides a chance to unplug from the pressures of the workplace and come back to daily life feeling renewed and more focused. A sense of adventure and anticipation can be created by the excitement of preparing and taking a trip, which can improve general well-being. Travel frequently evokes awe and the pleasure of discovering the unknown. By offering chances for rest, cultural inquiry, personal development, physical activity and the making of enduring connections and memories, travel has the potential to improve well-being. Travel, whether it's a quick weekend trip or a lengthy vacation, can improve one's general quality of life. To provide a responsible and sustainable travel experience for both persons and destinations, it is crucial to approach travel consciously while respecting local customs and the environment.

6. Religious/Pilgrimage Travel for Well-being/Wellness and Internal Healing

All over the world during endemic it was realized that being healthy is very much important and it has different healing dimensions. Good health does not include just the personal health protection measures but it encompassed bliss and solitude to some which can be fulfiled through travel. The concept of health has now expanded and it is much more than absence of any ailment/illness/disease. The concept of self-care has emerged during endemic and the definition of wellness has broadened. While we all faced lockdowns we realized a sudden absence of social interaction can be fulfiled or replaced by self-caring activities like Yoga, Meditation and staying in the lap of nature. Self-care concept is imbibed in the meaning of wellness and there are different dimensions of self-care. Self-care has been broadly defined to be a process of taking care of oneself body, mind and soul. All those behaviours which can promote health and give us an active life are the part of self-care. It also includes promoting various behaviours that promote health, well-being and active management of illness when it comes.

WHO defines self-care as 'the ability of individuals, families and communities to promote health, prevent disease, maintain health, and to cope up with illness and disability with or without the support of a health care provider'.

The concept of self-care can be satisfied by different types of travel and it can be a great healing strategy for Asian nations as these countries are rich in nature, flora and fauna spas, yoga centres, rejuvenation activities through meditation and other breathing exercises camps as well as fitness centres. Just living for few days in the lap of nature is itself quite healing. These destinations require projecting themselves to be destinations full of self-care activities which promote well-being. They have the ability of providing mystic experiences which can prove to be rich in psychological, emotional and spiritual self-care. The self-care is not only associated with physical dimensions but they help us in mental as well as social wellness (Arora & Sharma, 2021a).

Health, according to the WHO's definition, is not merely a disease-free body but also includes one's whole physical, mental and social well-being. Therefore, the significance of good health in the modern world goes beyond physical health to also include mental wellness.

Slade (2010) relates the idea of well-being with mental illness and emphasizes the significance of their general well-being. Not only the sickness should be taken into consideration while treating patients, but also the aspects of the everyday labour performed by those in employment, particularly women, as this has an impact on their health. Additionally, it is believed that good working conditions are crucial for psychological well-being. The importance of work affects how women's employment status is determined by their mental health. According to him, various work tasks performed by various professions might directly affect well-being. Researchers have also discovered a connection between religion, happiness and ultimately an individual's level of well-being (Lennon, 1994).

According to Cohen (2002), When assessing a person's level of life happiness, religion is a crucial factor. It also discusses the connection between happiness and spirituality. Other factors, such as religious coping strategies and religious convictions, have a significant role in determining people's level of well-being. As a result, travelling to religious locations that draw in good energy and can serve as a coping mechanism is a common choice in Asian countries.

Various authors have also covered a variety of spiritual aspects, which are crucial for life fulfilment and inspire optimism (Arora et al., 2021). Various authors have also covered a variety of spiritual aspects, which are crucial for life fulfilment and inspire optimism (Arora & Sharma, 2020; Ross, 1995). Travelling to the destinations which are projected as having spiritual or religious beliefs can act as a great resilience strategy not only for the economies suffering in tourism sector by COVID-19 but for the individuals as well (Arora & Sharma, 2023). One's true potential which is an important dimension of travel can also be influenced by travel behaviour. Travel has the potential of self-realization (Tiwari & Hashmi, 2022). COVID-19 affected deeply the emotional well-being of people across the world. One of the major push factors deriving to travel was found to be 'escape and relaxation' (Io & Luna Peralta, 2022). As per Global Wellness Institute (GWI), 'the global wellness economy was a $ 4.5 trillion market in 2018 and represented of all tourism expenditures'. Wellness tourism is on rise even more after endemic as the majority of the masses now are more concerned about their well-being and relaxation, escape from daily routine, visiting pollution fee spaces and find opportunities to stay close to nature is trending moreover well-being motivations are also related with seeking a healthy lifestyle while going on any kind of a trip whether it's for escape, spiritual or any other reasons. Well-being or wellness is a very subjective term and can have different meaning for different people. It includes elements of taking time, slowing down, escape, nutrition, self-care and self-development. It also includes travel for environment awareness that is being in nature, improving physical well-being focussing on emotional and mental well-being. It also has implications for relaxation and rejuvenation. The concept of well-being also includes certain intrinsic motivations like deriving inner pleasure and a state of bliss. It may have some satisfying experiences and may have the ability to experience certain mystic experiences beyond words for a traveller (Arora & Sharma, 2022b). The experiences may be quite meaningful which are quite rich in nature and are contributing to the overall personality of a traveller. The overall whole meaning of wellness and well-being has taken a wider perspective in this endemic. People are readjusting and recouping with opting various strategies where travel is one of them. There can be certain intrinsic experiences also which one traveller may achieve through mindful practices helping people to cope with greater stress and deep-down healthy isolation but not loneliness in this crowded world full of misery and pain.

Excessive work from home pressures, increased screen time and majority time spending online has shifted the thrust of travel to travel without work and overall wellness.

The concept of self-care is gaining a lot of importance. Asian countries especially India has been promoting itself to be Yogic destination which is again a very important form of self-care (Arora & Sharma, 2022a). Yoga is not just doing *asanas;* in fact, the ancient scriptures of Yoga also does not solely regard it as a practice of doing *asana* but it regards it a skill unifying mind and body. According to the survey travel is trending to become a self-care way. Travelling helps them to know new cultures, meeting new people, spending time with family and much more.

The GWI has categorized wellness into 11 sectors stated below and the share of each sector is quite overwhelming for endemic struck tourism sector.

The above Table 11.3 clearly shows that the market value of in particular wellness tourism is quite large and as per GWI the share of this sector is continuously increasing, the other related areas like Spas, Healthy Eating, Nutrition, & Weight Loss etc. is going.

The $639 billion wellness travel market's annual growth rate of 6.5% from 2015–2017 is more than double the 3.2% growth rate for tourism overall. World travelers made 830 million wellness trips in 2017, 139 million more than in 2015 – and these trips represent 17% of total tourism expenditures. Wellness tourism growth is very much a tale of developing markets, with Asia-Pacific, Latin America-Caribbean, Middle East-North Africa and Sub-Saharan Africa all clocking robust gains, and accounting for 57% of the increase in wellness trips since 2015. Over the past five years, Asia is the #1 gainer in both wellness tourism trips and revenues (trips grew a whopping 33% in two years, to 258 million annually). China and India rank #1 and #2 for growth worldwide, adding over 12 million and 17 million wellness trips respectively from 2015–2017.

Source: https://globalwellnessinstitute.org/industry-research/2018-global-wellness-economy-monitor/.

Table 11.3. The Wellness Economy.

The Wellness Sector	Value in Economy
Personal Care & Beauty	$955 billion
Healthy Eating, Nutrition, & Weight Loss	$946 billion
Physical Activity	$738 billion
Wellness Tourism	$436 billion
Traditional & Complementary Medicine	$413 billion
Public Health, Prevention, & Personalized Medicine	$375 billion
Wellness Real Estate	$275 billion
Mental Wellness	$131 billion
Spas	$68 billion
Workplace Wellness	$49 billion
Thermal/Mineral Springs	$39 billion

Source: The Global Wellness Institute (GWI), https://globalwellnessinstitute.org/press-room/statistics-and-facts/

7. Inbound Travel and Visiting Loved Ones as a Psychological Resilience Strategy

COVID-19 has raised a variety of issues that have caused societal disturbances, it poses a serious threat to the general well-being of families and children. COVID-19 will have long-lasting effects, and as a result, the results could be disastrous for one's health. Therefore, it becomes important to encourage resilience by taking into account factors like 'nurturing close relationships' and 'shared family beliefs'. Prime et al. (2020) hold the opinion that the current global crisis will have an impact on worldwide family well-being in addition to destabilising economies. In order to combat the risks posed by the endemic emotional anxieties, family leaders now have the responsibility to generate hope, nurture relationships and establish emotional security in the families.

Shaw (2020), found that individuals' mental health is now at stake due to COVID-19. The most significant factors affecting well-being that one should be aware of are hopelessness and burnout. In times of crisis, an open culture, trust and understanding between people – including family members, friends and others – are crucial.

The development of a positive outlook act as a buffer against distress (Babore et al., 2020). As a defence mechanism against distress, having a positive outlook has emerged as a coping strategy (Lazarus, 1995). In the end, it results in a higher quality of life. When someone is in a good frame of mind, they may interpret even bad circumstances with a positive outlook, which eventually helps them feel better. Babore et al. (2020) found 'Positive attitude' and accepting the situation as it is were shown to be the most effective coping mechanisms among healthcare personnel. Children helped professionals see the positive side of life, which made them feel less stressed. Gerhold (2020), in his survey done in Germany found that compared to younger people, those who were older felt COVID-19 to be more stressful. According to the emergency health corona virus professionals' stress coping mechanisms, media was a significant source of stress for the respondents (Munawar & Choudhry, 2020). Participants used coping mechanisms to reduce their media exposure in a similar way. The main coping mechanism was religion and a commitment to serving humanity. The study by Skapinakis et al. (2020); revealed that individuals who were overexposed to the media had very high levels of depression symptoms. However, those who had an optimistic outlook also shown greater levels of self-control. The COVID-19 incident had a significant emotional impact, especially on women and students.

The definition given by WHO in 1998 is much more comprehensive focussing on wellness by caring one's own self. This definition, which includes 'hygiene, i.e. general and personal nutrition (type and quality of food eaten), lifestyle, sporting activities, leisure activities, environmental factors, living conditions, social habits, socio-economic factors (income level, cultural beliefs and self-medication), has taken a broad look at the idea of self-care'. Therefore, Travel as a means of coping with physical and mental illness can add newer dimension to tourism sector.

8. Policy Implications of Enhancing Well-being Through Travel

Travelling for well-being has a number of policy ramifications that may contribute to social and individual well-being. These regulations seek to encourage and support travel in ways that maximize the good it can do for people's lives. The following significant policy ramifications can hold good in given circumstances.

8.1 Accessible Transportation Infrastructure

To make travel easier for people of all ages and abilities, governments should invest in current and accessible transportation infrastructure, such as highways, railroads, airports and public transportation systems. 'Accordingly, the institutional factors are considered as crucial for Sustainable Development effectivity' (Hunjra, Azam, et al., 2023). This policy implication holds good for developing, less developed and emerging nations.

Justification: Accessible transportation alternatives increase travel's well-being advantages by enabling a wider range of people to partake in travel.

8.2 Affordable Travel Options

Public transit subsidies, exemptions from taxes for travel-related expenses and partnership agreements with tourism organizations to provide cheaper airfare to those with low incomes are just a few examples of the policies that governments can enact to promote inexpensive travel options.

Justification: By lowering the financial obstacles to travel, more individuals will be able to benefit from its enhancement of well-being.

8.3 Sustainable Travel Methods

It's important in the current scenario to implement and execute sustainable tourism policies, such as those governing responsible tourism, environmental preservation and restriction of visitor numbers in sensitive habitats. Governments should take steps to introduce norms for promoting nature based or ecotourism in stricter sense. Further they can encourage use of green finance for achieving sustainable development (Hunjra, Hassan, et al., 2023), only then issues like sustainability and climate control could be taken care of.

Justification: Sustainable tourism or responsible tourism promotes the preservation of natural and cultural resources so that present-day travelers can continue to enjoy their travels in the future and the actions are sustainable in nature.

8.4 Promotion of Local and Rural Tourism

To extend the economic advantages of travel outside of major cities, the policy makers should encourage and support local and rural tourism activities. Government Grants, supporting advertising initiatives and infrastructural upgrades could all come in under this category.

Justification: Local and rural tourism may strengthen local economies, generate employment opportunities and provide real cultural experiences that are advantageous to both visitors and local communities.

8.5 Access to Opportunities for Cultural and Educational Travel

The governments or policy makers should stimulate educational and cultural exchange initiatives, particularly for students and young people, through partnering with educational institutions and providing scholarships and grants. Tourism related cultural exchanges and partnerships can be a major source of enhancing joy, peace and happiness promoting well-being.

Justification: Travelling can increase a person's understanding of the world, tolerance and personal development, leading way and opening gateways of well-being.

8.6 Support for Vulnerable Populations and Solo Travelers

To ensure safe and enjoyable travel, govt. or policy makers should provide services and support for single/solo travelers, families and vulnerable groups (such as the women, elderly and those with disabilities).

Justification: Everyone should be able to travel and enjoy it, regardless of their circumstances, a freedom to have a safe travel, for different motives enhances psychological well-being.

8.7 Crisis Management and Travel Security

There is a great need to do thorough crisis planning and prepare safety guidelines for visitors, including alternatives for information sharing, travel insurance and consular assistance at various levels. Policy interventions are required to have appropriate guidelines for travel in crises and ensuring travel security during crises or otherwise too.

Justification: Ensuring travelers' safety increases their willingness to travel, especially in tumultuous times.

9. Conclusion and Summative Remarks

All of us are badly hit by endemic psychologically, emotionally and many physically too. To achieve general well-being and excellent health, we must have a strong psychological foundation. The spiritual aspects of our well-being are equally important since they aid in overcoming our fascination with worldly things, help us prepare for a life of austerity with no expenses and alleviate feelings of hopelessness and distress (Braxton et al., 2007). Economic resilience is crucial because we must maintain our economies in the face of any obstacles. The importance of psychological resilience cannot be overstated. Travel for well-being seems like a workable strategy to resurrect the worst-hit industries, including tourism. Travel can be a potent resilience strategy for improving well-being, giving people a way to deal with obstacles in life and develop a more resilient mindset. Travel can offer many advantages to encourage resilience and general well-being in the face of adversity, such as stress, burnout or personal setbacks. By exposing people to novelty, relieving stress, creating fond memories and promoting personal growth, travel serves as a resilience strategy for improving well-being. It presents a special chance for people to grow in their ability to adapt, their capacity for coping with difficult situations and their level of self-assurance. A more balanced and resilient response to life's challenges can be achieved through embracing travel as a way to increase resilience, which ultimately improves overall well-being.

References

Arora, M., Kumar, J., & Valeri, M. (2023). Crises and resilience in the age of digitalization: Perspectivations of past, present and future for tourism industry. In *Tourism innovation in the digital era* (pp. 57–74). Emerald Publishing Limited. https://doi.org/10.1108/978-1-83797-166-420231004

Arora, S., & Sharma, A. (2020). Digital marketing for religious event of India for tourism sustainability and promotion. In *The Emerald handbook of ICT in tourism and hospitality* (pp. 453–465). Emerald Publishing Limited. https://doi.org/10.1108/978-1-83982-688-720201029

Arora, M., & Sharma, R. L. (2021a). *Post-pandemic psycho-social wellbeing in India: Challenges and the way ahead.* Alina COSTIN.

Arora, S., & Sharma, A. (2021b). Covid-19 impact on overtourism: Diversion from mass tourism to alternative tourism. In *Overtourism as destination risk* (pp. 275–283). Emerald Publishing Limited. https://doi.org/10.1108/978-1-83909-706-520211018

Arora, M., & Sharma, R. L. (2022a). Challenges of well being in and post-lockdown period: Exploring ayurveda, yoga and spirituality as A conduit for rejuvenation and overall wellness. *Prestige International Journal of Management & IT-Sanchayan, 11*(1), 1–17.

Arora, M., & Sharma, R. L. (2022b). Searching for a break from the drudgery of daily din: Analysing millennials' quest for spirituality and well-being during the COVID-19 pandemic. In *Crisis management, destination recovery and sustainability* (pp. 86–96). Routledge. https://doi.org/10.4324/9781003295839

Arora, M., & Sharma, R. L. (2023). Spirituality and yoga for well-being in a post-disaster scenario: Linking the qualitative facets of traditional Indian ways of life. In *Resilient and sustainable destinations after disaster* (pp. 227–239). Emerald Publishing Limited. https://doi.org/10.1108/978-1-80382-021-720231016

Arora, M., Sharma, R. L., & Walia, S. (2021). Revisiting the inner self in times of debilitating distress: Gateways for wellness through spiritual tourism. *International Journal of Religious Tourism and Pilgrimage*, *9*(5), 4. https://doi.org/10.21427/VY3H-Q158

Aydın, A. G., & Baykal, E. (2023). The impact of organizational strategic trends on individual-based resilience with a digital transformation perspective. *Human*

Resources Management and Services, 5(1), 3359. https://doi.org/10.18282/hrms.v5i1.3359

Babore, A., Lombardi, L., Viceconti, M. L., Pignataro, S., Marino, V., Crudele, M., Candelori, C., Bramanti, C., & Trumello, C. (2020). Psychological effects of the COVID-2019 endemic: Perceived stress and coping strategies among healthcare professionals. *Psychiatry Research, 293*, 113366. https://doi.org/10.1016/j.psychres.2020.113366

Bargain, O., & Aminjonov, U. (2020). Between a rock and a hard place: Poverty and COVID-19 in developing countries. https://ssrn.com/abstract=3614245

Bargain, O., & Ulugbek, A. (2020). *Poverty and Covid-19 in developing countries.* Bordeaux University. https://hal.science/hal-03258229

Braxton, N. D., Lang, D. L., Sales, J. M., Wingood, G. M., & DiClemente, R. J. (2007). The role of spirituality in sustaining the psychological well-being of HIV-positive black women. *Women & Health, 46*(2–3), 113–129. https://doi.org/10.1300/J013v46n02_08

Burke, M., Marlow, C., & Lento, T. (2010, April). Social network activity and social well-being. In *Proceedings of the SIGCHI conference on human factors in computing systems* (pp. 1909–1912). https://doi.org/10.1145/1753326.1753613

Caffo, E., Scandroglio, F., & Asta, L. (2020). Debate: COVID-19 and psychological well-being of children and adolescents in Italy. *Child and Adolescent Mental Health, 25*(3), 167–168. https://doi.org/10.1111/camh.12405

Caruana, E. J., Patel, A., Kendall, S., & Rathinam, S. (2020). Impact of Covid-19 on training and wellbeing in subspecialty surgery: A national survey of cardiothoracic trainees in the United Kingdom. *The Journal of Thoracic and Cardiovascular Surgery*. https://doi.org/10.1016/j.jtcvs.2020.05.052

Cohen, A. B. (2002). The importance of spirituality in well-being for Jews and Christians. *Journal of Happiness Studies, 3*(3), 287–310. https://doi.org/10.1023/A:1020656823365

Daniels, K. (2000). Measures of five aspects of affective well-being at work. *Human Relations, 53*(2), 275–294. https://doi.org/10.1177/a010564

De Ruyter, D. J. (2015). Well-being and education. In *Education, philosophy and well-being: New perspectives on the work of John White* (pp. 84–98). Routledge.

Estes, R. J., & Sirgy, M. J. (2019). Global advances in quality of life and well-being: Past, present, and future. *Social Indicators Research, 141*, 1137–1164. https://doi.org/10.1007/s11205-018-1869-4

Etheridge, B., & Spantig, L. (2020). *The gender gap in mental well-being during the Covid-19 outbreak: Evidence from the UK (No. 2020-08). ISER Working Paper Series.* http://hdl.handle.net/10419/227789

Field, J. (2009). *Well-being and happiness.* National Institute of Adult Continuing Education.

Finch, W. H., & Hernández Finch, M. E. (2020). Poverty and Covid-19: Rates of incidence and deaths in the United States during the first 10 weeks of the endemic. *Frontiers in Sociology, 5*, 47. https://doi.org/10.3389/fsoc.2020.00047

Gassman-Pines, A., Ananat, E. O., & Fitz-Henley, J. (2020). COVID-19 and parent-child psychological well-being. *Pediatrics, 146*(4). https://doi.org/10.1542/peds.2020-007294

Gerhold, L. (2020). *COVID-19: Risk perception and coping strategies.*

Giuntella, O., Hyde, K., Saccardo, S., & Sadoff, S. (2020). Lifestyle and mental health disruptions during COVID-19). Available at SSRN 3666985. https://doi.org/10.1073/pnas.2016632118

Hunjra, A. I., Azam, M., Bruna, M. G., & Bouri, E. (2023). A cross-regional investigation of institutional quality and sustainable development. *Journal of International Financial Markets, Institutions and Money*, *84*, 101758. https://doi.org/10.1016/j.intfin.2023.101758

Hunjra, A. I., Azam, M., Bruna, M. G., & Taskin, D. (2022). Role of financial development for sustainable economic development in low middle income countries. *Finance Research Letters*, *47*, 102793. https://doi.org/10.1016/j.frl.2022.102793

Hunjra, A. I., Azam, M., Bruna, M. G., Verhoeven, P., & Al-Faryan, M. A. S. (2022). Sustainable development: The impact of political risk, macroeconomic policy uncertainty and ethnic conflict. *International Review of Financial Analysis*, *84*, 102370. https://doi.org/10.1016/j.irfa.2022.102370

Hunjra, A. I., Hassan, M. K., Zaied, Y. B., & Managi, S. (2023). Nexus between green finance, environmental degradation, and sustainable development: Evidence from developing countries. *Resources Policy*, *81*, 103371. https://doi.org/10.1016/j.resourpol.2023.103371

Io, M.-U., & Luna Peralta, R. (2022). *Emotional well-being impact on travel motivation and intention of outbound vacationers during the COVID-19 endemic.* Leisure/Loisir. https://doi.org/10.1080/14927713.2022.2032809

Kim, J. H., & McKenzie, L. A. (2014). The impacts of physical exercise on stress coping and well-being in university students in the context of leisure. *Health*, *6*(19), 2570. http://www.scirp.org/journal/PaperInformation.aspx?PaperID=51245&#abstract

Knepple Carney, A., Graf, A. S., Hudson, G., & Wilson, E. (2020). Age moderates perceived COVID-19 disruption on well-being. *The Gerontologist*. https://doi.org/10.1093/geront/gnaa106

Krendl, A. C., & Perry, B. L. (2020). The impact of sheltering in place during the COVID-19 endemic on older adults' social and mental well-being. *Journals of Gerontology Series B: Psychological Sciences and Social Sciences*. https://doi.org/10.1093/geronb/gbaa110

Lazarus, R. S. (1995). Psychological stress in the workplace. In *Occupational stress: A handbook* (Vol. 1, pp. 3–14).

Lennon, M. C. (1994). Women, work, and well-being: The importance of work conditions. *Journal of Health and Social Behavior*, 235–247. https://doi.org/10.2307/2137278

Majumdar, P., Biswas, A., & Sahu, S. (2020). COVID-19 endemic and lockdown: Cause of sleep disruption, depression, somatic pain, and increased screen exposure of office workers and students of India. *Chronobiology International*, *37*(8), 1191–1200. https://doi.org/10.1080/07420528.2020.1786107

Marquez, D. X., Aguiñaga, S., Vásquez, P. M., Conroy, D. E., Erickson, K. I., Hillman, C., Stillman, C. M., Ballard, R. M., Sheppard, B. B., Petruzzello, S. J., King, A. C., & Powell, K. E. (2020). A systematic review of physical activity and quality of life and well-being. *Translational Behavioral Medicine*, *10*(5), 1098–1109.

McCunney, R. J. (2001). Health and productivity: A role for occupational health professionals. *Journal of Occupational and Environmental Medicine*, *43*(1), 30–35. http://www.jstor.org/stable/44995790

Moen, P., Pedtke, J. H., & Flood, S. (2020). Disparate disruptions: Intersectional COVID-19 employment effects by age, gender, education, and race/ethnicity. *Work, Aging and Retirement*, *6*(4), 207–228. https://doi.org/10.1093/workar/waaa013

Munawar, K., & Choudhry, F. R. (2020). Exploring stress coping strategies of frontline emergency health workers dealing Covid-19 in Pakistan: A qualitative inquiry. *American Journal of Infection Control*. https://doi.org/10.1016/j.ajic.2020.06.214

Popov, S., Sokić, J., & Stupar, D. (2021). Activity matters: Physical exercise and stress coping during the 2020 COVID-19 state of emergency. *Psihologija*, *54*(3), 307–322. https://www.ceeol.com/search/article-detail?id=959413

Prime, H., Wade, M., & Browne, D. T. (2020). Risk and resilience in family well-being during the COVID-19 endemic. *American Psychologist*. https://psycnet.apa.org/doi/10.1037/amp0000660

Rechtschaffen, D. (2014). *The way of mindful education: Cultivating well-being in teachers and students*. WW Norton & Company.

Ross, L. (1995). The spiritual dimension: Its importance to patients' health, well-being and quality of life and its implications for nursing practice. *International Journal of Nursing Studies*, *32*(5), 457–468. https://doi.org/10.1016/0020-7489(95)00007-K

Ryff, C. D. (1995). Psychological well-being in adult life. *Current Directions in Psychological Science*, *4*(4), 99–104. https://doi.org/10.1111/1467-8721.ep10772395

Ryff, C. D., & Singer, B. H. (2008). Know thyself and become what you are: A eudaimonic approach to psychological well-being. *Journal of Happiness Studies*, *9*(1), 13–39. https://doi.org/10.1007/s10902-006-9019-0

Shaw, S. C. (2020). Hopelessness, helplessness and resilience: The importance of safeguarding our trainees' mental wellbeing during the COVID-19 pandemic. *Nurse Education in Practice*, *44*, 102780. https://doi.org/10.1016%2Fj.nepr.2020.102780

Sirgy, M. J. (2012). *The psychology of quality of life: Hedonic well-being, life satisfaction, and eudaimonia* (Vol. 50). Springer Science & Business Media.

Skapinakis, P., Bellos, S., Oikonomou, A., Dimitriadis, G., Gkikas, P., Perdikari, E., & Mavreas, V. (2020). Depression and its relationship with coping strategies and illness perceptions during the COVID-19 lockdown in Greece: A cross-sectional survey of the population. *Depression Research and Treatment*. https://doi.org/10.1155/2020/3158954

Slade, M. (2010). Mental illness and well-being: The central importance of positive psychology and recovery approaches. *BMC Health Services Research*, *10*(1), 1–14. https://doi.org/10.1186/1472-6963-10-26

Steele, L. (2020). Holistic well-being: Mental, physical, and spiritual. *Good Health and Well-Being*, 373–382. https://doi.org/10.1007/978-3-319-95681-7_1

Stone, A. A., & Mackie, C. E. (2013). *Subjective well-being: Measuring happiness, suffering, and other dimensions of experience*. National Academies Press.

Tavares, F. F., & Betti, G. (2020). *Vulnerability, poverty and COVID-19: Risk factors and deprivations in Brazil*.

Tiwari, R., & Hashmi, H. (2022). Integrating concepts of destination image, travel motivations, expectation, and future behavior to create a model of wellness travel intentions. *International Journal of Spa and Wellness*, *5*(2), 185–206. https://doi.org/10.1080/24721735.2022.2057714

Tuzovic, S., & Kabadayi, S. (2020). The influence of social distancing on employee well-being: A conceptual framework and research agenda. *Journal of Service Management*. https://doi.org/10.1108/JOSM-05-2020-0140

Zach, S., Fernandez-Rio, J., Zeev, A., Ophir, M., & Eilat-Adar, S. (2021). Physical activity, resilience, emotions, moods, and weight control, during the COVID-19 global crisis. *Israel Journal of Health Policy Research*, *10*(1), 1–9. https://doi.org/10.1186/s13584-021-00473-x

Zautra, A. J., Arewasikporn, A., & Davis, M. C. (2010). Resilience: Promoting well-being through recovery, sustainability, and growth. *Research in Human Development*, *7*(3), 221–238. https://doi.org/10.1080/15427609.2010.504431

Zhang, S. X., Wang, Y., Rauch, A., & Wei, F. (2020). Unprecedented disruption of lives and work: Health, distress and life satisfaction of working adults in China one month into the COVID-19 outbreak. *Psychiatry Research*, 112958. https://doi.org/10.1016/j.psychres.2020.112958

Chapter 12

The Importance of Relational Marketing in Health and Wellness Tourism: A Study Applied to the Northern Euroregion of Portugal and Galicia

Diana Oliveira Marques[a], *Bruno Barbosa Sousa*[b] *and Ana Teresa Pedreiro*[c]

[a]Polytechnic Institute of Cávado and Ave (IPCA), Portugal
[b]Polytechnic Institute of Cávado and Ave (IPCA), CiTUR and UNIAG, Portugal
[c]University of Minho, Portugal

Abstract

Increased competition and the constant need for companies to differentiate make markets more competitive and consumers more demanding when making purchasing decisions. Consequently, companies recognized the importance of marketing and economics, which has gradually replaced its transactional practices, focused essentially on a single transaction, with relational practices, in which the customer is considered an integral part of the process. Creating long and profitable relationships has become a business priority and the use of tools such as customer relationship marketing (CRM) has become crucial. With the increasing importance of health and wellness tourism, thermalism has become a very competitive product wanted by tourists who need treatments that are not available in their country due to lack of availability or because they are quite expensive. Since tourism is integrated in the services sector, a set of relational marketing strategies can be applied to establish and maintain relationships with tourists. Therefore, it is important to understand how relational marketing can improve health and wellness tourism. Based on a combination of theoretical and practical research, this chapter explores the dynamic system and mechanism of innovation and development of relational marketing strategies in specific

The Emerald Handbook of Tourism Economics and Sustainable Development, 233–249

doi:10.1108/978-1-83753-708-220241012

contexts of tourism (i.e. health and wellness tourism) considering cross-border (i.e. Northern Euroregion of Portugal and Galicia). From an interdisciplinary perspective, the chapter presents insights for marketing and tourism (health and well-being) and for the economic development of territories (i.e. cross-border regions).

Keywords: Cross-border regions; economic development; relational marketing; health and wellness tourism; thermalism; Euroregion

1. Introduction

Commercial relations were associated with exchange activities produced between supply and demand. However, these relationships have undergone changes, elevating marketing as a tool that improves competitiveness and to which a social vision has been implemented, making the relationship the foundation for creating value to the organization and the customers (Arosa-Carrera & Chica-Mesa, 2020; Kottler et al., 2002). These changes were a consequence of fast progression of the world economy, which resulted in difficulties for organizations to survive in such a competitive environment. The changing role of marketing in organizations indicated a broad continuum of marketing relationships that emerged as a new paradigm of marketing functions along with transactional marketing (Kundu & Bhattacharya, 2019). The change in the direction and management of organizations, allowed companies to free themselves from the transactional process based solely on commercial exchanges and move towards an approach to building stronger and more lasting commercial relationships (Miquel-Romero et al., 2014; Grönroos, 2004).

Customer loyalty has become a prominent predictor in the business world due to its crucial role in business success (Paparoidami et al., 2019). It is necessary to meet customer satisfaction for such loyalty. When customers are satisfied, they demonstrate commitment to continually purchasing the same brand and become loyal customers. Consumer satisfaction becomes an important factor in creating loyalty, because when customers are satisfied with a product, they automatically return to buy and consume the same product (Lubis et al., 2020). In addition to satisfaction, there are other factors that influence and lead to customer loyalty, such as trust and commitment. The main objective of a company is to maintain consumer loyalty and demonstrate confidence in the services/products they provide (Hayati & Muddaststsir, 2019).

Tourism is a dynamic, economic and a social phenomenon. Health and wellness tourism such as thermalism are currently a popular and growing option where tourists seek to access procedures that are not available or because the costs are high in their home country (Sousa & Barros, 2021). Contemporary society has a broad and holistic perspective on what health and well-being are. Thermalism perfectly combines these two aspects. The practice of health care and wellness management depends on successfully informing potential clients about procedure options, service excellence, treatment facilities, tourism opportunities, travel benefits and destination choice. As such, there is a growing need to understand consumer behaviour in spa, wellness and health destinations (Sousa & Barros,

2021). Antunes (2012) states that spa institutions should seek to better identify their customers and develop individual and relevant relationships in order to understand their constant changing needs and thus enhance their loyalty. Although tourist loyalty is crucial and timely in the emerging wellness tourism industry, little research has addressed this issue of destination loyalty in the wellness spa tourism sector (Han et al., 2018).

In terms of natural resources and thermalism, the Northwest Peninsular (Galicia and Northern Portugal) constitutes the territory with the most thermal resources in Europe, which has allowed the construction of many thermal establishments, references in their sector. The tourist sector in the Northern Euroregion of Portugal and Galicia, thermal spas constitute a key activity for social, economic and environmental development in these regions. The Euro-region is one of the areas of the Iberian Peninsula with the greatest wealth of mineral-medicinal waters (Amboage et al., 2019). Therefore, it is important to understand how the Euro-region can evolve as a destination when developing a relational marketing strategy applied to a common product offer, such as thermal spas, to consequently be able to attract and retain tourists to the destination.

2. Literature Review

2.1 Relational Marketing

Changes in the marketing landscape, including the decline of the large bureaucratic organisation and the exponential increase of strategic alliances and business partnerships, served to reignite the interest in relationships as an appropriate lens through which business activity could be understood. Emerging ideas collided, creating knowledge about the phenomenon known as relationship marketing (O'Malley, 2014) (see Fig. 12.1).

With the emergence of technological advances, producers began to interact directly with many consumers and due to a variety of organizational development processes the direct relationship between producers and consumers returned to consumer and industrial markets, leading to a greater relational orientation. Among marketeers (Sheth et al., 2012). The hegemony of marketing management of mass-produced and distributed consumer goods was challenged and a relational approach to marketing was spread. Service companies and government organizations began to understand the need for marketing, service management and professionals to develope new strategies. This is how relational marketing emerges, dubbed as the new paradigm (Christopher et al., 2013) and represents the milestone of the transition from a traditional marketing approach oriented towards masses, shifting their central focus to building stable and lasting relationships with their customers (Antunes & Rita, 2007).

The monologs have now transitioned to dialogs with the consumers. There is an individualization of offers, services and messages for individual customers based on information about previous transactions, demographic and psychographic information and their preferences for media and distribution channels.

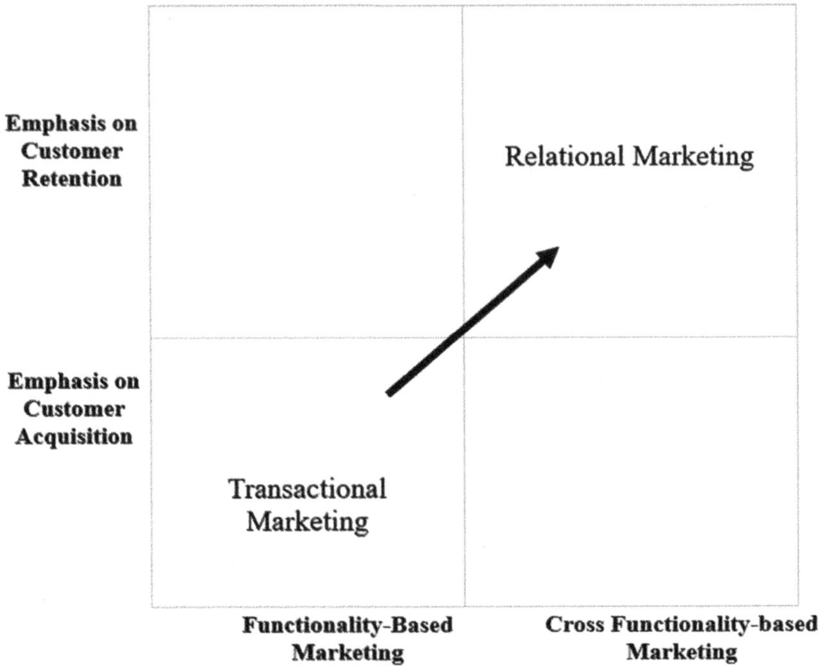

Fig. 12.1. Transition to Relationship Marketing.

With this type of procedures, companies hope to achieve greater growth, capturing more investments from customers by creating a strong relationship of loyalty (Kotler & Keller, 2012).

Also, the concept of relationship marketing emphasizes after-sales relationships. Its essence represents the establishment of a balance between marketing efforts to win new customers and maintain relationships with existing customers even after sales. The main theories of relationship marketing are not just about customer acquisition and sales transactions, but also about customer continuity and the importance of increasing the number of customers (Yegin, 2021).

Furthermore, through this marketing approach it was possible to recognize the importance of customer lifetime value, which in turn recognizes the customer as a revenue stream and a source of profits during a set of repeated long-term transactions and, therefore, favours the establishment of relationships throughout the customer's life to increase the company's profitability (Goldsmith & Tsiotsou, 2012). For relationships with customers to be long-term, it is necessary to adopt key relationship marketing strategies such as personalising the relationship with each customer, understanding the specific characteristics and requirements of each customer, placing all information in the database, retrieving it when necessary and carry out research on consumer tastes and habits (Edwards & Baker, 2020). It should be noted that for good management of relationship marketing, it is imperative to have a methodology that allows the design of medium and long-term goals, while short-term results relate to reducing the costs of relationship processes with the customer (Pereira, 2020).

The competitive environment that represents the current market situation requires companies to use relationship marketing if their focus is success (Austin-Campbell, 2021). This component of marketing benefits both consumers and companies. Without relationships, the company may suffer high costs to attract new consumers, and they may spend more time and money to purchase new products. Furthermore, the better the relationship between the company and client, the lower the risk perception. Therefore, relationship marketing assumes a central position in the financial situation and in increasing the company's competitiveness (Uğurlu, 2016).

2.2 Health and Wellness Tourism

Over time, the Tourism industry has increased exponentially and is currently considered one of the largest and most productive economic activities in the world that directly and indirectly affects other economic and cultural activities (Hadlan et al., 2021). Previous studies have expressed a growing concern about the applicability of digital marketing in promoting territories and promoting awareness of sustainability (Arora & Sharma, 2020) and also regarding alternatives to mass tourism (for example, in a niche logic) (Arora & Sharma, 2021; Gupta et al., 2022; Lubowiecki-Vikuk & Sousa, 2021; Santos et al., 2022). For instance, according to Sousa and Alves (2019), their study has identified that the global movement of tourism is seemingly showing an increased focus on the niche product or niche service. In this case, the question seems to be whether the further growth in demand for healthcare management and medical tourism – as a niche tourism example – products will continue until they take a form of mass tourism (Alves et al., 2021; Gupta et al., 2015) and health promotion (Loureiro et al., 2022). The new vogue of medical tourism forces to challenge and re-visit the power relationships that exist within contemporary tourism and the host–guest relationship. Currently, we are witnessing an increasingly accelerated and stressful pace of life, which has led to a clear tendency to search for environments and activities that promote well-being, wellness and relaxation. Consequently, interest in health and wellbeing tourism has seen rapid growth in recent years (Dillette et al., 2021). Health tourism represents a niche or market of special interest to the tourism industry and many countries are adopting its practice. This tourism practice contributes significantly to the economies of many countries, which increasingly offer health facilities and services to foreign clients (Heung & Kucukusta, 2013). 'Health tourism' does not only include medical tourism, but also takes on a special meaning associated with the well-being and happiness of tourists. In this way, we can define health tourism as a 'set of products that have health as their main motive and natural resources as support and aim to provide tourists with improved physical or mental well-being' (Cunha, 2006). Mourão (1992) distinguishes between two types of health tourism: 'transitory and medicinal', the first of which brings together those who are passing tourists and go 'to the spa to experience', while the second brings together 'the authentic Health Tourism [...] enthusiasts in hydromineral healing' (p. 676). Cunha (2006) defined 'two fundamental segments: those who move for primarily medical reasons and whose dominant motivation is healing or recovery and those who do so for reasons of prevention, well-being or recovery', concluding that 'probably the former do not reject welfare care, nor the latter reject the act or medical care'. Additionally, Cunha (2006) adds a third segment 'made of people who do not wish to have access to any form of private care, choose destinations or health resorts to enjoy the existing environmental conditions for reasons of rest, escape, or of contact with nature or another' (pp. 217–218).

Thus, the term health tourism is a comprehensive term covering two groups of tourists (medical and wellness), based on two different fundamental needs taking into account two main effects: curative and recuperative (prophylactic) (Romanova et al., 2015). Health systems use a pathogenic and reactive approach, focussing on the causes, consequences, diagnosis and treatment of illnesses and injuries. In contrast, wellness is a salutogenic and proactive approach focused on prevention, healthy lifestyles and the pursuit of optimal well-being (Global Wellness Institute, 2015).

One of the main reasons for the active development of health tourism is the significant difference in the level of health care and the cost of treatment in different countries. Thus, patients who do not have the opportunity to receive certain procedures or services in their own country, or do not want to wait, often find similar services abroad at lower prices and without delay (Veselova, 2017). Thus, a trip to a healthcare destination in another country opens access to services that could not be received at home, either due to a high cost or simply due to a lack of such services for legal reasons or due to the long waiting time to receive the services. desired (Romanova et al., 2015).

2.3 Thermalism in the Northern Euroregion of Portugal and Galicia

Thermal treatment is a tourist product highly recognized for its potential in exploring regional endogenous resources (Araujo et al., 2015). Thermalism also acts as a pillar to support other activities, is a net contributor to regional employment and wealth and offers a growth opportunity that has not yet been maximized. It is a heavily regulated activity at various stages of its production process, guided by multiple entities (in the spheres of health, tourism and geology) whose skills often overlap. Furthermore, it is subject to a variety of technical and administrative procedures, making it an activity that requires excessively demanding administrative processes (Ladeiras et al., 2015).

Thermalism constitutes a competitive offer contemplated in a new philosophy of occupation of free time and holidays, homoeostasis between health and well-being, leisure and tourism, contextualising itself as an active formula capable of satisfying the diverse arguments of tourism (Lourenço, 2012). Currently, spas are sought after for treatments and stress reduction, in other words, spa culture has transformed into health tourism, which begins in the traditional spa and ends in modern and luxurious spas (Turismo de Portugal, 2020).

The Euroregion represents cross-border micro-regions, with reduced geographic extension between contiguous spaces. Furthermore, cross-border regions may vary according to their degree of cooperation (high or low) or geographic extent (high or low). Depending on the perspectives, these can be interpreted differently. From a territorial point of view, they can represent political instruments for reconfiguring cross-border space (re-territorialization). From a policy and public perspective, these can be perceived as more pragmatic and functional policy-oriented organizations that can promote and stimulate joint cross-border initiatives and/or programs (Noferini et al., 2020).

The North of Portugal and Galicia share cultural, historical, geographic and economic ties, but have different historical, political and administrative realities (Pereira & Pereiro, 2014). Geographically and administratively, the Euroregion constitutes a cross-border space, located in the Northwest of the Iberian Peninsula (Pardellas & Padín, 2017; Pereira, 2020).

In the tourist sector of Galicia and the North of Portugal, thermal spa use is a fundamental activity for the social, economic and environmental development of the place. The Euroregion is one of the areas of the Iberian Peninsula with the greatest wealth of mineral-medicinal waters, in terms of quality and number of springs. For this reason, thermal tourism presents itself as an attractive tourist model, capable of generating employment and attracting tourists from different parts of Spain, Portugal and the rest of Europe (Amboage et al., 2019). It contributes to regional wealth and represents a growth opportunity that has not yet been maximized. Euroregion Galicia and Norther of Portugal makes up the territory with the most thermal resources in Europe. This fact is corroborated by the number of mineral water springs that appear in this territory, in addition to the quality and variety of their waters. This quantity of mineral water allows the construction of many spas, which are a reference in the tourism sector (Sánchez-Amboage et al., 2017).

Thermalism in this Euroregion is guided by several entities and is subject to different technical and administrative procedures. However, it is crucial that there is a strategic alignment of existing thermal laws throughout the Euro-Region to promote its sustainable growth. In this way, legislators could help promote a more favourable environment for the development and reinforcement of thermalism in the Euroregion (Ladeiras et al., 2015).

2.4 Cross-Border Tourist Demand Factors

Moral-Cuadra et al. (2019) define cross-border tourism as a temporary trip that involves, at least, an overnight stay in areas adjacent to the dividing line between countries, for different leisure reasons, where activities related to tourist activity are carried out. 'The search for border tourism seeks the same border and its sides as the central object of tourist motivation and experience' (Pereira & Pereiro, 2014). Cross-border tourism widens and stretches the border until it takes the tourist beyond the political-administrative border itself (Cardoso et al., 2023). Linked to new accessibility and centres of tourist interest further away from the political-administrative border itself, cross-border tourism extends to Porto, Espinho, Braga, Guimarães, Chaves, Santiago de Compostela, Vigo or Sanxenxo (Pereiro, 2013, pp. 75–82)

Tourists arriving in one region can benefit from travelling to other regions or neighbouring countries. There are cost advantages to cross-border cooperation when countries share the cost of promotion within the region, rather than being solely responsible for marketing expenses. The total cost can be shared by several countries. Another advantage is the creation and consolidation of peace processes between two neighbouring regions. Building economic activities and bringing together partners from both sides to co-create tourist experiences ultimately creates common economic interests and, eventually, trust and friendship. Revenues generated by cross-border tourist movements can be a catalyst for permanent peace solutions as both parties realize the benefits and risks involved. Such understanding can change the attitude and behaviour of people and local businesses over time, when the feeling of sharing from the same group becomes stronger (Kozak & Buhalis, 2019).

New competitiveness and spatial differentiation requirements are increasingly being demanded from destinations, according to the specific characteristics of each territory. However, it is crucial to understand what visitors to a given destination are willing to do and experience. The health and well-being tourist also seeks other experiences that create for destination management organization a need for integrated management of products and services to satisfy and, if possible, exceed the expectations of health and well-being tourists (Gonçalves & Guerra, 2019).

2.5 Relational Marketing in Health and Wellness Tourism

In the field of marketing, there is an interconnection between services and relationships. Service is based on relationships between parties and according to the research tradition of the Nordic School, relationship marketing cannot be implemented without service. The development of service logic in the literature emphasizes that adopting a service and marketing perspective allows service companies and product manufacturers to develop, maintain and improve relationships with customers (Grönroos, 2017).

Unlike products, services cannot be tried before purchasing, however, consumers have memories that they can share with others. In the tourism sector, many of the products sold are intangible experiences. To reduce rejection caused by the intangibility of services, consumers look for signs that generate trust in the service (Pereira, 2020).

As tourism is a dynamic, transversal and relational phenomenon and representative of a mainly service sector (Pereira, 2020), it is worth noting that in most tourist services, both the producer of the service as well as the consumer must be present for the operation to take place (Barros & Sousa, 2022). The human resources responsible for customer service are an integral part of the service. The inseparable nature of the service also means that consumers are an integral part of it. Both servers and consumers must understand the service delivery system, as they produce it together. The characteristic of inseparability requires that those who manage the tourism product organize both employees and consumers (Kotler et al., 2011).

With the growth of competition among tourist destinations and organizations, marketing tools are becoming increasingly important. This scenario of uncertainty in tourist destinations has led to new approaches to marketing, in which relational marketing stands out. Its way of operating is based on customer retention, with a high level of service and contacts, with a long-term vision. The value of customers assumes special importance, seeking to maintain high levels of commitment to the highest value customers (Antunes, 2012, pp. 218–219). Thus, and according to Sousa and Barros (2021) relationship marketing applied to the health and wellness tourism sector allows creating and mutually beneficial relationships.

3. Methodology

After reviewing the literature on the main themes, we developed a conceptual model based on the contribution of Moreira and Silva (2015) which presents a useful tool for an integrated approach to understanding loyalty of tourists towards the thermal spas they frequent, namely, through the analysis of theoretical and empirical evidence on the causal relationships between satisfaction, commitment, trust and loyalty. For this chapter, the method selected involved quantitative and qualitative methods (mixed methodology). The application of this methodology will allow investigating a greater extent of the research problem instead of a single, more focused method (Newman & Gough, 2020). Furthermore, mixed methodology allows a focused approach on process and results using qualitative and quantitative methods and data (Creswell & Tashakkori, 2007). Therefore, given the selected method, the data collection instruments to be used are interviews and questionnaire surveys. The adoption of both approaches aims to understand not only the perspective of consumers but also professionals on the research problem.

To analyse the quantitative data, the scale used to measure all variables was the original five-point Likert scale (1 = Strongly disagree and 5 = Strongly agree). Two surveys were made, one in Portuguese and the other in Spanish, in order to obtain insights from the population of both regions. In this sense, the sample is characterized as non-probabilistic, given the way data was collected (online), gathering a total of 329 responses considered valid. Then, the data obtained was analysed by the statistical software Statistical Package for the Social Sciences

(SPSS). Using SPSS, the following types of analyzes were explored: (1) Univariate analysis, to analyse the sociodemographic profile of respondents; (2) Bivariate analysis, to test and validate the hypotheses of this research through the study of associations between the variables of the proposed conceptual model. In this way, descriptive analysis techniques (generation of descriptive statistics such as mode and medians) and correlation (Pearson correlation coefficient) were used.

Regarding the analysis of the interviews (qualitative data), five dimensions were defined to cross-reference data from the different interviews on the theme of each dimension: (1) Identification of the Euroregion's key tourism products; (2) Identification of the target audience and thermal demand factors in the Euroregion; (3) Challenges faced and post-pandemic overcoming strategies; (4) Potential for creating a common product offering (thermalism); (5) Contribution to the structuring of a common relationship marketing strategy. The interviews were carried out with three entities related to the research topic, one located in the North of Portugal and two in Galicia.

4. Analysis and Discussion of Preliminary Results

Following the completion of data collection and all the processes for the statistical treatment of data, an analysis of the results obtained is now presented. Preliminarily, this analysis focuses on 329 surveys and 3 interviews carried out with public and private entities. In the data obtained by the portuguese survey, there is a strong dominance of a younger age group, being 84 respondents (40.8%) aged between 25 and 34 years, 50 respondents (24.3%) to the 35–44 age group and 32 respondents (15.5%) to the age group between 18 and 24 years old. Furthermore, this manuscript also included the participation of an older audience, with 23 respondents (6.8%) corresponding to the age group 45–54 years old, 23 respondents (6.8%) to the age group 55–64 years old and 2 respondents (1.5%) aged 65 or over. Similarly, the percentage of the various age groups of respondents to the Spanish survey are proportional to the age groups of the Portuguese survey, once again witnessing the predominance of the youngest age group in which 46 respondents (37.4%) are aged between 25 and 34 years old, 18 respondents (14.6%) between 18 and 24 years old and 36 (29.3%) between 35 and 44 years old. Although in a smaller percentage, there was also the presence of an older age group in which 16 respondents (13%) corresponded to people aged between 45 and 54, 4 respondents (3.3%) between 55 and 64 years old and 3 respondents (2.4%) among those aged 65 and over. Respondents centred between 18 and 25 years of age (14%), are characterized as being the least responsive age group to the questionnaire survey, as only 27 individuals in this age group did so. As for the gender category, it is possible to note the evident large percentage of female respondents compared to the totality of the male gender.

Regarding question 1 of the survey: 'What is your main reason for visiting thermal spaces?' as for the portuguese version of this survey, 116 respondents (56.3%) say they visit thermal spaces for their well-being. As for the remaining respondents, 29 (14.1%) frequent thermal spaces due to illness and 61 respondents

(29.6%) for prevention. As for participants in the Spanish survey, 50 respondents (40.7%) say they visit resorts for their well-being, 48 respondents (39%) for prevention and 25 respondents (20.3%) for illnesses. Through these results we can verify that there is a high concern with individual well-being, as evidenced by the literature of this chapter.

In the question 2: 'How did you take the initiative to visit thermal spaces?', regarding the portuguese survey, the majority of tourists, 72 respondents (corresponding to 35%), reported visiting thermal spaces on their own initiative. 64 respondents (31.1%) stated that advice from friends/family was the main factor in attending the spa. It should be noted that 47 tourists (22.8%) visit thermal spaces on medical recommendation, and 23 respondents (11.2%) were influenced by the media (social networks, television and advertising). Based on the Spanish survey, the factor that with the most influence on tourists in visiting spas was advice from friends and family, represented by 48 respondents (39%). 40 tourists (32.5%) visit on their own initiative, 28 tourists (22.8%) on medical recommendation and, finally and seven tourists (5.7%) visit spas due to the influence of the media.

On the other hand, in question 3 of the survey: 'Please indicate the number of times you have visited thermal spas in the Euro-region?' 117 participants (51.9%) in the first survey admitted having already visited thermal spas in the Euroregion 2–3 times, 46 respondents (22.3%) said they had already visited 4–5 times and 21 of the respondents (11.2%) are frequent thermal tourists in the Euroregion visiting thermal spas more than 5 times. On the other hand, 30 respondents (14.6%) had only visited the thermal spaces in this region once. Regarding the second survey, most tourists (corresponding to 41.5%) visited between 2 or 3 times or between 4 and 5 times (34.1%). 17 tourists (13.8%) attended only 1 time and 13 participants (10.6%) more than 5 times. In this way, most of respondents have already attended spas in the Euroregion more than once (85.4% of participants in the first survey and 86.2% in the second survey).

According to the descriptive analysis of the variables of the conceptual model, it was possible to observe that the mean and mode of the variables focused mainly on point 4 of the Likert Scale ('Agree') and in some cases on point 5 ('Strongly Agree') such as in the variable loyalty. Based on the results obtained from Pearson's correlation analysis it is notable that all correlations made between variables are positive and statistically significant. Also, there is a relationship between the study variables, with satisfaction being directly related to the commitment and trust of tourists and these two variables being directly related to loyalty to the spa. Even indirectly, satisfaction ends up influencing tourists' loyalty through trust and commitment.

From these results it is possible to affirm that tourists who show loyalty towards the spa they frequent are those who are more satisfied and consequently more committed and more confident with the spa. There is a great influence on tourists' commitment to loyalty, because the correlation of these two variables was presented as being the stronger correlation. There is thus a relational influence on the part of the indicator's satisfaction and commitment to loyalty to the spa frequented by tourists.

It was possible to determine through the interviews that there is great potential for creating a common product such as thermalism between the Euroregion, due to the large amount of thermal resources of both regions. The interviewees proposed the creation of a collaborative strategy between the two regions with the aim of creating a thermal circuit through joint promotion of thermal spas. Furthermore, interviewees highlight the importance of implementing a relationship marketing strategy in order to create synergies with tourists, taking into account their loyalty and proposed the creation of a customer loyalty programme so that it could be shared in the different thermal centres of the Euro-Region.

5. Conclusion

In this study, relational marketing is explored in the context of thermalism in the Northern Euroregion of Portugal and Galicia, where thermal waters are not only sources of well-being but also unique and rigorous experiences. Throughout this research, it became clear that relationship marketing plays a crucial role in the sustainability and growth of the thermal industry. A deep understanding of customers' needs and expectations proves to be fundamental to establishing solid and lasting bonds between thermal establishments and their visitors. The Euroregion, with its rich thermal heritage, offers a useful solid base for the application of relationship marketing strategies. The personalization of services, adaptation to cultural preferences and the promotion of different experiences are the pillars of this approach, where thermal waters are not only linked to their physical benefits but also to the history and identity of the region.

It is notable that the increase in satisfaction is entirely interconnected with the increase in other relational constructs such as commitment and trust, which, in turn, are related to the intensification of loyalty to the spa. This reflects the importance of spa centres working on a relational marketing strategy. The implementation of this strategy will increase the likelihood of making tourists more satisfied with the service provided, more committed and confident with the spa they visit and, consequently, more loyal, that is, more likely to revisit and recommend it.

The results of the survey data analyzes revealed that all correlations carried out for the analysis of the conceptual model are statistically significant, highlighting the correlation between trust and commitment between tourists and thermal centres, which suggests that marketers should be concerned about develop strategies that promote tourist satisfaction and trigger a feeling of pride and priority on tourists in choosing the thermal spa they visit compared to other possible choices. By applying this strategy, as the results demonstrate, the likelihood of revisiting and recommending of the spa is greater.

Although the defined methodology allowed the proposed objectives of this research to be achieved, some limitations were felt throughout the research. The first limitation is based on the fact that the sample of respondents is non-probabilistic in nature, which does not allow the generalization of the results across the population related to the study context. The second limitation corresponds to the exploratory interview method, which led to the interviewee presenting their own elements and opinions that do not meet the topic in question. We sought to overcome this situation, relating the results of the interviews with aspects that were considered interesting in this chapter. Another limitation is the fact that data collection was carried out through a survey made available online and had public access, which makes effective control of responses difficult, and the questionnaire may have been answered by people who do not meet the criteria defined for the representative sample of the study. This chapter presents some limitations of sampling nature and analysis and discussion of results. As for future research, it may lead to other interesting relationships by including different variables in the study, in addition to those that are already covered in this work, making the study more complete.

Based on a combination of theoretical and practical research, this chapter explores the dynamic system and mechanism of innovation and development of relational marketing strategies in specific contexts of tourism (i.e. health and wellness tourism) considering cross-border (i.e. Northern Euroregion of Portugal and Galicia). From an interdisciplinary perspective, the chapter presents insights for marketing and tourism (health and well-being) and for the economic development of territories (i.e. cross-border regions).

References

Alves, G. M., Sousa, B. B., & Belino, M. (2021). Understanding the brand management and rebranding processes in specific contexts of medical tourism. In *New techniques for brand management in the healthcare sector*. IGI Global.

Antunes, J. (2012). A importância do marketing relacional no turismo termal. Estudo da Região Dão-Lafões. In V. Joukes, L. Lourenço-Gomes, & A. Marta-Costa (Eds.), *Sustainable health and wellness destinations. Proceedings of the international conference held in Chaves* (pp. 219–236).

Antunes, J., & Rita, P. (2007). O marketing relacional e a fidelização de clientes – Estudo aplicado ao termalismo português. *Global Economics and Management Review*, *12*(2), 109–132.

Araujo, A. R., Paiva, T., Ribeiro, M. P., & Coutinho, P. (2015). Innovation in thermalism: An example in beira interior region of Portugal. In M. Peris-Ortiz & J. Álvarez-García (Eds.), *Health and wellness tourism* (pp. 165–180). Springer. https://doi.org/10.1007/978-3-319-11490-3_11

Arora, S., & Sharma, A. (2020). Digital marketing for religious event of India for tourism sustainability and promotion. In *The Emerald handbook of ICT in tourism and hospitality* (pp. 453–465). Emerald Publishing Limited. https://doi.org/10.1108/978-1-83982-688-720201029

Arora, S., & Sharma, A. (2021). Covid-19 impact on overtourism: Diversion from mass tourism to alternative tourism. In *Overtourism as destination risk* (pp. 275–283). Emerald Publishing Limited. https://doi.org/10.1108/978-1-83909-706-520211018

Arosa-Carrera, C., & Chica-Mesa, C. J. (2020). La innovación en el paradigma del marketing relacional. *Journal of Management and Economics for Iberoamerica, 36*(154), 114–122. https://doi.org/10.18046/j.estger.2020.154.3494

Austin-Campbell, S. (2021). Retail marketing industry comeback post pandemic. *Marketing: Theory and Research*, 1–11. http://doi.org/10.2139/ssrn.3895331

Barros, C. L., & Sousa, B. B. (2022). Relationship management in spa tourism: An approach to cross-border regions. In *Tourist behavior* (pp. 1–22). Apple Academic Press.

Cardoso, D., Sousa, B., Liberato, D., Liberato, P., Lopes, E., Gonçalves, F., & Figueira, V. (2023). Digital communication and the crisis management in hotel management: A perspective in the Euroregion North of Portugal and Galicia (ERNPG). *Administrative Sciences, 13*(8), 191.

Christopher, M., Payne, A., & Ballantyne, D. (2013). *Relationship marketing*. Taylor & Francis. https://doi.org/10.4324/9780080516042

Creswell, J. W., & Tashakkori, A. (2007). Differing perspectives on mixed methods research. *Journal of Mixed Methods Research, 1*(4), 303–308.

Cunha, L. (2006). Turismo de saúde – conceitos e mercados. *Revista Lusófona de Humandidades e Tecnologias, 10*, 79–84.

Dillette, A. K., Douglas, A. C., & Andrzejewskic, C. (2021). Dimensions of holistic wellness as a result of international wellness tourism experiences. *Current Issues in Tourism, 24*(6), 794–810. https://doi.org/10.1080/13683500.2020.1746247

Edwards, C. J., & Baker, B. L. (2020). Relational marketing throughout the history of commercial exchange: Blind spots in marketing's origin story. *Journal of Marketing Channels, 26*(3), 161–177. https://doi.org/10.1080/1046669X.2020.1757977

Global Wellness Institute. (2015). What is wellness. Global Wellness Institute. https://globalwellnessinstitute.org/what-is-wellness/

Goldsmith, R. E., & Tsiotsou, R. H. (2012). Implementing relationship marketing in hospitality and tourism management. In R. E. Goldsmith, & R. H. Tsiotsou (Eds.), *Strategic marketing in tourism services* (pp. 139–146). Emerald Group Publishing Ltd. https://doi.org/10.13140/RG.2.1.2327.6647

Gonçalves, E. C., & Guerra, R. J. (2019). O turismo de saúde e bem-estar como fator de desenvolvimento local: uma análise à oferta termal portuguesa. *Revista de Turismo y Patrimonio Cultural, 17*, 453–472. https://doi.org/10.25145/j.pasos.2019.17.030

Grönroos, C. (2004). The relationship marketing process: Communication, interaction, dialogue, value. *Journal of Business & Industrial Marketing, 19*(2), 99–113. https://doi.org/10.1108/08858620410523981

Grönroos, C. (2017). Relationship marketing and service: An update. *Journal of Global Scholars of Marketing Science, 27*(3), 201–208. https://doi.org/10.1080/21639159.2017.1318666

Gupta, A., Arora, N., Sharma, R., & Mishra, A. (2022). Determinants of tourists' site-specific environmentally responsible behavior: An eco-sensitive zone perspective. *Journal of Travel Research, 61*(6), 1267–1286. https://doi.org/10.1177/00472875211030328

Gupta, P., Pandey, B., Hussain, S., Kumar, A., Sharma, R., Yadav, M. K., Bhati, S., Khandelwal, A., Shyamal, S. Md., Khan, A., Mohata, V., Khan, F., Sharma, K., Arora, K., Gandhi, T., Kumar, P., & Kundu, R. (2015). Challenges and strengths of medical tourism in Delhi NCR. *DU Journal of Undergraduate Research and Innovation*, *1*(2), 230–251.

Hadlan, M., Jabban, A., Mousavi, S. H., & Sheikhbardsiri, H. (2021). Medical tourism development: A systematic review of economic aspects. *International Journal of Healthcare Management*, *14*(2), 576–582. https://doi.org/10.1080/20479700.2019.1677977

Han, H., Kiatkawsin, K., Jung, H., & Kim, W. (2018). The role of wellness spa tourism performance in building destination loyalty: The case of Thailand. *Journal of Travel & Tourism Marketing*, *35*(5), 595–610. https://doi.org/10.1080/10548408.2017.1376031

Hayati, I., & Muddaststsir, D. A. (2019). Effects of customer relationship management on customer loyalty. In *Proceedings of the 1st international conference on economics, management, accounting and business* (pp. 8–9). https://doi.org/10.4108/eai.8-10-2018.2288685

Heung, V. C., & Kucukusta, D. (2013). Wellness tourism in China: Resources, development and marketing. *International Journal of Tourism Research*, *15*(4), 346–359. https://doi.org/10.1002/jtr.1880

Kotler, P., Bowen, J. T., Makens, J., Garcia, J., & Flores, J. (2011). *Marketing Turístico*. Prentice Hall.

Kotler, P., & Keller, K. L. (2012). *Marketing management*. Pearson Education.

Kottler, P., Jain, D. C., & Maesincee, S. (2002). *El marketing se mueve: una nueva aproximación a los beneficios, el crecimiento y la renovación*. Paidós.

Kozak, M., & Buhalis, D. (2019). Cross–border tourism destination marketing: Prerequisites and critical success factors. *Journal of Destination Marketing & Management*, *14*, 1–9. https://doi.org/10.1016/j.jdmm.2019.100392

Kundu, A., & Bhattacharya, A. (2019). Dynamism in the marketing strategy: Transition from transactional to relationship marketing. *International Journal of Management Studies*, *6*(1), 3, 2231–2528. http://doi.org/10.18843/ijms/v6i1(3)/09

Ladeiras, A., Mota, A., & Pardo, M. C. (2015). A comparative study of thermal legislation in the Galicia–North Portugal Euroregion. In M. Peris-Ortiz, & J. Álvarez-García (Eds.), *Health and wellness: Emergence of a new market segment* (pp. 1–20). Springer. https://doi.org/10.1007/978-3-319-11490-3_1

Loureiro, H., Silva, M., Camarneiro, A. P., Pedreiro, A. T., & Mendes, A. (2022). REATIVA: An efficient health promotion program during retirement transition. *European Journal of Investigation in Health, Psychology and Education*, *12*(9). https://doi.org/10.3390/ejihpe12090095

Lourenço, F. M. (2012). *O posicionamento do turismo de saúde e bem-estar: O caso das Termas de São Pedro do Sul*. Instituto Politécnico de Viseu, Escola Superior de Tecnologia e Gestão de Viseu.

Lubis, A., Dalimunthe, R., Absah, Y., & Fawzeea, B. K. (2020). The influence of customer relationship management (CRM) indicators on customer loyalty of sharia based banking system. *Journal of Management and Marketing Review*, *5*(1), 84–92. https://doi.org/10.35609/jmmr.2020.5.1(8).

Lubowiecki-Vikuk, A., & Sousa, B. (2021). Tourism business in a vuca world: Marketing and management implications. *Journal of Environmental Management & Tourism*, *12*(4), 867–876. https://doi.org/10.14505//jemt.v12.4(52).01

Miquel-Romero, M. J., Caplliure-Giner, E. M., & Adame-Sánchez, C. (2014). Relationship marketing management: Its importance in private label extension.

Journal of Business Research, *67*, 667–672. https://doi.org/10.1016/j.jbusres.2013.11.025

Moral-Cuadra, S., Orgaz-Agüera, F., & Cañero-Morales, P. M. (2019). Attitudes towards border tourism and its relationship with visitor satisfaction and loyalty. *Geo Journal of Tourism and Geosites*, *25*(2), 609–622. https://doi.org/10.30892/gtg.25226-384

Moreira, A. C., & Silva, P. M. (2015). The trust-commitment challenge in service quality-loyalty relationships. *International Journal of Health Care Quality Assurance*, *28*, 253–266. https://doi.org/10.1108/IJHCQA-02-2014-0017

Mourão, B. M. (1992). *Medicina Hidrológica moderna terapêutica das Águas Minerais e Estâncias de Cura* (Vols. 1–732). Secretaria Municipal de Educação.

Newman, M., & Gough, D. (2020). Systematic reviews in educational research: Methodology, perspectives and application. In *Systematic reviews in educational research: Methodology, perspectives and application* (pp. 3–22).

Noferini, A., Berzi, M., Camonita, F., & Durà, A. (2020). Cross-border cooperation in the EU: Euroregions amid multilevel governance and re-territorialization. *European Planning Strudies*, *28*(1), 35–56. https://doi.org/10.1080/09654313.2019.1623973

O'Malley, L. (2014). Relational marketing: Development, debates and directions. *Journal of Marketing Management*, *30*(11–12), 1220–1238. https://doi.org/10.1080/0267257X.2014.939592

Paparoidami, N. G., Katsikeas, C. S., & Chumpitaz, R. (2019). The role of supplier performance in building customer trust and loyalty: A cross-country examination. *Industrial Marketing Management*, *78*, 183–197. https://doi.org/10.1016/j.indmarman.2017.02.005

Pardellas, X., & Padín, C. (2017). La Eurorregión Galicia-Norte de Portugal y los modelos de cooperación transfronteriza. *Polígonos. Revista de Geografía*, 29, 11–35. https://doi.org/10.18002/pol.v0i29.5199

Pereira, D. (2020). *Estratégias e ações de marketing para o destino Galiza e Norte de Portugal: um contributo para a sua valorização turística.* Instituto Politécnico de Viana do Castelo.

Pereira, V. D., & Pereiro, X. (2014). Turismo transfronteiriço na Euro-região Galiza-Norte de Portugal. *Revista Turismo & Desenvolvimento*, *2*(21/22), 285–294. https://doi.org/10.34624/rtd.v2i21/22.11585

Pereiro, X. (2013). *As Fronteiras e o Turismo. Cadernos Mateus DOC V · Fronteira* (pp. 75–82).

Romanova, G., Vetitnev, A., & Dimanche, F. (2015). Health and wellness tourism. In *Tourism in Russia: A management handbook* (pp. 231–287).

Sánchez-Amboage, E., Martínez-Fernández, V.-A., Juanatey-Boga, O., & Rodríguez-Fernández, M.-M. (2017). Modelos de Gestión de los Balnearios de la Eurorregión Galicia-Norte de Portugal. *Revista Portuguesa de Estudos Regionais*, *44*, 5–21. https://doi.org/10.59072/rper.vi44.455

Santos, V., Sousa, B., Ramos, P., & Valeri, M. (2022). Emotions and involvement in tourism settings. *Current Issues in Tourism*, *25*(10), 1526–1531. https://doi.org/10.1080/13683500.2021.1932769

Sheth, J. N., Parvatiyar, A., & Sinha, M. (2012). The conceptual foundations of relationship marketing: Review and synthesis. *The European Electronic Newsletter*, *13*(3), 4–26.

Sousa, B. M., & Alves, G. M. (2019). The role of relationship marketing in behavioural intentions of medical tourism services and guest experiences. *Journal of Hospitality and Tourism Insights*, *2*(3), 224–240. https://doi.org/10.1108/JHTI-05-2018-0032

Sousa, B. B., & Barros, C. (2021). Encouraging relational marketing in the wellness and thermalism segments. Економічний вісник Національного технічного університету України «Київський Політехнічний інститут», *28*, 98–102. https://doi.org/10.20535/2307-5651.18.2021.240453

Turismo de Portugal. (2020). *Dia do Património Termal.* http://www.turismodeportugal.pt/pt/Agenda/Paginas/dia-patrimonio-termal.aspx

Uğurlu, K. (2016). Relationship marketing in tourism sector. In C. Avcikurt, M. S. Dinu, N. Hacioglu, R. Efe, A. Soykan, & N. Tetik (Eds.), *Global issues and trends in tourism* (pp. 509–517).

Veselova, E. S. (2017). Medical tourism: Tourism with health benefits. *Problems of Economic Transition*, *59*(6), 480–500. https://doi.org/10.1080/10611991.2017.1352363

Yegin, T. (2021). Brand loyalty in creating relationship marketing practices: A study on GMS operators. *Electronic Journal of Social Sciences*, *20*(77), 201–216. https://doi.org/10.17755/esosder.661291

Chapter 13

The Economics of Market and Inclusive Growth in Tourism: A Narrative of Rural Artisans' Socio-economic Wellbeing

Sandhya H.[a], *Antony Davis*[a,b], *Bindi Varghese*[a] *and K.Lakshmypriya*[a]

[a]Christ University, India
[b]Christ Academy Institute for Advanced Studies, India

Abstract

Developing nations are starting to take centre stage and are becoming more conscious of their economic potential in the tourism sector. Earlier, these countries were largely left out of the tourist sector. Today, the foreign exchange reserves of these nations are mostly derived from tourism and a nation's balance of payments derived through tourism supports international trade and stabilizes the economy. The chapter adopts a case study method to evaluate the prospects of inclusive growth through socio economic well-being. In the dynamic and constantly evolving tourism industry, it is imperative to make the growth more inclusive so that there is a balanced approach to long term sustainable development. When it comes to using traditional skills and cultural heritage to participate in this economic paradigm, rural artisans face both opportunities and challenges as the global tourism landscape undergoes evolving transformations. The study commences by closely examining the current market obstacles that rural artisans in the tourism industry face, which encompass a range of issues such as shifting consumer preferences, market saturation and the aftermath of external shocks like pandemics. The chapter indicates major obstacles impeding the integration of rural artisans into the tourism value chain by means of a thorough empirical analysis. In order to overcome the obstacles, the study investigates inclusive growth and how it might act as a spur to improve the socio-economic circumstances of rural craftsmen. The findings in the chapter distil best practices that promote inclusivity, empower local

The Emerald Handbook of Tourism Economics and Sustainable Development, 251–272

doi:10.1108/978-1-83753-708-220241013

communities and support sustainable tourism development by looking at successful case studies and policy interventions.

Keywords: Inclusive growth; cultural tourism; arts and handicrafts; socio-economic wellbeing; community empowerment; community engagement; sustainable development

1. Introduction

The ongoing discourse among scholars and policymakers over how to address the high levels of poverty and income inequality can be enhanced by comprehending tourism's role in inclusive growth, particularly in emerging nations. The discussions around how tourism acts as one of the main drivers of inclusive growth can create opportunities for employment and economic participation (Adeniyi et al., 2023). Inclusive growth is coupled with equal opportunities (Rauniyar & Kanbur, 2010). The goal of inclusive growth is better opportunities for marginalized communities to actively engage in a nation's economic progress. Fairness, equal opportunity and protections for those impacted by changes in the labour market and employment landscape should be at the forefront of any successful growth strategy, according to the Commission on Growth and Development (2008). Sustainable growth is the mainstay of the inclusive growth framework. Sustainable growth is viewed as inclusive growth if it can (i) generate and increase economic possibilities and (ii) guarantee that these opportunities are widely accessible to all members of society so they may take part in and profit from growth (Bakker et al., 2023; Lee, 2019).

In many emerging economies, tourism is seen as a means of promoting employment growth and, in turn, enhancing the well-being of its stakeholders (Bakker et al., 2023; Snyman, 2012). But the theoretical underpinnings of the connection between inclusive growth and tourism development are still in their nascent stages. Bakker et al. (2023) argued that tourism should create productive employment opportunities, allowing entrepreneurs to participate in economic activities in the tourism sector. The sector has a large and diverse supply chain that includes artefacts, handicrafts, agri-products, transportation, art performances, etc. which has a multiplier effect to augment all economic sectors (Ashley & Mitchell, 2009). As tourism develops, it creates new forms of value. It reorganizes the internal distribution of capital, land usage and labour organization to meet the demands of tourism capital accumulation, resulting in the emergence of competitive models driven by digital platforms to the unscrupulous acquisition of land by real estate firms to create resorts and customized tourism experiences (Bakshi, 2023; Buhalis & Michopoulou, 2011). The complexity of financing and corporate ownership has brought the marginalized communities to a stage of poverty and exploitation, as a result of which many indigenous communities are forced to give up their ethnic workmanship, performing arts and folklore. They often end up taking underpaid jobs in the hospitality sector (Buhalis & Michopoulou, 2011). Many studies on the impact of tourism on the economy have focused on the macro perspectives of GDP (gross domestic product) of exports exchange rates and few studies have pointed out that an increase in tourism activity necessarily does not give equal participation for entrepreneurs to participate in wealth creation (Sánchez-Rivero et al., 2013) and many tourism growth studies do not include the marginalized groups whose participation in the economic activity in the sector can boost destination appeal and specifically cultural tourism (Cultural Tourism Market, 2023).

India stands out as a significant exporter of handicrafts, notably leading the handmade carpet market both in volume and value. From April to September 2023, handicraft exports reached Rs. 14,791 crore (US$ 1.77 billion), totalling Rs. 30,019 crore (US$ 3.60 billion) for the fiscal year 2023. In 2021–22, Indian handicraft exports surged 25.7% to US$ 4.35 billion. This growth, especially in items like carpets, has been remarkable over the past 3 years, with India contributing about 40% of the world's handmade carpet exports. In FY23, the overall value of handicraft exports spanned various categories, including wood wares at US$ 915.6 million, embroidered and crocheted goods at US$ 459.5 million, art metal wares at US$ 522.4 million, hand-printed textiles and scarves at US$ 303.3 million, imitation jewellery at US$ 186.5 million and other handicrafts at US$ 1.04 billion.

2. Tourism, Arts and Crafts

Cultural cities with diversity and resourcefulness can leverage the cultural component with the interconnectedness of cultural and economic vibrancy, which are influenced by interrelated dimensions of arts and crafts (Yan, 2023). Cultural destinations personified within historical landmarks, customary practices and socio-cultural heritage attract significant cultural tourists (Dugan et al., 2023). The predominant pattern of cultural tourism is high footfall of tourists to historical sites, museums and experiential engagement with local activities (Hamzah et al., 2023). Informal economies are constructed based on specific markets supported by artisans who curate handicrafts and traditional arts, which are integral to a region's identity and cultural legacy (Jamatia, 2023). A region's identity and cultural past are deeply ingrained in the handicrafts and traditional arts that craftsmen sustain, eventually becoming a foundation for informal economies (Jordan, 2023). Selling local handicrafts and other ethnic objects as souvenirs is a common way cultural sites interact with tourists (Darcy et al., 2011; Debbarma, 2020). The artisan markets contextualizes tourism by being key stakeholders through the local markets showcasing handmade goods, which provides a platform for artisans to sell their products directly to tourists (Zainuri & Huda, 2023). An experiential visitor engagement connects the local economies through craft workshops to give a social connection (Dey et al., 2020). Through artisans, perhaps an opportunity can be carved to enable an experience for tourist interaction and this flavour can add value to build the social connection between the local economy (Kanungo et al., 2021). Tourists may engage in craft workshops to learn traditional skills, fostering a deeper connection with the local culture (Bagnall & Southby, 2023).

Destinations can also capitalize through sustainable crafts, synergising the local essence of the destination (Custodio et al., 2023). Travellers can also be sensitized to the social fabric of the destinations. They can facilitate the host-guest relationship to be horned by acquiring the traditional craft-making skills, which strengthens their bond with the community (Baimuratova & Chhabra, 2023). Sustainable crafts can also help destinations make money by enhancing the local character of the area

(Bobo & Akhurst, 2023). Sustainable and environmentally friendly techniques are becoming increasingly important in the new work order, and there is a growing demand for crafts manufactured from recycled or locally sourced materials (Bone et al., 2023). A strategic alliance and partnership can foster benefits for the tourist and local communities, which can be established by integrating tourism with the arts and crafts (Boonyasurat et al., 2023). Perhaps the local artists and craftspeople have an opportunity to exhibit their talents and build an economic opportunity; tourists also get to see the cultural diversity of a place (Datta & Bhattacharyya, 2016; Karthik et al., 2023). Visitor engagement fosters a strong destination image with experiential aspects and the rich cultural diversity of the cultural cities (Deb, Mohanty et al., 2022; Khaledi Koure et al., 2023). When these elements work together, a region's cultural and economic development is personified (Kumari et al., 2019). The combined effect of regional integration, craftsmanship and arc-tician market boost the cultural image and advances the economic engagement positively in a region (Anaba & Cole-Adeife, 2023).

3. Tourism Arts and Crafts – A Traditional Perspective

The diversity and appeal of a destination are greatly enhanced by its local culture, which plays a crucial role in setting one place apart from another and enhancing the destination's competitiveness. Given the increased worldwide competition and shifting consumer partners that affect the supply and demand for tourism, this difference is becoming increasingly crucial for destinations looking to hold onto and expand their market share (Dibyendu & Bhattacharyya, 2016). Local legacies, such as arts and crafts, music, dance and cuisine, which are frequently unique to a particular nation, community, or social grouping, are among culture's most direct and tangible manifestations. It is often acknowledged that tourists have a strong desire for locally produced and handmade technologies. With the emergence of several sub-sectors, such as heritage tourism, arts tourism, gastronomy tourism, film tourism and creative tourism, the notion of cultural tourism gained popularity in the 1990s. This occurred during the height of globalization and technical development, which promoted more mobility through more affordable air travel, more accessible access to a broader range of destinations and cultural resources, the spread of media and the emergence of independent travel. By 2023, the revenue generated by cultural tourism worldwide is expected to reach US$5,931.2 million. Over the next 10 years, the demand for cultural tourism is expected to rise at a 14.4% CAGR on a global scale. The global market for cultural tourism is expected to be valued at around US$ 22,772.0 million by the end of 2033 (Cultural Tourism Outlook Report, 2022–23). Many nations view cultural tourism as a top priority to promote job development, particularly for young people, as it is another essential global employment pillar. Cultural tourism alone accounts for a substantial portion of tourism employment, which generates 40% of global tourism profits. The World Travel and Tourism Council (WTTC) estimates that the tourism sector contributes 330 million jobs, or one in 10 jobs worldwide (WTTC Report).

Micro, small and medium-sized enterprises (MSMEs) have proliferated due to the broad and diverse tourism industry, which attracts a steady stream of tourists interested in culture, history, adventure and world heritage. Many of these MSMEs are skill-based service businesses. These include providing culturally-related entertainment services, creating arts and crafts, regional cuisine and traditional crafts like basket weaving and hair braiding. Further, the tourism market presents countless sales opportunities for small business owners, such as cultural performers, vendors and manufacturers of handicrafts, furniture, soaps, speciality foods and other hand-made goods. There are a variety of sales venues and eager tourists looking for 'local', indigenous and culturally connected goods (Indian Handicrafts Industry and Exports Report, 2023).

The term crafts is defined as handmade items where the artisan pays attention to the material, design and workmanship of their product (Keane, 1996; Stoddard et al., 2012), and this is distinguished from souvenirs which can be anything like key chains, postcards, magnets etc.

3.1 Traditional Arts and Crafts in India

The majority of Indian crafts share an extensive lineage dating back hundreds of years. Production of pottery existed since the Harappan Civilization. The Mughal era was the pinnacle of development for India's most well-known handmade product, textiles. Favouritism from the government and nobles also allowed many other artisanal crafts to develop. Especially outside the nation, the Gupta era saw a great demand for crafts, including jewellery, metalwork and ivory carving. Silk spinning and weavers of all stripes reached their pinnacles during the Mughals.

Archaeology has confirmed that symbolic representations of social hierarchy have been used since the Palaeolithic. But there's a clear correlation between the rise in urbanization and the extraordinary rise in creating and consuming symbolic artefacts. Along with producing symbolic artefacts for religious and political purposes, the people of the river valley civilizations also developed unique ceramics, seals and bricks (Hunjra et al., 2022). The wealthy wore copper, bronze and gold jewellery, while the ordinary people used terracotta beads and bracelets. Ivory khol sticks, painted ceramics, semi-precious stones, bone arrowheads, copper utensils and ivory khol sticks have all been produced, according to archaeological discoveries, at numerous locations from various historical periods (Penhallurick, 2023). Craft manufacturing saw a resurgence in several nations when colonialism ended. For instance, to secure the survival of artisans and village craftspeople, handicrafts received renewed attention when India gained independence. It became a significant export for India in line with its new development paradigm. Handmade goods were being offered for a premium price all over the world (Khaledi Koure et al., 2023). Exporters had been interested in Indian handicrafts because of their high quality and variety. India's festivals presented the main markets' production of handicrafts created by talented artists. These national honours acknowledged the craftsmen's role in the country's economy.

The word 'handicraft' refers to skilled artisans' manual or minimal tool-based processing of raw materials to produce functional or ornamental goods made of clay, stone, paper, rock, etc. The Indian handicrafts are rich, and the country has a rich map of arts and crafts that dot the entire geographic spread of the country.

Hinduism, Jainism, Buddhism, Sikhism, Islam and other religions' aesthetic influences, as well as distinctive tribal aesthetics, are all represented in Indian handicrafts, which is a notable illustration of the country's syncretic culture. The wide range of craft products produced by traditional workmanship in historic settings results from centuries of continuous effort fuelled by the diverse contributions of many cultures. The handling techniques of local artisans have developed traditional handicrafts, resulting from the interactions between environmental elements, cultural expressions and practices (Karakul, 2015, 2018). The traditional knowledge and abilities of craftsmen passed down through the master-apprentice relationship from generation to generation have been the primary source of the diversity of traditional handicrafts. Customary workmanship in historically significant settings results from an intricate process wherein artisans apply their knowledge, abilities and inventiveness. India and the rest of the world have impacted Indian crafts and global aesthetics. Not by accident, but rather because of the Indian artisans' dedication to their profession, this age-old relationship has endured for thousands of years. There has always been a profound reverence and spirituality associated with the craftsman's creations, making their work for Indian artisans more than just a means of subsistence. The crafts are categorized as.

- Crafts linked to construction include working with stone, timber frames and mud bricks and crafting using stone, wood, metal, glass, glazed tile and wall painting (Karakul, 2018).
- Meanwhile, culturally rich traditional crafts encompass various skills such as working with stone, wood, metal, glazed tile, glass, making carpets, rugs, cloth, writing tools, tiles, ceramics, pottery, embroidery, leatherwork, crafting musical instruments, copper artistry, basket weaving, felt making and different weaving techniques (Matias et al., 2012).

There has been a tremendous interest in cultural tourism and a shift in travellers to appreciating ethnic and traditional art forms, folklore and crafts. There is a growing emphasis on inclusive growth and sustainability. This has immensely benefitted the rejuvenation of various handicrafts in the country. The unorganized handicraft industry can contribute to the local community's economic growth. Numerous studies offer differing recommendations and fixes for strengthening the regional economy. Concentrating attractions and activities in less developed areas fosters partnerships and collaborations between local and neighbouring communities and catalyses tourism-induced economic stimulus.

4. Challenges Faced by Arts and Craftsmen

The second largest employer in India, after agriculture, is the craft industry, which employs over 3,000 different craftspeople, numbering about 200 million. The handicraft industry is vulnerable to the pandemic because it relies on unofficial and diversified markets and is highly disorganized (Rao & Yadav, 2022). The concept of inclusive growth has been of great interest to economists, government and NGOs, yet the possibilities of the tourism sector to contribute to this by involving people across all sections of economic status remains scanty. The major factors that contribute to the inclusive approach of the tourism sector has been identified by (Bakker, 2020) as.

- Increase in productive employment opportunities.
- Equitable access to this job.
- Equal outcome of these opportunities.

To improve productive employment avenues in tourism, the sector needs to increase the volume and value of direct and indirect employment opportunities. The artisans and craftsmen are confronted with multiple challenges in terms of opportunities for self-employment and access to job opportunities.

4.1 The Current Crafts Sector in India Is Categorized Under Four Verticals

(1) The organization – The artisans and craftsmen still belong to the unorganized sector, with very few associated with formal structures. They still work in informal groups under traders and master craftsmen, which often results in exploitation. This sector has four central units: Self-help groups (SHGs), cooperatives, producer companies and private limited companies.
(2) Procurement and processing of raw materials -Historical linkages between the development of crafts and locally sourced materials meant that raw materials were generally accessible to craftspeople. Craftspeople also had access to communal resources thanks to the jajmani system, which involved a reciprocal interaction between artisanal castes and the larger village community to supply goods and services. Unfortunately, artists now find it challenging to obtain high-quality raw materials at reasonable costs due to the collapse of these old frameworks and corporate industry competition. If raw material banks are not there, they frequently depend on local vendors who supply them with raw materials on demand – albeit at exorbitant costs – or use non-conventional options (Tidd & Bessant, 2015).
(3) Production – Crafts are typically produced in households, with several family members participating in various stages. Craftsmen work in community environments, even where formalized artisan systems are present. Since most craftspeople also work in agriculture to augment their income, production is typically seasonal, with a pause during the harvest season (Raya et al., 2021).
(4) Intermediary trade – India's poor infrastructure and connectivity make it challenging to aggregate items, contributing to many of the current supply chain bottlenecks in the crafts industry. There are insufficient incentives for buyers and merchants to resolve supply-side problems that arise upstream, which prevents artists from taking advantage of market opportunities.

Inclusive and Sustainable development places a high priority on the preservation of cultural resources. Preserving our cultural heritage and reviving old buildings can bring financial benefits. It boosts property values, creates jobs for local craftsmen and promotes heritage tourism. Sadly, crafts have lost their importance due to industrialization. People now see this industry as 'decorative, peripheral and elitist', considering it backwards and inferior (Rasool et al., 2021). The lack of proper laws and support from the government worsens the situation, making craftsmen feel like their trade is dying. At this juncture, The Living Human Treasures System was created by UNESCO as part of the 2003 Convention to protect traditional handicrafts, organize craftspeople and pass on their skills to future generations. The system encourages state governments to set up their national systems. According to UNESCO, the 2003 Convention primarily addresses the knowledge and skills required for craftsmanship rather than the actual craft products (Vlahos, 2023). Safeguarding was only addressed in relation to the activities of craftsmen who produce crafts and teach others how to do the same, not in relation to the preservation of craft objects.

4.2 The Significant Challenges Faced by Artisans and Craftsmen in India Are

- Low productivity and lack of information – Artisans and craftsmen lack access to sophisticated production techniques, equipment and raw materials. Furthermore, many artists operate out of small workshops or residences that do not have the necessary infrastructure or current tools (Bhakshi, 2023). They also lack access to information on funding and market trends; the exclusion from mainstream information results in heavy debts and poor livelihoods for the artisans.
- Disintegrated value chain – The sector is highly unorganized with unscrupulous exploitation by middlemen. (Verma & Shekhawat, 2023) in her study on chikankari workers, she points out that This sector is rife with middlemen who unfairly take advantage of the circumstances and continue to exploit craftsmen. They take direct contracts from the factory owners and marketers and pay them menial wages. Not even 5% of the sale price is given to the women artisans. An artisan makes about 700 INR monthly, whereas a middleman makes about 20,000 INR monthly.
- Lack of successors: Certain crafts are in danger of extinction as the next generation is not interested in taking up these skills due to poor economic advantage; Aranmula Kannandi, a unique mirror made of metal, is almost disappearing as the family is not willing to part with the knowhow to outsiders and the younger generation in the family has not been keen on carrying on the tradition of making these mirrors. Their inclination to pursue other trades stems from their parents' struggles to find markets and reasonable product pricing. Secondly, they are being pushed towards lower-paying white-collar jobs rather than teaching and skilling them on the value of crafts this needs to be incorporated into the curriculum. Further, in India, specific castes are closely related to certain crafts. Rural youth are discouraged from entering the family trade because they face social exclusion and stigma, like bamboo products, leatherwork, pottery, etc., where they hail from the lowest caste (Deb, Nafi et al., 2022).

- Poor Marketing and pricing strategies: The current state of fragmented data on the production of handicrafts and artisans in rural villages in India is a significant hurdle to developing a streamlined marketing strategy with an organized system. The market and lifestyle integration of crafts and arts is yet unexplored to its fullest potential; hence, market research is very scanty. New designs, trends, technology and product positioning beyond a collector's item or a showpiece, a gifting item should be replaced to get more reach for handicrafts produced (Hossain et al., 2022). End-to-end value chain integration involving Marketing research, product development, distribution, pricing, Packaging, advertising, sales promotion and branding are tasks involved in marketing (Abirami et al., 2017). Handicraft businesses cannot afford to establish a distinct, well-organized marketing system since they are typically conducted by either artisans or the community of artisans and craftsmen using a proprietary type of organization or else middlemen who exploit them. Craftsmen have very little knowledge of pricing strategies and fix prices on cost-plus pricing strategy where creative elements, novelty and unique designs, which are the intellectual property of the artisan, are not considered while pricing the products, thereby fetching poor prices even in the export market.

Numerous emerging nations understand that the growth of artisan crafts can significantly impact regional development. Nevertheless, they do not have a clear policy framework to direct the development activity. Moreover, there is typically no explicit decision regarding the division of ministerial duties to promote artisan craft. Governments in developing nations frequently struggle to set up an effective ministerial coordination mechanism because of the wide range of development issues (Varghese & Chennattuserry, 2022), which include but are not limited to the eradication of poverty, rural development, preservation of traditional cultural references, promotion of export, human resource development and technological innovation.

5. Pricing, Market Strategies and Institutional Support to Promote Handicrafts

In a world where machines are taking over, the crafts industry is the only one that recognizes the value of human engagement. Its reliance on skilled craftsmanship, creativity and human interaction has allowed it to remain relevant and distinctive in the contemporary world. Since the beginning of Indian civilization, the country's rural market has developed organically with community participation.

While natural and material cultural heritage has always piqued tourists' interest, living cultures, lifestyles and traditional craftsmanship have also captured attention recently. Traveller mobility to experience local customs, celebrations, traditional craftsmanship and cultural practices has increased recently, coinciding with a rise in concern for intangible cultural heritage elements (Karakul, 2019).

The emergence of international online retailers presents a huge opportunity to support the resurgence of Indian design, crafts and arts. It will not only help karigars become more tech-savvy but also open a worldwide market for the very last karigar.

He will not need an intermediary since he can provide direct customer service to online shoppers. Nonetheless, the government must also make sure that individual craftsmen are aware of the standards, laws and regulations about exports and provide them with information and training on online marketplaces (Varghese et al., 2024) The international positioning of Indian arts, crafts and design will be aided by access to commercial and urban areas, exhibitions and fairs organized according to international standards, stimulating global interest globally through associations and supporting community companies.

A growing number of historical villages are being developed to attract tourists who want to see the production of goods and the local culture. A few heritage villages are already effective models of the communal spirit of the craftsman. The governmental initiatives to support artisans have improved with a few interventions like.

- Ambedkar Hastshilp Vikas Yojana: This government scheme to aid in crafts people's infrastructure, technological and human resource development requirements. To facilitate bulk production and economies in acquiring raw materials, the goal is to mobilize artisans into self-help organizations and societies.
- Mega Cluster Scheme: Creating jobs and raising the living standards of craftsmen are two of the programme's goals. This programme uses a cluster-based methodology to scale infrastructure and production chains at handicraft centres, particularly in rural areas.
- Marketing Support and Services Scheme: Through financial support, this programme helps craftsmen organize and participate in trade shows and exhibitions domestically and internationally. It also offers interventions for domestic marketing events.
- Research and Development Scheme is initiated to collect feedback on social, economic, aesthetic and promotional aspects of handicrafts, artisans and craftsmen.
- Comprehensive Handicrafts Cluster Development Scheme: This plan expands handicraft clusters' production chain and infrastructure. This plan also attempts to provide sufficient infrastructure for value addition, production and quality control.

Public policies that promote cooperation should be viewed as a strategic element that enhances the activity by improving social inclusion, job creation, income and the artisan's standard of living (Sehnem et al., 2020). Intangible cultural heritage, including traditional crafts in particular, can provide income and jobs for living people, even the impoverished. Craftsmen and their families, as well as others engaged in manufacturing raw materials needed for craft production and transporting and selling finished goods, all benefit financially from traditional craftsmanship. Given the significance of climate change and the need for environmentally friendly products, it is now appropriate for Indian arts, crafts and design to have a central role. Today's market is filled with toxic produce, whereas Indian handcrafted goods are entirely recyclable, renewable, biodegradable and reasonably priced. Transitioning to handcrafted products is necessary to build an inclusive society and ethically satisfy customer needs (World Tourism Organization, 2021).

6. Wellbeing Spectrum and Quality of Life of Artisans

The socio-economic well-being of artisans requires favourable working conditions, a fair standard of living, good pay and job prospects that advance socio-economic well-being (Irvansyah et al., 2023). Handcrafted goods are manufactured by skilled labourers and artisans, using traditional methods or craft skills (Chatterjee et al., 2017). The fieldwork indicated that the absence of unprepared and uninterested next-generation heirs does not demonstrate a wholehearted passion for continuing the expert workmanship and, thereby, causes the experienced craftsmen to become demotivated (Ahamed & Karim, 2019). The socio-economic well-being of craftsmen depends on the particular environment, locality and kind of artisan labour engaged (Ma & Mohame, 2023). Improving the well-being of craftsmen requires tackling several socio-economic issues, including income, education, working conditions, market access and cultural preservation (Medeiros et al., 2023). The roots of artisan engagement are deeply rooted and found in cultural customs (Afriyie et al., 2023). The preservation and upkeep of traditional practices may positively impact the well-being of craftspeople (Liu et al., 2023). Acknowledging and valuing traditional handicrafts can also boost a sense of pride and identity, thereby improving the destination's reputation and helping it carve out a niche (Aula & Masoodian, 2023).

Artisanal products and crafting methods can become more efficient less expensive and offer new opportunities by embracing innovation and technology (Dolapoglu et al., 2023). Productivity and competitiveness can be increased by teaching craftspeople how to use contemporary technologies and innovative methods (Brillet et al., 2023). Collaboratively, governments, non-governmental organizations, corporations and consumers may guarantee a more equitable and sustainable future for artisanal communities (Afrifah & Mensah, 2023). Achieving socio-economic well-being encourages a multi-faceted, integrated strategy involving partnerships and cooperation between the public sector, business organization, local communities and other related groups. It is imperative to consistently evaluate and modify tactics to effectively tackle changing obstacles and prospects within the socio-economic context (Ahamed & Karim, 2019). Nonetheless, an inclusive engagement approach can drive economic status by integrating standard procedures and approaches to improve socio-economic well-being (Irvansyah et al., 2023). These engagements extend a fair and equitable development while empowering the small-time local communities. Fig. 13.1 shows an Inclusive Socio-Economic Wellbeing Process.

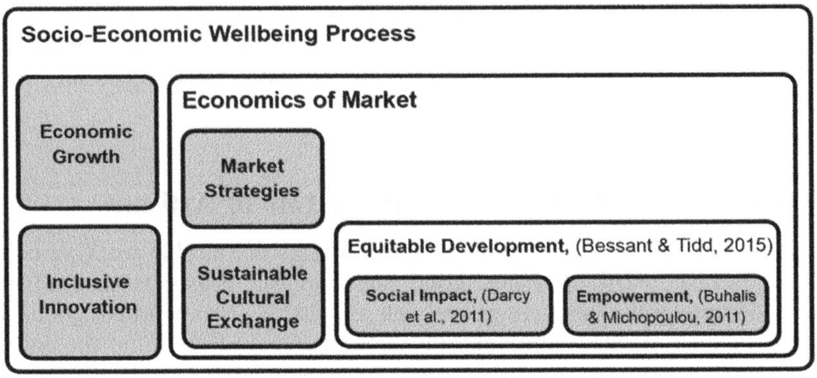

Fig. 13.1. An Inclusive Socio-Economic Wellbeing Process. *Source*: Complied by authors.

7. Case Study – 1: Communities and Social Wellbeing: Fostering Socio-Economic

Upliftment With Inclusive Engagement – A Case Study on Channapatna, Karnataka, India

Through boosting economic opportunities, tourism can assist localities in achieving financial stability. The demands of the sector for a means of subsistence may be addressed through the implementation of a framework for inclusive involvement. Being financially independent can benefit the artisans in the local community and facilitate social and economic empowerment.

Karnataka, India's one state, encourages the local communities to encourage and create a market for themselves with locally sourced wooden toys. The approach personifies a socially conscious and responsible mechanism to boost its economy. It thrives when it participates maximally in inclusive business environments with an entrepreneurial focus on financial independence and economic survival (Varghese et al., 2022). Channapatna, a small town in Karnataka, fosters inclusive participation via toy manufacturing units which are community-led and channelised by empowered teams laying its operations which guarantee business dealings with a broad spectrum of visitors while advancing regional culture, upholding community values and improving general well-being. This is how the location accomplishes inclusivity.

- Economic opportunity and Inclusivity are a philosophy ensured by businesses that are also reachable by individuals with reachability and locational advantages uphold the economic opportunity and inclusivity ideology. The services are rendered with outstanding professionalism and competence when access is granted through aggregators such as networked collaborators and suppliers. The need for informal economies drives the local communities to engage in inclusion.

- Local economies can inculcate polite interactions to foster cultural sensitivity, providing visitors with information about regional customs, traditions and etiquette.
- Local community engagement facilitates interconnectedness while working with local artists, performers and organisations to offer authentic cultural experiences to the guests, such as workshops, local markets, or traditional performances. Locally sourced cuisines, handcrafted crafts and other commodities from nearby places can aid and support small local businesses and boost the local economy.

Community-based organisations (CBOs) and SHGs can drive the community's social and cultural framework while generating some cash for businesses. The Channapatna community with artisans provides travellers with experiential tourism with life-changing experiences while improving the neighbourhood, promoting cross-cultural exchange and developing morally and environmentally responsible tourism practices by emphasising inclusive involvement.

The Ideology of Inclusivity

The local communities, often marginalised or from excluded populations, are among the many people who are to be guided as empowered teams for supporting and improving the quality of life. The local market with local communities whose needs are addressed by a fundamental aspect of the inclusive innovation approach is pivotal. As new ideas, products, services and policies emerge, the community gains financial independence and empowerment. The ultimate goal of inclusive innovation is to guarantee that everyone can profit from innovation, regardless of their financial situation, gender, age, ethnicity, physical limitations, or any other attribute.

Empowerment and Community Engagement

Informal economies can generate revenue for local artisans, markets, restaurants and other small businesses in addition to themselves. These businesses create job

(*Continued*)

(*Continued*)

opportunities that are both direct and indirect, enabling the local community to work in the tourism industry and raise their level of living. Small-time local businesses can be aided to provide opportunities for visitors and locals to engage culturally.

While tourists can learn about local customs, habits and traditions, localities can get insights into different civilisations. The preservation and promotion of cultural heritage are facilitated by tourism businesses, which often showcase local architecture, crafts and rituals.

Source: Compiled by the authors

8. Discussion

Strategic planning and execution are necessary for rejuvenating, resuscitating and repositioning arts and crafts through pricing (Maliao et al., 2023). It can be helpful to identify the target audience and their preferences to understand better the current market trends and demands for arts and crafts (Metaragakusuma et al., 2023). To serve a broad spectrum of customers, provide various products at various price points (Mensah-Abludo et al., 2023). Offer bundled packages to entice buyers to purchase more than one item. To raise the perceived value of some crafts, emphasis should be given to their uniqueness and rarity (Ngonso & Egielewa, 2023). Craftsmen can emphasize the skill and creativity behind each piece by closely examining the pricing strategies of competitors in the market and concentrating on distinctive and superior products that stand out in the marketplace (Prince, 2023). A fair profit margin must be considered when carefully evaluating the production costs, which include labour, materials and overhead, as well as the time and expertise required for each craft (Singh et al., 2023). Establish pricing tiers according to the size or complexity of the craft (Su et al., 2023). In order to boost sales volume, provide discounts on large purchases (Slak Valek & Mura, 2023). Introduce limited-edition merchandize to establish a feeling of uniqueness (Skibins et al., 2023). Incorporating contemporary styles into regional arts and crafts has the potential to stimulate economic development in local communities. Cross-cultural influences can give traditional arts new life and stimulate a lively exchange of ideas (Kaur et al., 2023). This promotes cultural exchange and appreciation by opening up new markets for these traditional crafts and broadening the creative horizons of local artisans.

8.1 Outcome and Implications

Traditional crafts create new markets and opportunities for economic growth as they adapt to meet contemporary demands. The concept of Craft village has been getting popular in India. Kerala Arts and crafts Village is a centre for art, culture and entertainment designed and built with a number of goals, including the improvement of indigenous technology and artisan capabilities, product diversification and the guarantee of a respectable living for the craftsmen through market expansion. Nettoorpetti, glass and stone crafts, dry flowers, artistic paper-to-metal creations, horn carvings, hammocks and jewellery and pooram crafts date back hundreds of years. Teak, rosewood and white wood are used to make toys, furniture and laminated wooden crafts. The only locations where one can purchase Kerala's traditional crafts with geographical indicator status are Kerala Arts and Craft Village and Sargalaya. The unique products are presented along with live demos of production offering the visitors a special opportunity to view all in a single place and appreciate not just the artefacts but the whole process of making which is a testimony of creativity, craftsman ship and perseverance. Similar artisans' village like Cholamandala artist village in Tamil Nadu, Pilikula artisans village in Karnataka create experiential space for people to appreciate artefacts and the whole process of making handicrafts. Thereby reviving dying craft forms and creating employment opportunities for artisans. These projects will create financial inclusion as they connect artisans with the markets aiding with better technology, trends and designs to improve the marketability and reach of the artefacts and crafts.

The nation's creative artisans showcase their expertise via stunning designs of permanent sculptures, such as 3D Holograms, Pattachitra paintings of Bengal and Odisha, Kerala murals and the vibrant paper quilling crafts that date back thousands of years to ancient Egypt. Craftsmen and artisans can reach a wider audience, boost sales and create a more sustainable lifestyle using modern designs, materials and marketing techniques. This increase in revenue benefits local artisans directly as well as the community's general well-being. Digital marketing and developing a visible presence online are excellent marketing strategies to attract potential customers globally (Tedjokoesoemo et al., 2023). Create a robust online presence on social media and with a website (Wilson-Nash et al., 2023). E-commerce and digital platforms allow artisans to present their creations to a worldwide clientele (Afriyie et al., 2023). Technology can also improve production, increasing efficiency while preserving the authenticity of the craft (Singh et al., 2023). Modern tools and methods can expedite production, guaranteeing a better product while requiring less time and effort. Present the creative process and the history of the crafts. Use digital marketing techniques to connect with more people (Boonyasurat et al., 2023). Create a community around your brand by interacting with clients via newsletters, events and social media (Maliao et al., 2023). Promote client endorsements and reviews to foster trust. Keep abreast of consumer preferences and industry developments. Use consumer feedback to enhance your offerings and pricing plan (Ohara et al., 2023).

9. Conclusion and Future Directions

Multi-faceted initiatives are necessary to improve the socio-economic status of rural artisans. Implementing strategic interventions that include skill development, equitable market access and a sensible pricing structure, in addition to brushstrokes and chisel marks, is imperative. Subsequent investigations ought to concentrate on cultivating inventiveness, safeguarding cultural legacy and advocating for sustainable growth in artisanal communities. Education plays a pivotal role in ensuring the longevity of local arts and crafts. Integrating new trends involves educating artisans and the broader community about adapting to contemporary demands while preserving cultural heritage. Through the examination of skill development, market accessibility, financial inclusion and the ecological consequences of artisanal practices, scholars can enhance our comprehensive comprehension of the forthcoming obstacles and prospects. Workshops, training programs and educational campaigns can bridge the gap between tradition and modernity, empowering artisans with the skills and knowledge needed to navigate the changing landscape. With this information, policies and initiatives that support rural artisans can be developed, guaranteeing a prosperous and long-lasting future for this essential part of India's cultural fabric. Beyond helping the artisans financially, empowering them also involves protecting their cultural identities, building community resilience and appreciating their priceless contributions as stewards of tradition. By incorporating new trends, traditional craftsmen and modern designers can work together to create a dynamic synergy revitalising craft.

References

Abirami, P., Velavan, M., Arunkumar, S., Anand, V. V., Sivasumbramanian, J., & Abirami, G. (2017). Indian handicrafts and its challenges faced by artisan community. *International Journal of Economic Research*, *14*(7), 431–438.

Adeniyi, O., Kumeka, T. T., Orekoya, S., & Adekunle, W. (2023). Impact of tourism development on inclusive growth: A panel vector autoregression analysis for African economies. *Tourism Economics*, *29*(3), 612–642. https://doi.org/10.1177/13548166211061154

Afrifah, M., & Mensah, J. (2023). Diaspora tourism and homeland development: Exploring the impacts of African American tourists on the livelihoods of local traders in Southern Ghana. *African Geographical Review*, *42*(2), 125–140. https://doi.org/10.1080/19376812.2021.1997612

Afriyie, R. Z., Arthur, E. K., Gikunoo, E., Baah, D. S., & Dziafa, E. (2023). Potential health risk of heavy metals in some selected vegetable crops at an artisanal gold mining site: A case study at Moseaso in the Wassa Amenfi West District of Ghana. *Journal of Trace Elements and Minerals*, 100075. https://doi.org/10.1016/j.jtemin.2023.100075

Ahamed, D. M., & Karim, R. (2019). Impact of rural tourism product on the socio-economic condition of Artisans in Birbhum District: A case study on Kantha Stitch handicraft. *Research Directions*, Special Issue-June, 153–162.

Anaba, E., & Cole-Adeife, O. (2023). Keloid disease: Prospcetive cross-sectional study of clinicoepidemiological and health related quality of life at a tertiary center. *Annals of Clinical Sciences*, *8*(1), 20–27.

Ashley, C., & Mitchell, J. (2009). *Tourism and poverty reduction: Pathways to prosperity*. Taylor & Francis.

Aula, I., & Masoodian, M. (2023). Creativity and healthy ageing: Future research directions. *Journal of Population Ageing*, 1–21. https://doi.org/10.1007/s12062-023-09431-8

Bagnall, A. M., & Southby, K. (2023). *Systematic review of community infrastructure (place and space) to boost social relations and community wellbeing: Five year refresh*.

Baimuratova, S., & Chhabra, D. (2023). Slow food tourism and quality of life: The social capital perspective. In *Handbook of tourism and quality-of-life research II: Enhancing the lives of tourists, residents of host communities and service providers* (pp. 383–395). Springer International Publishing. https://doi.org/10.1007/978-3-031-31513-8_26

Bakker, M. (2020). A conceptual framework for identifying the binding constraints to tourism-driven inclusive growth. *Tourism Planning & Development*, *16*(5), 575–590. https://doi.org/10.1080/21568316.2018.1541817

Bakker, M., van der Duim, R., Peters, K., & Klomp, J. (2023). Tourism and inclusive growth: Evaluating a diagnostic framework. *Tourism Planning & Development*, *20*(3), 416–439. https://doi.org/10.1080/21568316.2020.1850517

Bakshi, R. (2023). *Empowering Indian artisans for an inclusive & sustainable economic growth*. https://www.businessworld.in/article/Empowering-Indian-Artisans-For-An-Inclusive-Sustainable-Economic-Growth-/05-05-2023-475564/

Bhakshi, R. (2023). Indian artisans are in trouble. *Assistthem*. https://www.hindustantimes.com/opinion/indian-artisans-are-in-trouble-assist-them-101688136788139.html

Bobo, B., & Akhurst, J. (2023). "Okokuqala ngokuya ndandiqala kwakungekho easy": Feeling empowered to take collective action through community engagement. *Journal of Community & Applied Social Psychology*, *33*(2), 425–435. https://doi.org/10.1002/casp.2659

Bone, J. K., Fancourt, D., Fluharty, M. E., Paul, E., Sonke, J. K., & Bu, F. (2023). Associations between participation in community arts groups and aspects of wellbeing in older adults in the United States: A propensity score matching analysis. *Aging & Mental Health*, *27*(6), 1163–1172. https://doi.org/10.1080/13607863.2022.2068129

Boonyasurat, W., Piriyatachagul, G., Thammawong, P., & Niratsayaputi, N. (2023). Crafts and folk art tourism development in Chiang Mai towards UNESCO creative cities network membership. *Journal of Sustainable Development*, *15*(4), 1–84. https://ideas.repec.org/a/ibn/jsd123/v15y2023i4p84.html

Brillet, L., Peries, M., Vernhet, C., Rattaz, C., Michelon, C., & Baghdadli, A. (2023). Dyadic effect of coping on the perceived impact of ASD of children on parental quality of life: Report from the ELENA cohort. *Journal of Autism and Developmental Disorders*, *53*(1), 38–49. https://doi.org/10.1007/s10803-021-05414-w

Buhalis, D., & Michopoulou, E. (2011). Information-enabled tourism destination marketing: Addressing the accessibility market. *Current Issues in Tourism*, *14*(2), 145–168. https://doi.org/10.1080/13683501003653361

Chatterjee, R., Mukherjee, D., Chakraborty, G., & Deb, A. K. (2017). Factors influencing growth of rural entrepreneurship in Tripura: A socio-economic perspective. *The Journal of Innovations*, *1*, 47–57.

Commission on Growth. (2008). *The growth report: Strategies for sustained growth and inclusive development*. World Bank Publications.

Cultural tourism market Outlook. (2023–2033). https://www.futuremarketinsights.com/reports/cultural-tourism-market

Custodio, H. M., Hadjikakou, M., & Bryan, B. A. (2023). A review of socioeconomic indicators of sustainability and wellbeing building on the social foundations framework. *Ecological Economics*, *203*, 107608. https://doi.org/10.1016/j.ecolecon.2022.107608

Darcy, S., Cameron, B., & Pegg, S. (2011). Developing a business case for accessible tourism. *Accessible Tourism: Concepts and Issues*, 241–259.

Datta, D. B., & Bhattacharyya, S. (2016). An Analysis on problems and prospects of Indian handicraft sector. *Asian Journal of Management*, *7*(1), 5–16. https://doi.org/10.5958/2321-5763.2016.00002.0

Deb, S. K., Mohanty, P. P., & Valeri, M. (2022). Promoting family business in handicrafts through local tradition and culture: An innovative approach. *Journal of Family Business Management*. https://doi.org/10.1108/JFBM-10-2021-0131

Deb, S. K., Nafi, S. M., & Valeri, M. (2022). Promoting tourism business through digital marketing in the new normal era: A sustainable approach. *European Journal of Innovation Management*.

Debbarma, S. (2020). *A study on the socio economic status of rural handicraft artisans in Hezamara RD Block West Tripura District, Tripura*. https://repository.tribal.gov.in/handle/123456789/74224

Dey, P., Ray, N., Dash, C. K., Das, D. K., & Das, P. K. (2020). *Mapping of socio-economic condition and occupational structure of patuas in the context of Pattachitra (scroll painting): A study from Bharatpur Village in Bankura West Bengal, India*. ISSN: 1746 1278 (153).

Dibyendu, B. D., & Bhattacharyya, S. (2016). An analysis on problems and prospects of Indian handicraft sector. *Asian Journal of Management*, *7*(1), 5–16. https://doi.org/10.5958/2321-5763.2016.00002.0

Dolapoglu, N., Ozcan, D., Tulaci, R. G., & Tulacı, R. G. (2023). Is orthorexia nervosa a non-specific eating disorder or a disease in the spectrum of obsessive-compulsive disorder? *Cureus*, *15*(5).

Dugan, R., Ubal, V. O., & Scott, M. L. (2023). Sales well-being: A salesperson-focused framework for individual, organizational, and societal well-being. *Journal of Personal Selling and Sales Management*, *43*(1), 65–83. https://doi.org/10.1080/08853134.2022.2093733

Hamzah, R., Meliza, J., Adelia, I., Siagian, R., & Zainuddin, Z. (2023). Acceleration of tourism development in the process of achieving an increase in the number of tourists. *Jurnal*, *1*(1), 26–29. https://ijosthes.org/index.php/j/article/view/12

Hossain, M. R., Akhter, F., Sharma, A., & Hassan, A. (2022). Thirty years of research on application of technology in tourism and hospitality industry: A systematic literature review. *Technology Application in Tourism in Asia: Innovations, Theories and Practices*, 3–36. https://doi.org/10.1007/978-981-16-5461-9_1

Hunjra, A. I., Azam, M., Bruna, M. G., & Taskin, D. (2022). Role of financial development for sustainable economic development in low middle income countries. *Finance Research Letters*, *47*, 102793.

Indian handicrafts industry and exports report. (2023). https://www.ibef.org/exports/handicrafts-industry-india

Irvansyah, A., Muljono, P., Fatchiya, A., & Sadono, D. (2023). Social entrepreneurship competence of non-formal education managers for the empowerment of learning communities. *Journal of Nonformal Education, 9*(1), 179–189. https://doi.org/10.15294/jne.v9i1.42727

Jamatia, P. L. (2023). The role of youth in combating social inequality: Empowering the next generation. *International Journal of Social Science, Educational, Economics, Agriculture Research and Technology, 2*(8), 229–238.

Jordan, M. (2023). The power of connection: Self-care strategies of social wellbeing. *Journal of Interprofessional Education & Practice, 31*, 100586. https://doi.org/10.1016/j.xjep.2022.100586

Kanungo, P., Sethi, N., & Biswal, P. (2021). Socio-economic condition, welfare schemes, and occupational structure of 'pattachitra' artisans in Odisha, India. *Creative Industries Journal, 14*(1), 81–106. https://doi.org/10.1080/17510694.2020.1775029

Karakul, Ö. (2015). An integrated methodology for the conservation of traditional craftsmanship in historic buildings. *International Journal of Intangible Heritage, 10*, 135–144.

Karakul, Ö. (2018). Designing the museum of turkey's intangible cultural heritage: A studio experience of interior architectural design. *Milli Folklor, 30*(120).

Karakul, Ö. (2019). The effects of tourism on traditional craftsmanship for the sustainable development of historic environments. *European Journal of Sustainable Development, 8*(4), 380. https://doi.org/10.14207/ejsd.2019.v8n4p380

Karthik, A., Jeniffer, M., & Chandran, R. A. (2023). Multi-dimensional strands of rural tourism in AlankuppamA stimulus for destination branding. *International Research Journal of Economics and Management Studies IRJEMS, 2*(1). https://doi.org/10.56472/25835238/IRJEMS-V2I1P113

Kaur, J., Kaur, K., & Singh, R. (2023). Adopting technology for sustainable development: Reflections on innovative ecosystem. In *Fostering sustainable development in the age of technologies* (pp. 93–111). Emerald Publishing Limited.

Keane, M. J. (1996). Sustaining quality in tourism destinations: An economic model with an application. *Applied Economics, 28*(12), 1545–1553. https://doi.org/10.1080/000368496327525

Khaledi Koure, F., Hajjarian, M., Hossein Zadeh, O., Alijanpour, A., & Mosadeghi, R. (2023). Ecotourism development strategies and the importance of local community engagement. *Environment, Development and Sustainability, 25*(7), 6849–6877. https://doi.org/10.1007/s10668-022-02338-y

Kumari, A., Brahmi, M. K., & Thakur, K. S. (2019). Socio-economic status of forest-based rural artisans in mid-hill zone of Himachal Pradesh. *International Journal of Farm Sciences, 9*(2), 103–108. http://doi.org/10.5958/2250-0499.2019.00053.3

Lee, N. (2019). Inclusive growth in cities: A sympathetic critique. *Regional Studies, 53*(3), 424–434. https://doi.org/10.1080/00343404.2018.1476753

Liu, Y., Groot, B., de Kock, L., Abma, T., & Dedding, C. (2023). How participatory arts can contribute to Dutch older adults' wellbeing – Revisiting a taxonomy of arts interventions for people with dementia. *Arts & Health, 15*(2), 153–168. https://doi.org/10.1080/17533015.2022.2035417

Ma, S. N. Z., & Mohame, B. B. (2023). Heritage tourism and cultural identity in China: Challenges and opportunities for development. *International Journal of Professional Business Review, 8*(7), 47.

Maliao, R. J., Fernandez, P. R., & Subade, R. F. (2023). Well-being of artisanal fishing communities and children's engagement in fisheries amidst the COVID-19 pandemic: A case in Aklan, Philippines. *Humanities and Social Sciences Communications, 10*(1), 1–12. https://doi.org/10.1057/s41599-023-01716-9

Matias, Á., Nijkamp, P., & Sarmento, M. (Eds.). (2012). *Quantitative methods in tourism economics*. Springer Science & Business Media.

Medeiros, T., Costa, S. G., Mendes, J., Tomás, L., & Ferreira, J. A. (2023). Senior entrepreneurship, sustainable tourism, and preservation of collective memory. In *Advances in Tourism, Technology and Systems: Selected Papers from ICOTTS 2022* (Vol. 2, pp. 193–205). Springer Nature Singapore. https://doi.org/10.1007/978-981-19-9960-4_18

Mensah-Abludo, E., Peprah, K., & Baddianaah, I. (2023). Application of triple bottom line concept to artisanal and small-scale mining activities in Wa East District, Ghana. *Journal of Degraded & Mining Lands Management, 10*(3). https://doi.org/10.15243/jdmlm.2023.103.4547

Metaragakusuma, A. P., Sakakibara, M., Arifin, Y. I., Pateda, S. M., & Jahja, M. (2023). Rural knowledge transformation in terms of mercury used in artisanal small-scale gold mining (ASGM) – A case study in Gorontalo, Indonesia. *International Journal of Environmental Research and Public Health, 20*(17), 6640. https://doi.org/10.3390/ijerph20176640

Ngonso, B. F., & Egielewa, P. E. (2023). Socio-economic power of social media: An examination of social media usage by rural artisans in selected villages in Nigeria. *International Journal of Economics, Commerce, and Management, 11*(7), 178–192.

Ohara, E. E., Ogar, I. F., Felix, A. C., & Okon, I. O. (2023). Effect of women participation in artisanal fishing on their socio-economic well-being in coastal communities in cross river state, Nigeria. *Global Journal of Agricultural Sciences, 22*(1), 93–100. https://www.ajol.info/index.php/gjass/article/view/257703

Penhallurick, R. D. (2023). *Tin in antiquity: Its mining and trade throughout the ancient world with particular reference to Cornwall*. Taylor & Francis.

Prince, S. (2023). 3 Rural authenticity and agency on of contemporary craft-artists on Bornholm, Denmark. *Islandscapes and Tourism: An Anthology, 36*.

Rao, R. H. S., & Yadav, C. K. (2022). *Grama Shilpi – The 'Haat' of North Karnataka*. https://www.iimb.ac.in/turn_turn/grama-shilpi%E2%80%93haat-north-karnataka.php

Rasool, H., Maqbool, S., & Tarique, M. (2021). The relationship between tourism and economic growth among BRICS countries: A panel cointegration analysis. *Future Business Journal, 7*. https://doi.org/10.1186/s43093-020-00048-3

Rauniyar, G., & Kanbur, R. (2010). Inclusive growth and inclusive development: A review and synthesis of Asian Development Bank literature. *Journal of the Asia Pacific Economy, 15*(4), 455–469. https://doi.org/10.1080/13547860.2010.517680

Raya, A. B., Andiani, R., Siregar, A. P., Prasada, I. Y., Indana, F., Simbolon, T. G. Y., ... Nugroho, A. D. (2021). Challenges, open innovation, and engagement theory at craft smes: Evidence from Indonesian batik. *Journal of Open Innovation: Technology, Market, and Complexity, 7*(2), 121. https://doi.org/10.3390/joitmc7020121

Sánchez-Rivero, M., Pulido-Fernández, J. I., & Cárdenas-García, P. J. (2013). In A. Matias (Ed.), *Tourism growth versuseconomic development: An analysis by multivariate techniques*. https://doi.org/10.1007/978-3-7908-2879-5_13

Sehnem, S., Piekas, A., Dal Magro, C. B., Fabris, J., & Leite, A. (2020). Public policies, management strategies, and the sustainable and competitive management model in handicrafts. *Journal of Cleaner Production*, *266*, 121695. https://doi.org/10.1016/j.jclepro.2020.121695

Singh, A. S., Parahoo, S. K., Ayyagari, M., & Juwaheer, T. D. (2023). Introduction: How could rural tourism provide better support for wellbeing and socio-economic development? *Worldwide Hospitality and Tourism Themes*, *15*(1), 3–7. https://doi.org/10.1108/WHATT-08-2022-0100

Skibins, J. C., Das, B. M., & Schuler, G. (2023). Digital modalities, nature, and quality of life: Mental health and conservation benefits of watching bear cams. *Human Dimensions of Wildlife*, *28*(3), 218–232. https://doi.org/10.1080/10871209.2021.2024629

Slak Valek, N., & Mura, P. (2023). Art and tourism – A systematic review of the literature. *Tourism Review*, *78*(1), 273–290. https://doi.org/10.1108/TR-05-2022-0214

Snyman, S. L. (2012). The role of tourism employment in poverty reduction and community perceptions of conservation and tourism in southern Africa. *Journal of Sustainable Tourism*, *20*(3), 395–416. https://doi.org/10.1080/09669582.2012.657202

Stoddard, J. E., Evans, M. R., & Shao, X. (2012). Marketing arts and crafts: Exploring the connection between hedonic consumption, distribution channels, and tourism. *International Journal of Hospitality & Tourism Administration*, *13*(2), 95–108. https://doi.org/10.1080/15256480.2012.669308

Su, M. M., Wall, G., Ma, J., Notarianni, M., & Wang, S. (2023). Empowerment of women through cultural tourism: Perspectives of Hui minority embroiderers in Ningxia, China. *Journal of Sustainable Tourism*, *31*(2), 307–328. https://doi.org/10.1080/09669582.2020.1841217

Tedjokoesoemo, P. E. D., Nilasari, P. F., & Sari, S. M. (2023). *Addressing the independent learning curriculum (Kurikulum Merdeka Belajar) as a form of positive disruption to empower the community*. Doctoral dissertation, SciTePress. https://www.scitepress.org/PublicationsDetail.aspx?ID=hYKD2sApP7g=&t=1

Tidd, J., & Bessant, J. (2015). *Gestão da inovação-5*. Bookman Editora.

Varghese, B., & Chennattuserry, J. C. (2022). Community resilience and crisis management: Stakeholders perspective of the tourism industry. In A. Hassan, A. Sharma, J. Kennell, & P. Mohanty (Eds.), *Tourism and hospitality in Asia: Crisis, resilience and recovery*. Springer. https://doi.org/10.1007/978-981-19-5763-5_2

Varghese, B., Chennattuserry, J. C., & Kureethara, J. V. (2022). Sinking houseboats and swaying home stays: Community resilience and local impacts of COVID-19 in managing tourism crisis in Kerala. *International Journal of Tourism Cities*, *8*(4), 805–819. https://doi.org/10.1108/IJTC-01-2021-0004

Varghese, B., Joseph, E. K., Lakshmypriya, K., Kallarakal, T. K., & Mehta, H. (2024). Analogy of social entrepreneurship and community empowerment: An inclusive tourism approach with technological intervention. In A. Sharma (Ed.), *International handbook of skill, education, learning, and research development in tourism and hospitality*. Springer International Handbooks of Education. Springer. https://doi.org/10.1007/978-981-99-3895-7_11-1

Verma, V., & Shekhawat, K. (2023). Wellness tourism motivates tourists to participate. *Journal of Mines, Metals and Fuels*, *71*(2).

Vlahos, N. (2023). *Reclaiming participatory governance: Social movements and the reinvention of democratic innovation*. Routledge.

Wilson-Nash, C., Pavlopoulou, I., & Wang, Z. (2023). Selecting, optimizing, and compensating during lockdown: How older consumers use social networking services to improve social well-being. *Journal of Interactive Marketing*, *58*(2–3), 301–320. https://doi.org/10.1177/10949968231155156

World Tourism Organization. (2021, April 9). https://www.unwto.org/news/international-tourism-arrivals-could-fall-in-2020. Accessed on April 10, 2021.

Yan, M. (2023). *Protection and development of arts and crafts from the perspective of environment and public health*. https://doi.org/10.38007/IJAID.2023.040102

Zainuri, A., & Huda, M. (2023). Empowering cooperative teamwork for community service sustainability: Insights from service learning. *Sustainability*, *15*(5), 4551. https://doi.org/10.3390/su15054551

Chapter 14

Halal Tourism as a New Strategy for Sustainable Tourism

Fathimah Nachia Syed and Narentheren Kaliappen

Universiti Utara Malaysia, Malaysia

Abstract

The research about the concept of Halal tourism has emerged recently. It becomes the vital factor in determining the tourists's satisfaction or their loyalty. Thus, this concept needs to be developed in certain areas because it still not acknowledged. The purpose of this study is to align the theoretical foundations of Halal tourism to conventional tourism paradigms. It investigates the value of Muslim tourists' perception in the context of Malaysian tourist destination. The six variables of Muslim tourist perceived value (MTPV) are examined such as quality, price, emotional, social, physical and non-physical attributes. The respondents of the research are a total of 205 Muslim tourists in Langkawi, Malaysia during October 2021 (Langkawi Tourism Bubble). The results indicated four (4) variables have impacts on tourist satisfaction. Practical implications will impact towards cognitive, affective and Islamic (i.e. Halal) values on tourist satisfaction. It was examined as one of the priotitize destination experience on tourist satisfaction. The findings provide Malaysian tourism with significant managerial implications. It also impacted the Halal tourism as a new approach specifically in the post COVID-19 era. Hence, in improving Muslim tourist satisfaction, destination marketers should consider the Halal tourism. Tour agencies also should scrutinize the product and services value including destination attributes that they offer. The destination's competitiveness will be strengthened with the right destination attraction, facilities, accommodation and activities that suit Muslim tourists.

Keywords: Halal tourism; Muslim tourist perceived value; Muslim-friendly; sustainable tourism; tourist satisfaction; Malaysia

The Emerald Handbook of Tourism Economics and Sustainable Development, 273–292

doi:10.1108/978-1-83753-708-220241014

1. Introduction

Halal means 'permissible' in Arabic. Halal applies to everything in compliance with Islamic law and injection (Sharia) acceptable. Each Muslim, for example, is required to eat only what is Halal and reject what is not. This goes to food and beverages, products and services that Muslims often use must comply with Halal standards and Sharia legislation. World Travel Market (2007) described the Halal tourism as a sort of sacred tourism that is compliant with and obedience to Islamic teaching. It providing destinations appropriate for Muslim tourists that can provide Halal services and products. Halal tourism provides Muslim tourists with an alternative option as it keeps reminding them that Sharia law enforcement concerns and questions are addressed. Islamic tourism's main components are identical to traditional tourism components such as hotels, restaurants, logistics, finance and vacation packages. The aims of this study is to examine whether the Muslim tourist perceived value (MTPV) affects Muslim tourist satisfaction in the context of 'Halal' or 'Islamic' tourism. This MTPV has been created by Isa et al. (2018) who defined MTPV is Tourist perceived value involves tourist responses which vary depending on differences in culture and the sense in which they exist.

Muslim Population Worldwide (2017) indicated that muslims represented one of the world's largest tourism destinations, as its community forms an international market with more than 2.1 billion potential customers. The Malaysian tourism industry has been contributing more than 13% to the country's gross domestic product (GDP) since 2013 due to the government effort in diversifying its revenue sources. In the same vein, an estimated RM 80.1 billion was inbound from the tourism expenditure in 2014 (Department of Statistics Malaysia, 2017). The sector has also contributed to job creations for the local population.

The tourism industry can be a nation's economic driver for a developing country like Malaysia. There is potential in Malaysia's Tourism industry when we are looking to expand and diversify the revenue. Over 13% GDP of Malaysia has been contributed by the tourism industry since 2013 and there is an increase in inbound tourism since 2010. The tourism industry recorded a significantly improved performance for tourists that have visited Malaysia in 2018. The portion of gross value added of tourism industries (GVATI) to GDP is 15.2% in 2018 if compared with 2017 which only 14.6%. A value of RM220.6 billion has been recorded for GVATI, sustained a significant growth of 10.0% (2017: 10.3%). (Department of Statistics Malaysia, 2019).

In addition, to boosting the income of a nation, the tourism industry also generates job opportunities for the local people. World Travel and Tourism Council (2015) disclosed that In 2014, Travel & Tourism's overall employment contribution, as well as for jobs implicitly reinforced by the industry, amounted to 13.0% of the overall employment (1,770,000 jobs). This is predictable to rise to 1,824,000 jobs by 3.1% in 2015 and to increase to 2,489,000 jobs by 3.2% in 2025 (14.4% of that same total).

The Malaysian government has repeatedly provided assistance and subsidies to the industry of local tourism and travel because of its global influence. This comprises supporting the tourism industry to explore new and undiscovered zones to create an appealing business model as a travel destination for Malaysia and this including Halal tourism which aimed at Muslim travellers from around the world.

Halal tourism offers an alternative to Muslim tourists who love to travel without worrying about breaking Sharia law compared to conventional travel. Malaysia is well-positioned as a predominantly Muslim country to provide desirable destinations for Muslim tourists without ignoring their religious needs or traditions. Nowadays, the growth of Muslim consumer society around the world are fit with the efforts to implementing Halal tourism. In reality, Muslims are said to slowly to be a part of a strong consumer community in the tourism sector. Middle East people are keenly interested in looking for Muslim-friendly travel and leisure destinations. For an example is United Arab Emirates, Saudi Arabia and Bahrain (Eid & El-Gohary, 2015b). Consequently, it has been proposed that a clearer picture of their particular attitudes, desires and expectations is required (Prayag & Hosany, 2014).

Currently, the development of the tourism market that offers by hotels with the Shari'ah-compliant concept is growing especially in countries like Malaysia, UAE and Turkey. The potential growth of the hotels can help to raise the Muslim Tourists all over the world although hotels with Shari'ah-compliant have existed in certain particular countries like Kuwait and Saudi Arabia. That is why more factors need to be examined to paint clearer pictures.

Thus, in the background of 'Halal' or Islamic travel, this study intends to do so which is to paint a clearer picture of their particular attitudes, desires and expectations are required. Given that research is fairly new in such an area, it is expected that study's result will contribute useful insights to business organisations in the tourism sectors and the Malaysian Government. This study mainly explores the influence of the two constructs. The first is Eid and El-Gohary (2015a) established Muslim tourist perceived quality (MTPV); MTPV measures the understanding of Muslim tourists in rapports of their travelling impressions. The subsequent is to explore the impact of expectation of the destination on MTPV. This is also significant as it is said that the destination encounter significantly impacts visitor happiness, decision-making process and desire to return (Bonn et al., 2005).

2. The Unfamiliarity of Halal Tourism

Although its acceptance has increased significantly in the past decade, the concept of halal tourism remains imprecise. The halal tourism which used terminology is sometimes inaccurate, and there are many likely causes for it; a comparatively recent trend alien to the standard language of tourism might well be one and missing total institutionalisation (Neveu, 2010). Diversity of Islamic societies, differences of standard religious, diverse Islamic interfaces with socio-economic and cultural factors and the diversified cultures of Islam, are the reasons for ambiguity in the concept of halal tourism. (Henderson, 2003; Zamani-Farahani & Henderson, 2010). Henderson (2003) also claimed that the idea of halal tourism is distinct from the other in Islam, which affects diplomacy, economies, cultures and societies. Halal can only provide tourism and remain in tune with Muslim beliefs, as halal tourism must cover many perceptible, undeniable aspects of Shariah (Islamic teachings). This is very distinctive from just assure the halal food. Halal tourism will lead to inquiry about the Islamic trampling environment. For instance, dress code, exclusive worship-and prayer areas, the type of allowable fun, hotel personnel's demographic profile, including the gender and faith, as well as the code of interactions among the opposing genders (Battour et al., 2011; Henderson, 2011).

Zeithaml (1988) and Bitner (1990) view quality in service as very important to the business profit and growth of the service industry as good service quality attracts more customers and generates more revenues. According to Han and Ryu (2009), customers frequently consider price as an important factor in their expectations of the service performance. It shapes their attitude and behaviour that lead to either negative disconfirmation, positive disconfirmation or simple disconfirmation. Emotional is a feeling that triggered by products or services and thus affective feelings affects tourist satisfaction. That's why it is important the tourist feels comfortable and satisfied with the quality of services provided at the destination and the price offered is affordable (Yoon & Uysal, 2005). According to Sweeney and Soutar (2001) and Cengiz and Kirkbir (2007), the variable of social states that perceived utility derived from the products and services ability in enhancing their self-image and become a well-known products and services provider. The tourism industry offered the value of products and services which is become the variable of physical attributes. For instance, all Halal food and drinks, utensils and equipment used by Muslims. Meanwhile, non-physical attributes are the intangible attributes that comply to Sharia regulations. For an example, space of prayer, separate of Muslim's swimming pool, spa and hair saloon for men and women (Eid & El-Gohary, 2015a).

The principles of shari'a are the requirements for all Muslims, and their sensitivity to the implementation of these philosophies is essential because all the Islamic deed are not tolerable unless they are properly performed such as daily routine prayers and must be in clean conditions. Islamic education required a Muslim to abstain from violent and indulgent eating. Shari'ah principles not only forbid adultery, gambling and eating pork but as well as other (forbidden) food consumption, inappropriate sale or drinking of liquor and dressing. Consequently, conformity with shari'a should be a requirement for Muslim clients' high-satisfaction on tourism experiences. Analysis 7 of the factor affecting customer satisfaction and tourist satisfaction is limited (Isa et al., 2018).

For instance, the Forozia et al. (2013b) examined customer satisfaction in Malaysia's 3-star hotels. Another study was done by Isa et al. (2018) only focused on Pahang, Penang and Terengganu. However, minimal research has been conducted to observe factors affecting customer satisfaction in terms of these six variables. Thus shortcomings in consumer satisfaction need to be strengthened (Forozia et al., 2013a) and that is why the researcher conducted this study.

This study focused on Langkawi, Malaysia because Langkawi has been the main attraction for destinations and lots of local and foreign tourists went there. Langkawi is also known as The Jewel of Kedah. On 1 June 2007, the UNESCO has awards the Langkawi Island as a World Geopark status (Naturallylangkawi.my, 2019).

According to the latest global survey conducted by Crescent Rating, Malaysia was ranked first in the Global Muslim Travel Index in 2018. Meanwhile, Langkawi has been named as a top destination for Muslim travellers eight years. This Malaysian archipelago offers a variety of halal options. The trip advisor listed these four hotels (The Adya Hotel, The Danna Hotel, The Datai Hotel and The Andaman) that are famous for its Muslim friendly facilities and services. For instance, The Andaman has provides the guest with the halal-friendly menus and suites which is a private pool or in-suite hot tub. In Friday prayers, Masjid al Istiqua is a nearest mosque from the resort and the hotel can provide transportation to go for Friday prayer.

Nonetheless, it is very complex and dynamic how Halal evolves in specific markets and consumer groups. In the case of participation in Islamic tourism, evaluating the value of tourism products and satisfaction requires an entirely different method due to Islamic shari'a requirements. Muslims require appropriate products, facilities and conditions to engage in tourism activities. Therefore, the physical and non-physical attributes, quality, price, emotional and social were not only represent the satisfaction of the customer alone, but also the attributes that comply with the shari'a.

Moreover, all the researchers proposed to call the hotels that deliver goods and services in the industry of halal tourism as 'halal hotels' and 'Muslim tourists' to the tourists. Lack of halal standards has become one of practical challenges in halal tourism. The physical and non-physical attributes, quality, price, emotional and social must not only represent the satisfaction of the customer alone, but also the attributes that comply with the shari'a, especially for hotels. The market is not standardized when Muslims comprehend and have quite different in adapt their lives to Islam. This research implication: This study is presenting the literature reviews about the physical and non-physical attributes, quality, price, emotional and social. The halal certificates of hotel companies will assure the customers. The hotel needs to inform its objective visitors about the hotel's location before they arrive at the hotel. This is because, the hotels with halal concept and other hotels are usually located in a resort destination, specifically in Malaysia.

This research work is one of the foremost papers that differentiates Islamic tourism and the concept of halal tourism. It provides several issues of conceptual and practical in the industry of halal tourism especially in term of quality, price, emotional, social and physical and non-physical attributes.

3. The Theory of Expectancy Disconfirmation (EDT)

This research based on the expectancy disconfirmation theory (EDT). Oliver (1977, 1980) introduced the expectancy-disconfirmation theory (EDT) as the most effective customer satisfaction measurement. The model suggests buyers buy services and good with perceptions of the expected performance in anticipation of purchasing. Then the rate of standards becomes a benchmark against which the consumer is evaluated. That is, the outcomes will be evaluated to standards once the product or service has been used. If the outcome suits the verification of assumption should occur.

There is disconfirmation when there are different beliefs and consequences. A consumer is either happy or disappointed as a consequence of a positive or negative difference between desires and experiences. Therefore, when service quality is better than expected by the customer, there is a positive disconfirmation resulting in satisfaction with both expectations and performance. Meanwhile, when customer receive the service performance as expected, it leads to the satisfaction with a confirmation between expectations and perceptions. The items measured under this expectancy are quality, price, emotional, social, physical attributes and non-physical attributes. All these variables will lead to either positive, negative or simple disconfirmation.

In contrast, there is a detrimental disconfirmation between standards and assumptions which creates disappointment when the quality of the services or the products is not as high as the expectations of the customer.

In this study, factors in the EDT model is used which are the MTPV by providing perceived quality, moderate prices, emotional, social, physical attributes and non-physical attributes to measure the satisfaction (Al-Sabbahy et al., 2004).

4. Six Parameters of Muslim Tourist Perceived Values (MTPV)

The perceived value is different from the satisfaction of customers. The former offers consumers with quality, while the latter has to do with meeting the needs of customers. At different stages of procurement process, especially pre-and post-purchase (Woodruff, 1997) the perceived value can happen. It is a system that is arbitrary. Customer perceived value involves customer responses which vary depending on differences in culture and the sense in which they exist (Prebensen et al., 2013; Sweeney & Soutar, 2001). It has several qualities, including perceived quality, consumer satisfaction and price elasticity (Al-Sabbahy et al., 2004).

The research began to conceptualize the perceived value of the customer of Zeithaml (1988), which described as 'the consumer's ultimate estimation of the effectiveness of a product based on the assessment of what is obtained and what is provided'. (Khan & Kadir, 2011). Stated that the perceived value idea was multi-dimensional and split into two sections, mental and affective. Eid and El-Gohary (2015a) developed a new measuring scale in relation to Islamic tourism in order to assess perceived value for Muslim tourists. The researcher named this MTPV and it aligned with the amount of consumer experience after using the services and products.

Six parameters of MTPV, including quality, price, emotional, social, physical attribute and non-physical attribute are proposed by the scholars. The mental interest applies to tourists logical and economic valuations of the price, products and services quality they purchase. Sweeney and Soutar (2001) define price as a benefit extracted from products and services because of its potential short-term and long-term cost reduction. The value parameter is a vital effect on the level of customer service (Bolton & Lemon, 1999; Parasuraman & Grewal, 2000; Varki & Colgate, 2001). Sweeney and Soutar (2001) also defined value as the usefulness extracted from the perceived quality and planned consumer results. In terms of customer loyalty, the principles of value of partnership, product quality and quality of service and customer interest were a major driver of company satisfaction (Anderson & Mittal, 2000; Cooil et al., 2007).

The social parameter relates to the presumed utility resulting from the willingness of products and services to boost their self-image and become a well-known supplier of products and services (Cengiz & Kirkbir, 2007; Sweeney & Soutar, 2001). The emotional variable is an important factor in increasing tourist satisfaction. For an example, stress relief, family unity, enthusiasm, fun and security (Sweeney & Soutar, 2001; Yoon & Uysal, 2005). The utility of a single item is also a key factor in raising tourism satisfaction. Islamic value is differentiate into a vector with physical and non-physical attributes.

The variable of physical attributes resulting from the quality of the tourism industry's products and services. This contains all the food and drinks Halal used by Muslims, utensils and appliances (Eid & El-Gohary, 2015a). Nonphysical attributes are intangible qualities that conform with Sharia laws such as praying room, segregated Muslim swimming pool, segregated men's and women's spa and beauty salon (Eid & El-Gohary, 2015a).

Customer-perceived value has been debated in the literature in recent years as a factors predicting consumer behaviour (Hellier et al., 2003). It is constructive and direct background to customer satisfaction (McDougall & Levesque, 2000; Patterson & Spreng, 1997). Subsequently, in order to attract more Muslim visitors than ever before, the Malaysian tourism industry will concentrate on Islamic qualities that are expected to meet the standards of tourists. In a study by Henderson (2011), several Muslim countries have been proposed to be more tolerant in order to attract more visitors to their country. A nation that plans to provide Halal tourism, however, should first consider Halal's meaning. Thus, it must take into consideration aspects such as halal food and beverages, halal facilities and culture, Islamic Codes of dress and services of worship (Hashim et al., 2010; Zamani-Farahani & Henderson, 2010; Battour et al., 2010). In this analysis, the MTPV level was used from the viewpoint of Muslim customers to analyse brand interest.

4.1 Quality

According to Awad (2012) Quality of service is the customer's assessment on the level or standard. Quality of service is the core of the service industry's business growth Better quality of service brings more customers and creates more income. Photo and Tourist Satisfaction with increasing the quality of their facilities are in the interest of the management of each tourist destination. Managers will please their visitors and keep tourists faithful to the destination by enhancing their operation. Poor customer service reliability hampers the business' capacity and causes consumers to change service providers, according to Sparks and Westgate (2002). That idea relates to the tourism sector. If there is no quality service offered by any single tourist destination, tourists will choose other destinations.

The previous studies about travel services quality are trying to facilitate the factors that affect customers' perceived quality of travel services. It also determine the differences between expectation of customers and actual service experience (Dires & Anteneh, 2016). Hence, in order to get better understanding about the dimensions of travel service quality, there is crucial to conduct study on the travel services business. Researchers must consider several common dimensions and combine them for research purpose. By providing with modern and latest facilities and services at tourism destination visitors are found to be more satisfied (Dires & Anteneh, 2016).

4.2 Price

Many marketing studies show that service performance variation increases customer hesitation. In this circumstance, customers frequently see price as a cue in their service performance expectations, which also influences their behaviour and attitude (Han & Ryu, 2009). In fact, cost is also has been identified as a determinant of quality expectations (Varki & Colgate, 2001), which is why a number of works by scholars have researched some facets of price within advertising literature. For example, Jiang and Rosenbloom (2005) analysed consumer perceptions of prices in the digital environment and found that perceptions of prices have a clear and beneficial impact on customer satisfaction overall.

In fact, Bolton and Lemon (1999) explored effect of cost equality on the use of cell phone and television products by consumers. Their results showed that their consumer performance and behavioural goals in both sectors were significantly affected by customer views of cost fairness/unfairness. In yet another research, Han and Ryu (2009) explored the function of price as a moderating in the relationships between perceived food quality, service quality and restaurant business customer satisfaction and physical environment. They also claimed that the understanding of a reasonable price by consumers intervenes as a parameter moderator to maximize the effect on their experience with performance.

All these research highlight different aspects of price, but little study on the theory of price acceptance has been carried out. In the theory, the definition of market tolerance is similar to one which the theorists term the reservation cost within microeconomics, corresponding to the extreme price a consumer is willing to pay for the product or service in question (Huber et al., 2001; Martín-Consuegra et al., 2007).

Price is known as an essential determinant of satisfaction (Jiang et al., 2019). According to Lockyer's study (2005), prices are one of the major factors affecting hotel selection and satisfaction. A hotel's medium and affordable price can achieve maximum satisfaction (Olga, 2018). Therefore, if the consumer discovers that the hotel's pricing system is unfair, the loyalty will be diminished and future customers will not frequent the hotel (Bassey, 2014).

4.3 Emotional

The emotional parameter refers to the usefulness extracted from the emotions or affective state created by a consumer and is a crucial factor in the tourist satisfaction. Hence, factors such as happiness, family unity, anticipation, fun and security are significant (Bigné et al., 2008; Lin & Liang, 2011; Sweeney & Soutar, 2001; Yoon & Uysal, 2005). Bagozzi et al. (1999) defined emotion as an emotional state of readiness that is the product of mental interpretations of events or thoughts; phenomenological tune; followed by physiological processes; is often conveyed externally and could contribute to specific actions to validate or control the emotion, obviously depending on its existence and the person with it.

4.4 Social

The variable of social (i.e. affective value) refers to the presumed usefulness arising from the potential of products and services. It helps to improve their self-image and become a well-known supplier of products and services (Chiu et al., 2014). The affective quality comprises the feelings and emotions produced by the use of the products and services. The social parameter relates to the presumed utility resulting from the willingness of products and services to boost their self-image and to become a well-known supplier of products and services (Cengiz & Kirkbir, 2007; Sweeney & Soutar, 2001). The parameter of emotion implies the utility extracted from a product's emotions or affective condition and is a foundational factor in that tourist satisfaction. Thus, factors such as comfort, unity of family, anticipation, fun and security are essential (Chiu et al., 2014).

4.5 Physical Attributes

Islamic quality is split into physical and non-physical quality variable. The physical quality parameter (i.e. Islamic cost) extracted from the significance of the tourism industry's products and services. It consists of all the Muslim's Halal food and beverages, cookware and appliances (Eid & El-Gohary, 2015b).

4.6 Non-Physical Attributes

Nonphysical (i.e. Islamic value) attributes. These are intangible attributes which comply with Sharia laws. For instance, space of prayer, separation of Muslim swimming pool, men's and women's spa and hair saloon (Eid & El-Gohary, 2015a). The literature review's purpose is to describe and clarify the workings of a theoretical framework that could be used as a foundation for creating a visitor experience measurement method. Hence, for empirical research purposes, we suggest the destination with creation of Islamic physical and non-physical qualities. Ozdemir and Met (2012) also stated that several hotels in Turkey offer separate swimming pools and recreational facilities. It is because, they know that Muslims typically analyse a dress code and avoid from mixing with different gender. Therefore, Muslim consumers with Islamic religiosity reduces the impact to Islamic physical attributes and non-physical attributes on customer satisfaction.

5. Muslim Tourist Satisfaction

Satisfaction is definite as an overall performance assessment which related to all previous company experiences. Satisfaction can be impacted by objective factors such as features of products and services, whereas the subjective factors such as feelings and customer needs. Satisfaction genuinely plays a main role in creating value to consumers (Tanaskovic et al., 2013).

Customer satisfaction is one of the key factors of performance in any company (Lovelock & Patterson, 2015). Satisfaction level is the result of the expected and real performance discrepancy. Of instance, if the product or service works better than expected, positive disconfirmation as recognized as satisfaction would occur (Oliver, 1980). In general, customer satisfaction is the product of the understanding of the interest received by the customer in a partnership or service relative to the competitors (Blanchard & Galloway, 1994).

Tourism includes the selling of memorable experiences. It is different from the sale of other products and services. It needs the satisfaction of the perceptions of tourists to be successful. Consequently, the quality of its execution is arbitrary in many ways; based on how visitors view it. Therefore, tourism is a user-centric sector. Therefore, it is an essential component in assuring its sustainability. It is important in creating a competitive offer with growth potential. It is not an easy task, though, to gain customer satisfaction.

Customer satisfaction is a subjective evaluation that provides businesses with major challenges in recognising, obtaining and sustaining customer satisfaction (Su, 2004). Hence, it requires the emotional state of the consumer because it is harder to manage.

Accumulative phenomenon also called customer satisfaction and is comprehended as a customer's reaction to delivery. Customer positively measures their purchase experience after using a service or product. Tourist satisfaction in regards to tourism relates to the tourist's pleasure after having visited a specific destination. Literature from the past has indicated consumer purpose, engagement and actions (Chi & Qu, 2008; Hutchinson et al., 2009; Su & Hsu, 2013; Su & Swanson, 2017). According to Bitner (1990), Bolton and Drew (1991), the quality, environment, price and personal differences were related with customer satisfaction. Zeithaml and Bitner (2003) claimed that customer satisfaction receive after-purchase feedback. That is, the post-purchase interaction's evaluation of consumers relative with their desires.

Furthermore, if the assessment of the products and services by visitors meets their standards, their happiness will be improved. The positive impact leads by good experience of visitors receiving. It is because the satisfaction and happy of the tourists will make them revisit and also will suggest the location to their friends, family and acquaintances. However, the unsatisfied customer are not expected to do the same Word-of-mouth has been recognized as a significant factor in motivating visitors to visit that destination.

Overall satisfaction of this study is adopted, highlighting various comparison and recommendations standards of essence of satisfaction's initial service performance, the emotional executed by describing the pattern of consumer satisfaction (Eid & El-Gohary, 2015a). Broadly speaking, the indicator of the profitability of the market performance is satisfaction. It increase the interest income and market share.

6. Methodology

The research was conducted using on a basis of quantitative research technique. Descriptive study structure is used in this research to identify the six defined factors that affect Muslim tourist perceived interest against Muslim tourist satisfaction as it explains the circumstance and does not specify causes and effects.

Self-administered questionnaire has been used in this data collection process. The questionnaire has been distributed to 205 Muslim tourists in Langkawi, Malaysia during October 2021 (Langkawi Tourism Bubble). The respondents answered the online questionnaire using the Google Form on the researcher's tablet where the respondents can answer the questionnaire on the spot. The data analysed in this study is a primary data. It is because the data was collected only for the uses of this study.

This research used the technique of convenience sampling for non-probability sampling. A maximum of 205 questionnaires distributed equally to the target population. It is based on the standard sample size principle at least five times the amount of indicators if the population is unknown (Hair et al., 2003).

7. Effects of MTPV on Muslim Tourists Satisfaction

The relationship between the variables have been investigated in this study. The regression results presents that four variables (quality, price, social and physical attribute) are an important factors of tourist satisfaction. While the remaining two factors (emotional and nonphysical attribute) shows insignificant towards tourist satisfaction.

This study used the correlation analysis to analysed the relationship between the independent variables which are quality, price, emotional, social, physical attributes and nonphysical attributes. The result shows that four variables (quality, price, emotional and physical attribute) have significant correlation between tourist satisfactions. Meanwhile, social and nonphysical attribute have no significant correlation towards tourist satisfaction. Hence, the researcher believe that the respondents do not view social and nonphysical attribute to be the important factors that can contribute to tourist satisfaction.

7.1 Effects of Quality on Tourist Satisfaction

The definition of service quality is the gap between customer's expectations and perception based on disconfirmation paradigm (Parasuraman et al., 1990). Quality plays a very important role to contribute to the tourist satisfaction. Quality is believed to be critical for Muslims tourists to observe satisfaction. In this case it is believed that when the goods and services offered are in good quality, the tourists will be satisfied. As the objective is to look into the relationship between quality and tourist satisfaction, researcher has come up with a hypothesis that predicts quality will be a significant influence that will effects tourist satisfaction. The result from previous chapter stated that quality and tourist satisfaction have a significant relationship. The regression beta value for quality is at 0.345. This result shows positive relationship which shows that the higher the quality of the products and services the higher the tourist satisfaction. The findings of quality shows similar outcome as previous study by Isa et al. (2018) and Eid and El-Gohary (2015a) that quality effects tourist satisfaction.

7.2 Effects of Price on Tourist Satisfaction

Price is the primary factors that effects the decision to buy or use the products and services (Campo & Yagüe, 2008) thus lead to the satisfaction of the tourists. Price determine the tourist satisfaction as when the price is too high, less tourists will buy or use the products and services thus lead to lower satisfaction of the tourist. The constructed hypothesis is to examine if price will have significant impact on the tourist satisfaction.

To test the stated hypothesis, the Pearson Correlation test has been done. The results stated that there is positive correlation between price and tourist satisfaction as the correlation coefficient is at 0.666 and the regression value is at $b = 0.141$. The hypothesis H2 is accepted as the p-value is smaller than 0.05.

7.3 Effects of Emotional on Tourist Satisfaction

The variable of emotions derived from the feelings or affective state. It influenced by a products and services that usually increase tourist satisfaction. Therefore, factors such as family togetherness, fun, excitement, relaxation and safety are important (Sweeney & Soutar, 2001; Yoon & Uysal, 2005). The constructed hypothesis is to examine if emotional have significant impact on the tourist satisfaction. Based on the result in previous chapter, the regression value of emotional variables is at $b = 0.152$ and this signify the hypothesis is accepted. The Pearson Correlation Test also suggest that there is moderately high correlation between emotional and tourist satisfaction as the correlation coefficient is the second highest at 0.682. The regression value is at 0.152. This result align with previous study by Isa et al. (2018) that shows significant effects of emotional on tourist satisfaction.

7.4 Effects of Social on Tourist Satisfaction

The variable of social refers to the perceived utility. It derived from the products and services capability in self-image enhancement and became well known products and services provider (Cengiz & Kirkbir, 2007; Sweeney & Soutar, 2001). The hypothesis is constructed to test if the variable of social has an effects on tourist satisfaction and the length of the influences. As per stated in previous chapter, the regression value of social variable is at $b = 0.001$, thus suggested that the proposed hypothesis is rejected. This means social does not play important role in tourist satisfaction. This findings shows similar result with previous study by Eid and El-Gohary (2015b). The social aspect are not significant role in tourist satisfaction. The reason why the result of this study is insignificant is because respondents which are the tourist, went to Langkawi for their own pleasure and comfort not to get any social approval. The respondents feel like just because a lot of people are using or chose that tourism package or particular hotels, that mean they have to choose that as well. Mostly people will match their desire and budget. Based on the questionnaire 50% of the respondent are singles and 28% respondents get their monthly income below RM999 which can indicate they are on low budget vacation. Thus they usually stay at budget hotel or any hotel that is within their budget rather than 4–5stars hotel to get social approval or trying to make good impression on society.

7.5 Effects of Physical Attribute on Tourist Satisfaction

Physical attributes variable came from the value of the products and services offered. In inclusive with the Muslims Halal food and drinks, equipment and utensils used by them (Eid & El-Gohary, 2015a). This is very important as Muslims due to the fact that halal not only mean food but also in term of the preparation all of these will effects Muslim's belief and thus will directly affect the tourist's satisfaction. The term 'Halal' goes deeper than just the word. Halal is a food that is hygienic and fulfils religious requirement (Soong, 2007). Halal food not 'haram' are clarified as unused equipment that produced any haram food before and also the ingredients of making halal foods must not be mixed with the haram ingredient or equipment (Nordin, 2007). The hypothesis constructed is to test if physical attribute has relationship with tourist satisfaction and the result shows that the hypothesis is accepted as the regression value of physical attribute is at $b = 0.203$. This means that physical attribute is important to tourist satisfaction. This results shares the same outcome with previous study by Isa et al. (2018) that shows physical attribute does effect tourist satisfaction.

7.6 Effects of Non-physical Attribute on Tourist Satisfaction

Nonphysical attributes are the intangible attributes that follows the Sharia regulations. For example, separation of swimming pool, hair saloon, prayer room and spa for men and women. The hypothesis is constructed to test if non-physical attribute has a relationship with tourist satisfaction and the length of the influences. The results shows that the proposed hypothesis is rejected as the regression value of nonphysical attribute variable is at $b = 0.085$. This means non-physical attribute does not play important role in tourist satisfaction. This finding support the result from previous study by Eid and El-Gohary (2015b) which stated that non-physical attribute does not affect tourist satisfaction as the result is not significant. Non-physical attribute is not significant because most tourist went for a holiday and Langkawi offered variety of fun adventure sight-seeing along with physical activities so they don't pay attention to all this non-physical matter. Hair Salon and spa is part of non-physical attribute but most hotels did not provide hair salon and as for spa mostly only available at hotel that is four stars and above. Since Langkawi have beach so they will go there instead of swimming pool. Furthermore there are many swimming attire that is sharia compliance, so the Muslim tourist usually will wear proper Islamic compliance attire to swimming pool and that's why they can share the public swimming pool with men and woman. Thus, explains why non-physical attribute does not affect tourist satisfaction.

8. Practical Contributions

The creation of more Halal products and services should be implement by Malaysian government such as hotels and other infrastructures. It is one the way to attract more Muslim tourists from other countries. Since the study shows that the Islamic quality and tourist satisfaction are related.

Encouragement of Halal awareness to Malaysian is needed. This can be done by Malaysian Government primarily through education in school. In order to become a 'Halal tourism' hub in Asia continent, the awareness can also be done through advertisement or broadcasting. Halal awareness to non-Muslim is crucial to guarantee the success of Halal tourism because Malaysia is a multiracial country.

Furthermore, this study provides the hoteliers with the useful information about these influencing factors. It can be implement in the high competitive field in order to fulfil and satisfy the needs of tourist. Besides that, this study also give advantages to academics and educators. They may have more detail information and knowledge about the factors that can affect the tourist satisfaction in related field.

Tourism sectors primarily travel agents and suppliers from local products and services should be on alert on both internal and external factors as why Muslim tourists select specific destination over the others. With this, it is safe to say that to attract more Muslim Tourist, the Malaysian tourist attraction destination providers should prepare various options of Halal products and services. Thus, it will increase the tourism revenue in Malaysia. Nevertheless, the quantity and quality of the attraction's destination's 64 infrastructure must parallel with the concept of Halal even some of the Muslim tourist doesn't matter is they are Malaysian or foreigners, focus more on Halal products and services available instead of the environment and the culture itself but it is also important to preserve every factors. The satisfaction of tourist is derived from the good impression of the destination.

9. Post-Pandemic: The Rise of Halal Tourism in Malaysia

As the country recovers from the Covid-19 outbreak, Malaysia stays committed to developing more Muslim-friendly tourism categories. Datuk Seri Nancy Shukri, Minister of Tourism, Arts and Culture, said the country would continue to establish the segments of Muslim-friendly travel that have become crucial part towards the global tourism market (Yusof, 2021).

The ministry has been working with its agency, the Islamic Tourism Centre (ITC) since 2009 to establish Muslim-friendly travel programmes that include local experiences. In 2019, an increase of 20% in Muslim tourists entering the nation was observed. As a result, while planning for the future after the pandemic, this market sector will be more focused.

The Muslim consumer market is projected to be a vital strategy for boosting the tourism industry and economies with 1.8 billion Muslims worldwide. Although, devastated by the pandemic, with Islamic Tourism possibly leading the way. As borders gradually reopened, Muslim travellers would be able to meet their travel needs. Muslim tourist expenditure reached USD194 billion at the peak of pre-pandemic tourism and is anticipated to reach USD208 billion in 2024 when COVID-19 is factored in ('Islamic tourism paves the way forward for economic and tourism recovery', 2021).

Despite the recent travel disruption, a significant figure is still projected. The economy sectors from all the industry players are encouraged to participate in and benefit from the Islamic Tourism economy. It inclusive of travel, finance, medical, media and entertainment, medical, food and beverage, technology, telecommunications, transportation, education and retail.

Hence, this research on the effects of MTPV on Muslim tourists' satisfaction is crucial to be considered as guidance for a new strategy to penetrate the Halal tourism market in the post-pandemic world. All industry players should consider these values to improve the satisfaction of Muslim tourists.

10. Conclusion

The study comes to conclusion with several statements based on the problem statement generated, research questions, objectives, relevant literature reviewed, analysis conducted on the data, generated results as well as the discussion made. As said by Hernández-Lobato et al. (2006), tourism services and products which provide identity credibility and social prestige are the key factor which contributes quality to tourists. Pricing strategies are, however, also an essential element for suppliers of destinations. Price does not only imply value inferences; it can affect real performance.

Quality is recorded to be the biggest contributing factors that effects the tourist satisfaction while physical attribute comes in second and emotional variable comes in third. These three (3) variables deemed to be the highest factors that contribute to tourist satisfaction. While the other three (3) variables (price, non-physical attribute and social) comes in fourth, fifth and sixth respectively.

The last two (2) variables which are social and non-physical attribute are not significant. This means these two variables does not affect tourist satisfaction.

Thus, Muslim tourist market should be taken into consideration in order to understand of its needs. It is better to make the necessary preparations to accommodate them, opens the door to a new source of revenue or means of income from a potentially lucrative market in this post-pandemic world.

References

Al-Sabbahy, H. Z., Ekinci, Y., & Riley, M. (2004). An investigation of perceived value dimensions: Implications for hospitality research. *Journal of Travel Research*, *42*(3), 226–234. https://doi.org/10.1177/0047287503258841

Anderson, E. W., & Mittal, V. (2000). Strengthening the satisfaction-profit chain. *Journal of Service Research*, *3*(2), 107–120. https://doi.org/10.1177/109467050032001

Awad, B. A. A. G. (2012). The relationships between service quality, satisfaction, and behavioral intentions of Malaysian spa center customers. *International Journal of Business and Social Science*, *3*(1).

Bagozzi, R. P., Gopinath, M., & Nyer, P. U. (1999). The role of emotions in marketing. *Journal of the Academy of Marketing Science*, *27*(2), 184–206. https://doi.org/10.1177/0092070399272005

Bassey, F. O. (2014). *The effect of perceived price fairness on customer satisfaction and loyalty*. Doctoral dissertation. Eastern Mediterranean University (EMU)- Doğu Akdeniz Üniversitesi (DAÜ).

Battour, M. M., Ismail, M. N., & Battor, M. (2010). Toward a halal tourism market. *Tourism Analysis*, *15*(4), 461–470. https://doi.org/10.3727/108354210X12864727453304

Battour, M., Ismail, M. N., & Battor, M. (2011). The impact of destination attributes on Education. Muslim tourist's choice. *International Journal of Tourism Research*, *13*(6), 527–540.

Bigné, E. J., Mattila, A. S., & Andreu, L. (2008). The impact of experiential consumption cognitions and emotions on behavioral intentions. *Journal of Services Marketing*, *22*(4), 303–315. https://doi.org/10.1108/08876040810881704

Bitner, M. J. (1990). Evaluating service encounters: The effects of physical surroundings and employee responses. *Journal of Marketing*, *54*(2), 69–82. https://doi.org/10.1177/002224299005400206

Blanchard, R. F., & Galloway, R. L. (1994). Quality in retail banking. *International Journal of Service Industry Management*, *5*(4), 5–23. https://doi.org/10.1108/09564239410068670

Bolton, R. N., & Drew, J. H. (1991). A multistage model of customers' assessments of service quality and value. *Journal of Consumer Research*, *17*(4), 375–384. https://doi.org/10.1086/208564

Bolton, R. N., & Lemon, K. N. (1999). A dynamic model of customers' usage of services: Usage as an antecedent and consequence of satisfaction. *Journal of Marketing Research*, *36*(2), 171–186. https://doi.org/10.1177/002224379903600203

Bonn, M. A., Joseph, S. M., & Dai, M. (2005). International versus domestic visitors: An examination of destination image perceptions. *Journal of Travel Research*, *43*(3), 294–301. https://doi.org/10.1177/0047287504272033

Campo, S., & Yagüe, M. J. (2008). Research note: Effects of price on tourist satisfaction. *Tourism Economics*, *14*(3), 657–661. https://doi.org/10.5367/000000008785633596

Cengiz, E., & Kirkbir, F. (2007). Customer perceived value: The development of a multiple item scale in hospitals. *Problems and Perspectives in Management*, *5*(3), 252–268.

Chi, C. G. Q., & Qu, H. (2008). Examining the structural relationships of destination image, tourist satisfaction and destination loyalty: An integrated approach. *Tourism Management*, *29*(4), 624–636. https://doi.org/10.1016/j.tourman.2007.06.007

Chiu, C. M., Wang, E. T., Fang, Y. H., & Huang, H. Y. (2014). Understanding customers' repeat purchase intentions in B2C e-commerce: The roles of utilitarian value, hedonic value and perceived risk. *Information Systems Journal*, *24*(1), 85–114. https://doi.org/10.1111/j.1365-2575.2012.00407.x

Cooil, B., Keiningham, T. L., Aksoy, L., & Hsu, M. (2007). A longitudinal analysis of customer satisfaction and share of wallet: Investigating the moderating effect of customer characteristics. *Journal of Marketing*, *71*(1), 67–83. https://doi.org/10.1509/jmkg.71.1.067

Department of Statistics Malaysia. (2017). *Inbound from the tourism expenditure in 2014*. https://v1.dosm.gov.my/v1/index.php?r=column/cthemeByCat&cat=111&bul_id=ekpONHlVZllyN001YzJNYjBGVmtzQT09&menu_id=TE5CRUZCblh4ZTZMODZIbmk2aWRRQT09

Department of Statistics Malaysia. (2019). *Tourism satellite account 2018*. https://v1.dosm.gov.my/v1/index.php?r=column/cthemeByCat&cat=111&bul_id=Wk1KWlpxZTRDWnVhVWNMV21ZVVY3Zz09&menu_id=TE5CRUZCblh4ZTZMODZIbmk2aWRRQT09

Dires, A., & Anteneh, M. (2016). The effect of service quality on tourist satisfaction/loyalty at Gondar and its vicinity. *International Journal in Management and Social Science*, *4*(3), 445–458.

Eid, R., & El-Gohary, H. (2015a). Muslim tourist perceived value in the hospitality and tourism industry. *Journal of Travel Research*, *54*(6), 774–787. https://doi.org/10.1177/0047287514532367

Eid, R., & El-Gohary, H. (2015b). The role of Islamic religiosity on the relationship between perceived value and tourist satisfaction. *Tourism Management*, *46*, 477–488. https://doi.org/10.1016/j.tourman.2014.08.003

Forozia, A., Zadeh, M. S., & Gilani, M. H. (2013a). Customer satisfaction in hospitality industry: Middle East tourists at 3star hotels in Malaysia. *Research Journal of Applied Sciences, Engineering and Technology*, *5*, 4329–4335.

Forozia, M., Zadeh, M. S., & Hemmati, M. (2013b). Customer satisfaction in Malaysia's 3-star hotels: A study among Middle East tourists. *Research Journal of Applied Sciences, Engineering and Technology*, *5*(12), 4329–4335.

Hair, J. F., Babin, B., Money, A. H., & Samouel, P. (2003). *Essentials of business methods research*.

Han, H., & Ryu, K. (2009). The roles of the physical environment, price perception, and customer satisfaction in determining customer loyalty in the restaurant industry. *Journal of Hospitality & Tourism Research*, *33*(4), 487–510. https://doi.org/10.1177/1096348009344212

Hashim, N. H., Murphy, J., Purchase, S., & O'Connor, P. (2010). Website and email adoption by Malaysian hotels. *International Journal of Hospitality Management*, *29*(1). https://doi.org/10.1016/j.ijhm.2009.05.003

Hellier, P. K., Geursen, G. M., Carr, R. A., & Rickard, J. A. (2003). Customer repurchase intention: A general structural equation model. *European Journal of Marketing*, *37*(11/12), 1762–1800. https://doi.org/10.1108/03090560310495456

Henderson, V. (2003). The urbanization process and economic growth: The so-what question. *Journal of Economic Growth*, *8*(1), 47–71. https://doi.org/10.1023/A:1022860800744

Henderson, J. C. (2011). Religious tourism and its management: The hajj in Saudi Arabia. *International Journal of Tourism Research*, *13*(6), 541–552. https://doi.org/10.1002/jtr.825

Hernández-Lobato, L., Solis-Radilla, M. M., Moliner-Tena, M. A., & Sánchez-García, J. (2006). Tourism destination image, satisfaction and loyalty: A study in ixtapa-zihuatanejo, Mexico. *Tourism Geographies*, *8*(4), 343–358. https://doi.org/10.1080/14616680600922039

Huber, F., Herrmann, A., & Wricke, M. (2001). Customer satisfaction as an antecedent of price acceptance: Results of an empirical study. *The Journal of Product and Brand Management*, *10*(3), 160–169. https://doi.org/10.1108/10610420110395403

Hutchinson, J., Lai, F., & Wang, Y. (2009). Understanding the relationships of quality, value, equity, satisfaction, and behavioral intentions among golf travelers. *Tourism Management*, *30*(2), 298–308. https://doi.org/10.1016/j.tourman.2008.07.010

Isa, S. M., Chin, P. N., & Mohammad, N. U. (2018). Muslim tourist perceived value: A study on Malaysia Halal tourism. *Journal of Islamic Marketing*, *9*(2), 402–420. https://doi.org/10.1108/JIMA-11-2016-0083

Islamic tourism paves the way forward for economic and tourism recovery. (2021, November 15). *Islamic Tourism Centre of Malaysia*. ITC. https://itc.gov.my/islamic-tourism-paves-the-way-forward-for-economic-and-tourism-recovery/

Jiang, S., Feng, D., & Lu, C. (2019). A sustainable innovation – Additional services for products based on personalised customer value. *Sustainability*, *11*(6), 1763. https://doi.org/10.3390/su11061763

Jiang, P., & Rosenbloom, B. (2005). Customer intention to return online: Price perception, attribute-level performance, and satisfaction unfolding over time. *European Journal of Marketing*, *39*(1/2), 150–174. https://doi.org/10.1108/03090560510572061

Khan, N., & Kadir, S. L. S. A. (2011). The impact of perceived value dimension on satisfaction and behaviour intention: Young-adult consumers in banking industry. *African Journal of Business Management*, *5*(11), 4087–4099.

Lin, J. S. C., & Liang, H. Y. (2011). The influence of service environments on customer emotion and service outcomes. Managing service quality. *International Journal*, *21*(4), 350–372. https://doi.org/10.1108/09604521111146243

Lockyer, T. (2005). Understanding the dynamics of the hotel accommodation purchase decision. *International Journal of Contemporary Hospitality Management*, *17*(6), 481–492. https://doi.org/10.1108/09596110510612121

Lovelock, C., & Patterson, P. (2015). *Services marketing*. Pearson Australia.

Martín-Consuegra, D., Molina, A., & Esteban, Á. (2007). An integrated model of price, satisfaction and loyalty: An empirical analysis in the service sector. *The*

Journal of Product and Brand Management, *16*(7), 459–468. https://doi.org/10.1108/10610420710834913
McDougall, G. H., & Levesque, T. (2000). Customer satisfaction with services: Putting perceived value into the equation. *Journal of Services Marketing*, *14*(5), 392–410. https://doi.org/10.1108/08876040010340937
Naturallylangkawi.my (2019). About Langkawi – Naturally Langkawi [online]. Available at: https://naturallylangkawi.my/about-langkawi/
Neveu, E. (2010). *Sociologie du journalisme*. La Découverte.
Nordin, A. (2007). *Perception of consumers in Malaysia towards Halal certification from JAKIM*. University Malaya.
Olga, S. (2018). *Determinants of customer satisfaction on Chinese and Russian M commerce markets*.
Oliver, R. L. (1977). Effect of expectation and disconfirmation on post exposure product evaluations: An alternative interpretation. *Journal of Applied Psychology*, *62*(4), 480. https://psycnet.apa.org/doi/10.1037/0021-9010.62.4.480
Oliver, R. L. (1980). A cognitive model of the antecedents and consequences of satisfaction decisions. *Journal of Marketing Research*, *17*(4), 460–469. https://doi.org/10.1177/002224378001700405
Ozdemir, I., & Met, O. (2012). The expectations of Muslim religious customers in the lodging industry: The case of Turkey. *Current Issues in Hospitality and Tourism Research and Innovation*, 323–328.
Parasuraman, A., & Grewal, D. (2000). The impact of technology on the quality-value-loyalty chain: A research agenda. *Journal of the Academy of Marketing Science*, *28*(1), 168–174. https://doi.org/10.1177/0092070300281015
Parasuraman, A., Zeithaml, V. A., & Berry, L. L. (1990). *Delivering service quality*. The Free Press.
Patterson, P. G., & Spreng, R. A. (1997). Modelling the relationship between perceived value, satisfaction and repurchase intentions in a business-tobusiness, services context: An empirical examination. *International Journal of Service Industry Management*, *8*(5), 414–434. https://doi.org/10.1108/09564239710189835
Prayag, G., & Hosany, S. (2014). When Middle East meets West: Understanding the motives and perceptions of young tourists from United Arab Emirates. *Tourism Management*, *40*, 35–45. https://doi.org/10.1016/j.tourman.2013.05.003
Prebensen, N. K., Woo, E., Chen, J. S., & Uysal, M. (2013). Motivation and involvement as antecedents of the perceived value of the destination experience. *Journal of Travel Research*, *52*(2), 253–264. https://doi.org/10.1177/0047287512461181
Soong, S. F. V. (2007). *Managing halal quality in food service industry*.
Sparks, R., & Westgate, M. (2002). Broad-based and targeted sponsorship strategies in Canadian women's ice hockey. *International Journal of Sports Marketing & Sponsorship*, *4*(1), 48–73. https://doi.org/10.1108/IJSMS-04-01-2002-B006
Su, A. Y. (2004). Customer satisfaction measurement practice in Taiwan hotels. *International Journal of Hospitality Management*, *23*(4), 397–408. https://doi.org/10.1016/j.ijhm.2004.02.002
Su, L., & Hsu, M. K. (2013). Service fairness, consumption emotions, satisfaction, and behavioral intentions: The experience of Chinese heritage tourists. *Journal of Travel & Tourism Marketing*, *30*(8), 786–805. https://doi.org/10.1080/10548408.2013.835228
Su, L., & Swanson, S. R. (2017). The effect of destination social responsibility on tourist environmentally responsible behavior: Compared analysis of first-time and

repeat tourists. *Tourism Management*, *60*, 308–321. https://doi.org/10.1016/j.tourman.2016.12.011

Sweeney, J. C., & Soutar, G. N. (2001). Consumer perceived value: The development of a multiple item scale. *Journal of Retailing*, *77*(2), 203–220. https://doi.org/10.1016/S0022-4359(01)00041-0

Tanaskovic, M., Fagiano, L., Smith, R., Goulart, P., & Morari, M. (2013). Adaptive model predictive control for constrained linear systems. *European Control*. https://doi.org/10.23919/ECC.2013.6669544

Varki, S., & Colgate, M. (2001). The role of price perceptions in an integrated model of behavioral intentions. *Journal of Service Research*, *3*(3), 232–240. https://doi.org/10.1177/109467050133004

Woodruff, R. B. (1997). Customer value: The next source for competitive advantage. *Journal of the Academy of Marketing Science*, *25*(2), 139. https://doi.org/10.1007/BF02894350

Yoon, Y., & Uysal, M. (2005). An examination of the effects of motivation and satisfaction on destination loyalty: A structural model. *Tourism Management*, *26*(1), 45–56. https://doi.org/10.1016/j.tourman.2003.08.016

Yusof, T. A. (2021, July 14). *Malaysia committed to creating more Muslim-friendly tourism segments, says Nancy*. New Straits Times. https://www.nst.com.my/news/nation/2021/07/708318/malaysia-committed-creating-more-muslim-friendly-tourism-segments-says

Zamani-Farahani, H., & Henderson, J. C. (2010). Islamic tourism and managing tourism development in Islamic societies: The cases of Iran and Saudi Arabia. *International Journal of Tourism Research*, *12*(1), 79–89. https://doi.org/10.1002/jtr.741

Zeithaml, V. A. (1988). Consumer perceptions of price, quality, and value: A mean-send model and synthesis of evidence. *Journal of Marketing*, *52*(3), 2–22. https://doi.org/10.1177/002224298805200302

Zeithaml, V. A., & Bitner, M. J. (2003). *Services marketing: Integrating customer focus across the firm* (2nd ed.). McGraw-Hill/Irwin.

Chapter 15

Economic Contribution and Employment Opportunities of Tourism and Hospitality Sectors

Ravi Kant Modi

Nirwan University Jaipur, India

Abstract

A lot of research and studies have focused on how tourism might help a country's economy growth in recent years. Year after year, governments throughout the world pump a ton of money into tourist infrastructure because everyone knows it boosts economies. These days, tourism is a major force in the growth and prosperity of nations. According to data from the World Travel and Tourism Council and the World Tourism Organization, the tourism sector supported 320 million jobs, or 10% of all employment, and generated 10.4% of the world's GDP (gross domestic product) in 2018–2019. This sector is vital to the growth and economy of approximately 191 countries and 25 regions. The leisure industry received a disproportionate share of total expenditure (79% vs 22% for businesses), and it was responsible for 7% of all exports worldwide and 27.5% of all service exports. Domestic tourism accounted for 71.2% of all tourist expenditure in 2018–2019, with emerging nations witnessing the fastest rise. By fostering a sense of national pride and disseminating development and regional economic benefits, domestic tourism promotes possibilities. Without a doubt, novel advancements are possible atop these preexisting elements. As a means of economic progress, cultural exchange and the cultivation of mutual love, travel is becoming increasingly important as some countries turn inward.

Keywords: Economic contribution; employment opportunities; employment challenges; tourism; hospitality; visitor exports

The Emerald Handbook of Tourism Economics and Sustainable Development, 293–306

doi:10.1108/978-1-83753-708-220241015

1. Introduction

Worldwide, the tourism industry is among the most prominent and fast expanding markets. Three main sectors have emerged as engines of economic growth in the modern world: technology, telecommunications (including software and cyber-space) and the hospitality and tourist industries.

With a growth rate of 4%, the travel and tourism industry has outpaced the world economy for eight years running. This growth is attributed to the rise of middle-class families and strong worldwide consumer spending. In 2018, the Travel and Tourism gross domestic product (GDP) of the United States, China, Japan, the United Kingdom, the European Union, the Middle East and the Eastern region collectively constituted 47% of the global total. While the Asia-Pacific area continued to outperform the rest of the world in 2018, expanding by 6.4%, the 8.6% increase in North Africa stands out. This growth demonstrates not just the sector's international influence and advantages, but also its resiliency and ability to overcome security concerns and the present worldwide terrorist environment. With that said, Singapore, Macau, Maldives and Ethiopia In terms of travel and tourism's impact on GDP growth in 2018, the top four countries were Egypt, Turkey, St. Kitts and Nevis and Ecuador (Travel and Tourism Economic Impact 2019–2023). Job creation is a direct result of this growth. The travel and tourism industry is a powerhouse when it comes to job creation; it supports one out of every 10 jobs globally and has been responsible for one out of every five jobs generated in the last five years. Work in the tourism industry is especially helpful for young people, women and other underrepresented groups. Destination beauty, peace, nature, security, and the strength of its currency are all factors that influence the flow of passengers throughout the globe.

To put it simply, tourism is the sum of all the activities, services and businesses that come together to give a vacation. This includes things like transportation, lodging, dining, shopping, entertainment, business and general hospitality.

2. Literature Review

One study that looked at the financial effects of tourist growth was 'Impact of Tourism on Employment: An Econometric Model of 50 Central and East European areas' by Aguayo (2011). By use of an econometric model, he demonstrated that tourism had a beneficial effect on work opportunities and market services. The model presented in this study demonstrates how tourism might positively affect the services industry. The tourism industry has the potential to spur quicker job growth and regional development because to the strong ties between different sectors in the area (Aguayo, 2011).

In 2010 book 'Tourism Development Principles and Practices', by Bhatia A.K. paints a comprehensive picture of the economic and historical dimensions of tourism. Infrastructural development, regional growth, job creation, etc. all see a multiplier effect from tourist revenue. Also covered in depth in this book are the many tourist organizations and the part played by travel agents in the industry's growth (Bhatia, 2010).

In 2008 study titled 'Employment and Income Potentiality of Tourism Development in Amhara area Ethiopia,' by Ajala. O.A. investigated how the industry may boost the local economy, create jobs and raise people's living standards in the Amhara region of Ethiopia. Hotels, souvenir shops, tour guides and water transportation companies have all been the author's primary areas of research (Ajala, 2008).

'Tourist: A Modern Synthesis' by Page and Connell (2006) analysed many factors that influence the development of jobs in the tourism sector, with a focus on the many kinds of tourism. While resort development will certainly increase employment opportunities, locals may not always reap the most rewards (Page & Connell, 2020).

Regarding the 'Challenges of Sustainability of Tourism Development in Developing World', Tosun (2001) wrote a chapter. Although it may conflict with the ideas of long-term sustainable development, governments in poor nations do not have many choices but to encourage present tourist development, according to this research. There is a mismatch between the requirements of developing nations and the tourist development methods designed for wealthy nations (Tosun, 2001).

3. Summarized Results

The entire contribution of the travel and tourism industry to the global economy reached 10% of GDP in 2019. In 2013, the travel and tourism industry provided around 266 million jobs, or 8.9% of the total workforce (Aynalem et al., 2016). In 2021, this number increased to 2 in 11 out of all occupations in the globe. Despite the global economic and social crisis, the travel and tourism sector of the global economy is expected to generate 296 million jobs by the year 2023. This indicates that tourism has the potential to become a major job generator, as stated in the report that was published by the International Labour Organization (ILO)'s world Dialogue Forum on New Developments and Challenge (International Labour Organization, 2010).

One major source of employment is the tourism industry, which requires a large amount of human work. It allows young people, women and migrant workers quick access to the workforce and is one of the world's leading employers of varying skill levels. For example, in 2019 alone, tourism in India created 25 million jobs and that number is expected to rise to 40 million in 2023, making it the second-largest employer in the travel and tourist industry worldwide (Dayananda, 2014). Numerous job openings are created by the tourism and friendly reception sectors in various sectors, such as lodging, transportation, attractions, safari sites, etc. As a result, every strategy or programme for the growth of tourism must take into account the accessibility of competent and educated labour. However, working in the tourist business isn't without its challenges, such as poor income, long, antisocial hours and precarious employment status. High employee turnover and the difficulty in finding qualified candidates are the most noticeable and immediate effects of this predicament and they have a negative impact on the industry's performance.

Indeed, employees working in the tourist and friendly greeting industries are receiving less than enough attention. To put it another way, individuals are the essential component of a service delivery system that is both great and punctual. These workers are among the most important resources that the firm has as a consequence of this. There are two key methodologies that will be used in order to express the link that exists between Human Resources and Industry.

The first criterion for a successful tourist industry is the availability of a sufficient supply of high-quality employees or property force.

The second point is that all parties involved, including managers and consumers, should treat employees fairly and with respect. One of the most important factors that determines the perceptions and experiences that visitors have of a website is the amount of professionalism that is shown by its staff members.

As a result, the goal of this review paper is to give a succinct overview of the information about the potential and difficulties associated with the job market in the travel and hospitality sector. This research also aims to draw attention to the significance of the results for pertinent stakeholders, including scholars, legislators and industry owners, so that they can take appropriate action to develop a sustainable workforce.

4. Economic Contribution of Travel and Tourism in Countries GDP

For the majority of nations, the travel and tourism industry is a key driver of national GDP growth. The sector has substantial indirect and induced effects on the economy in addition to its direct effects. Travel and tourism's direct contribution is the only one that can be quantified using the technique authorized by the World Organization's Statistics Division for tourism satellite accounting. However, Travel & Tourism's overall effect is significantly greater and World Travel and Tourism Council (WTTC) intends to capture this, as well as any indirect or induced consequences, via its yearly report (WTTC, 2014, 2015).

These two types of travel expenditures – the 'internal' (the total expenditure within a country on travel by both citizens and visitors) and the 'individual' (the expenditure by the government on travel services directly linked to guests) both contribute directly to the value that travel and tourism create.

Hotels, airlines, airports, travel agencies and recreation services are just a few of the industries that are specific to tourism and deal directly with travelers. This approach can be used to evaluate the direct contribution of travel to value. The value is calculated based on the output that is recorded in national accounting. One way to determine the direct contribution of travel to value is to 'net out' the purchases made by the various tourist sectors from the total internal disbursal. This allows for the determination of the value of travel. This is in accordance with the 2008 tourism satellite account: advised method framework's concept of tourist value (TSA: RMF 2008).

The 'broader impacts', or indirect and induced effects, on the economy are a part of travel's overall contribution. The 'indirect' contribution includes the value and jobs supported by:

- The investment expenditure in the travel and tourism industry is a crucial part of both the present and the future of the industry. This includes investments like new aircraft purchases and hotel building.
- Spending by the government on behalf of the 'society at large' has numerous positive effects on the travel and tourist industry, including marketing and sales, airports, administration, security, cleanliness and more.
- Expenditures made inside the country by industries directly serving the tourism industry, such as hotels for food and cleaning services, aeroplanes for fuel and catering and travel agencies for information technology services.

The 'induced' contribution reflects the worth and employment opportunities created by the distribution of these World Health Organization funds, which are used either directly or indirectly by the travel industry.

5. Opportunities for Employment in the Tourism and Hospitality Sectors

The ability to create as many jobs as possible is the tourist industry's greatest strength. Regional and economic growth are both aided by tourism. One of India's most important sources of foreign currency, according to a recent report, is the tourist industry, which has flourished thanks to globalization and the open economy. With it comes the chance to build up the necessary infrastructure. Improving a country's roads, automobiles, communication, water supply, airports and railway stations is crucial to any country's overall prosperity. The growth of a nation is the only determinant of its economic development and industrial advancement. In addition, the tourist industry is crucial to the improvement of these infrastructures.

Job openings in the business and hospitality industries are often generated indirectly or directly depending on the level of engagement or contribution offered by commercial enterprises. The sum of all job openings directly supported by the travel and tourist industry is known as direct employment opportunities. Restaurants, hotels, tour guides, scouts, local transportation, protected areas, resorts and shopping centres, souvenir shops, photography, sightseeing tours and other occupations are all examples of professions that fall under the aegis of the hospitality industry.

There are a great number of companies that derive their revenue from the tourism and hospitality industry. These enterprises include those that provide supplies to eateries, construction firms and aircraft manufacturers, producers of various handicrafts, marketing firms and accountancy services.

There was a direct impact of $2,750.7 billion from travel and tourism on GDP in 2022. (3.2% of GDP). In 2023, this is expected to increase by 3.6% to reach

USD 2,849.2 billion. The hospitality, tourism, airline and other passenger transportation sectors are mostly responsible for this economic activity. But it also include things like the restaurant and recreation sectors, which rely on visitors for direct support. By 2029, the travel and tourism industry is projected to account for 3.5% of GDP, or 4,065.0 billion USD, an increase of 3.6% a year (Caribbean Hotel & Tourism Association, 2011) (Fig. 15.1).

When evaluating the impact of tourism on the economy, the economic indicators that are taken into consideration include income, employment, investment and development. The wages and salaries that are paid to individuals who are employed in a manner that either directly fulfils the demands of visitors or indirectly benefits from their distribution are likely to form the majority of the financial gain in a labour-intensive enterprise such as commerce and cordial welcome. The tourism industry stands to gain a lot from places that draw in a lot of people. When people visit a place, they tend to spend a lot of money since there are so many things to do and see.

Globally, the popularity of tourism and international travel has increased, with many people viewing it as a fundamental human right. As a result, destinations are working hard to meet the demands of tourists and provide them with high-quality services. Travelers also have a lot of holiday options to choose from. International organizations in the tourism and hospitality industries are setting standards for quality customer service, which helps to ensure sustainable development. Lastly, the ongoing decline in oil prices has made travel more affordable. The increasing influx of tourists to these locations is mostly due to the aforementioned factors. Consequently, a large staff is needed to cater to the needs of the visitors. This is why there is a wide range of job possibilities in the tourist and hospitality industries and its ancillary fields.

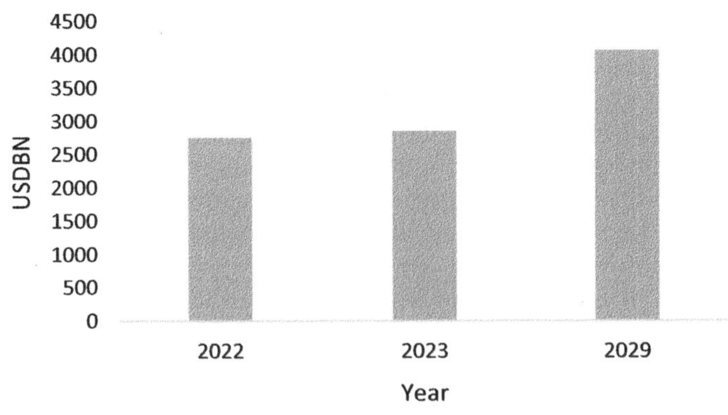

Fig. 15.1. Direct Impact of Travel and Tourism on GDP. *Source:* Produced by author.

Expected growth of 3.6% to USD 9,126.7 billion (10.4% of GDP) in 2023 from \$8,811.0 billion in 2022, bringing the overall contribution of travel to GDP to 10.4%.

Forecasts indicate that it will reach USD 13,085.7 billion by 2029, an increase of 3.7% per year (11.5% of GDP) (Fig. 15.2).

6. Employments Challenges in Tourist Industries and Hospitality

There were 122,891,000 direct jobs in the travel industry in 2022 (or 3.8% of the total workforce), and experts predict that number will rise to 125,595,000 in 2023 – a 0.2% increase (3.9% of total employment) (Fig. 15.3). Employers in this sector include lodging establishments, tour operators, airlines and rideshares (excluding commuter services). Included in this category are the building and leisure sectors, which rely on visitors for revenue.

When larger consequences like as investment, the provision of resources and iatrogenic financial gain are taken into consideration, travel was responsible for contributing to the employment market in 2022 to the tune of 318,811,000 jobs (10.0% of total employment). This is expected to reach 328,208,000 jobs in 2023, an increase of 2.9% (10.1% of total employment).

With an expected annual growth rate of 2.5%, the travel industry is projected to provide 420,659,000 jobs by 2029, accounting for 11.7% of total employment.

With a 2.1% annual growth rate over the next decade, the travel and tourism industry will directly support 154,060,000 employment by 2029. The next

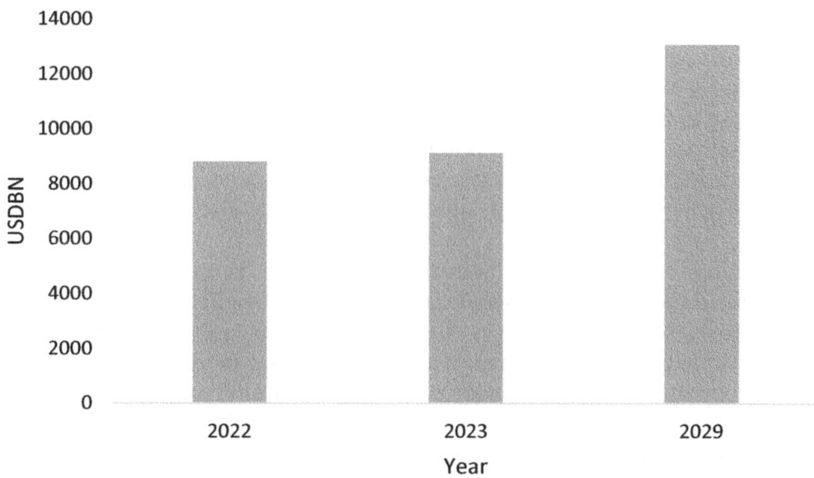

Fig. 15.2. Overall Contribution of Travel and Tourism to GDP.
Source: Produced by author.

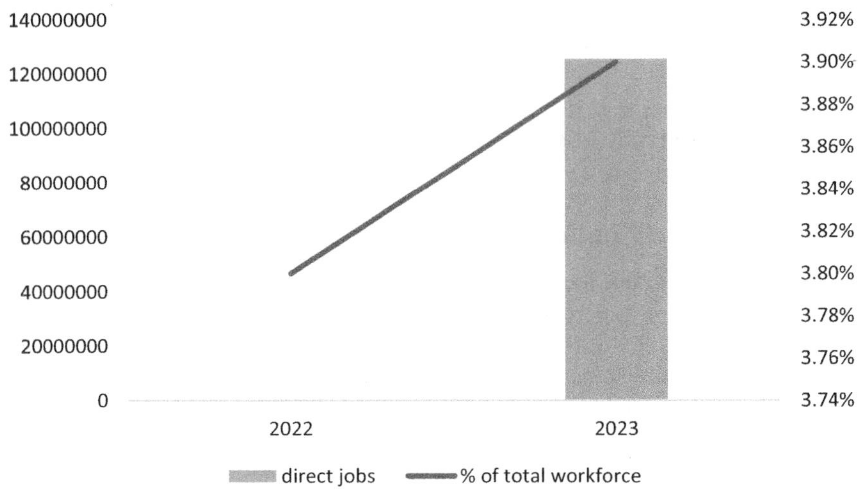

Fig. 15.3. Direct Impact of Travel and Tourism to Employment.
Source: Produced by author.

paragraphs address the key difficulties, as recognized by the majority of studies, of working in the tourist industry and providing friendly greeting.

Unfair treatment: One kind of worker is not handled in the same manner as another. For example, according to the International Labour Organization and the United Nations Development Programme (ILO-UNDP), which cited Thomas (2013), there is a gender imbalance in the industries. Despite the fact that they only own one percent of the property and provide 10% of the revenue, women are responsible for producing 50% of the food and doing 66% of the labour in the globe. These conditions exist because women are less able than men to obtain land, capital and education. Working from home or in family companies puts women at a higher risk of not only confronting bias and an overabundance of labour in both professional and familial settings, but also of not being legally protected.

The United Nations World Tourism Organization released a report in 2011 that evaluated the opportunities and difficulties faced by girls seeking employment in the tourism sector. The report found that insecure jobs are more common among unskilled or semi-skilled girls and that they are more likely to experience violence, exploitation, stress, sexual harassment and unfavourable working conditions. There are several reasons why women in underdeveloped countries have not been able to gain economic independence and take fully in the economy. Cultural norms, low levels of education and training, poverty, poor maternal health and a lack of sexuality education are a few of these causes. Businesses may occasionally be situated distant from residential areas, particularly in developing countries and communities, which will cost women money in terms of time and travel because they often have less access to resources that allow them to be flexible with their schedules and finances (United Nations World Tourism Organization [UNWTO], 2011).

The second issue is that women are more likely to be employed in lower-level positions and vocations that provide less opportunities for professional growth, while males are more likely to be employed in crucial management jobs. The number of women who hold senior and managerial positions is consistently low. In Europe, for example, women make up just 32% of managers in enterprises established in the European Union.

Family responsibilities and caregiving are still not allocated in an equitable manner. Girls are primarily responsible for providing care for members of the family who are reliant on them. Girls are far more likely than men to make the decision to take parental leave. As a result of this, as well as the scarcity of resources for elder and child care, women are often forced to quit the workforce. Women who have dependent children have a lower employment rate than males who have dependent children, with a rate of 62.4% compared to 91.4% (UNWTO, 2011). Vergas had also made the observation that, in contrast to occupations in other sectors, careers in the hospitality and tourist industries are often aimed at those who are younger than 35 years old. 43.4% of workers in this sector in Spain are between the ages of 25 and 34. It was stressed by Swarbrooke that companies typically reject female candidates who are beyond a particular age (about 35) and prefer resumes that include photographs of women. The tourism industry's dismal record of hiring and promoting persons with disabilities only rubs salt in the wound.

Inadequate pay and working conditions: Studies conducted by the European Foundation for the Improvement of Living and Working Conditions (Eurofound) reveal that, in comparison to average salaries in almost every EU member state, the average payment rate in hotels and restaurants is comparatively low. There are a number of features that are associated with the tourism and hospitality sectors. These include low hourly earnings, overtime without extra compensation, lengthy workweeks (up to 50 hours) and inadequate or non-existent breaks during peak season.

A lot of repetitive tasks like booking computers and running theme park rides are among the other problems, along with an inappropriate corporate philosophy, management style and leadership style, unplanned hiring, the use of foreign labour, seeing workers as expenses rather than assets, inflexible leadership that goes against democracy and inadequate training. These are but a handful of the issues that have been noted. The tourist and hospitality sectors commonly experience difficulties in terms of employment due to one or more of the following issues, which is the reason why this is the case: and seasonality (during some months of the year, there are relatively few tourists, forcing these sectors to reduce workforce). One very clear illustration of this is the fact that travellers only visit Ethiopia between the months of September and February. Other examples include part-time work that requires an excessive amount of hours, family work that is either low-paying or unpaid and informal, often illegal, labour that is considerably more difficult to quantify.

In addition, the multitude of difficult conditions that the tourist and hospitality sectors have been confronted with have an effect on the employment opportunities that are available. For example, there are numerous terrorist attacks occurring in popular tourist destinations; there are emerging infectious diseases; there are global economic crises; and the tourism and hospitality industries are experiencing technological advancements like the use of e-business, e-marketing and virtual tourism. As a consequence of these reasons, a great number of enterprises operating in the tourism and hospitality industries are experiencing financial losses. The major reason why businesses fire their workers, which in turn leads to an increase in the unemployment rate, is because of this.

7. Best Practices

The hotel, catering and tourism industries in the Euro Zone and several Western countries are considered best practices due to their size and rapid growth, with an average of 50%–70% female participation. More significantly, the food, catering and lodging sectors – rather than the economy at large – have larger percentages of women employees with competitive wages and working conditions in the Euro zone, Croatia, Serbia, Norway and Switzerland. In some nations, such the Bahamas, Canada and Australia, over 55% of the workforce is made up of women. Jamaica, Canada, the Dominican Republic and India are some of the countries that have managed their human resources in the tourist industry in the most effective manner. To put that into perspective, women only make up roughly 25% of the workforce in the tourist business in Egypt.

8. Implications of the Results and Future Directions

According to a 2010 UNWTO research, tourism can help reduce extreme poverty and empower impoverished women to escape the cycle of poverty through training, entrepreneurship, formal and informal jobs and community improvement. But not every woman is profiting from company development in the same way. The poorest women may not always be able to take advantage of the growth of the tourism industry due to a lack of resources and education. While commerce promotes women's empowerment in some areas, tourism has a detrimental impact on women's lives and reinforces existing gender and economic disparities in other areas.

Thomas claims that social discourse and discussions with all relevant parties – governments and intergovernmental bodies, local government sectors, trade unions, communities and their diverse member groups, non-governmental organizations and community-based tourism initiatives – can advance equity in terms of opportunities and treatment, close the pay and salary gaps for jobs of equal value, encourage more women to participate in decision-making to increase the number of women in managerial positions. These are just some of the benefits that can be achieved through these strategies. Through consistent meetings and conversations with the ILO, in order to achieve equitable pay and work environments free from sexual harassment, discrimination and exploitation, we must fight towards. We also need to ensure and enforce compliance with labour laws that uphold fundamental human rights, prohibit forced labour and the exploitation of child labour and put an end to discrimination. In addition, we need to create jobs that are stable, full-time and offer fair salaries and benefits (International Labour Office [ILO], 2013).

The creation of democratic, participatory management cultures is crucial for creating productive workforces in the tourism and hospitality sectors. The management of human resources is also emphasized, with a particular emphasis on bringing quality, recognising employees as assets, rewarding top performers and enhancing ties with academic institutions to provide courses that are more closely suited to the needs of the travel and tourism sector. For most employees in the field, this is the only method of accessing educational possibilities.

In order to meet consumer demand for responsible travel and to promote environmental and social performance, the United Nations Environment Programme and the ILO pay proper attention to standardization and certification. Enterprises such as lodging facilities, resorts, marinas, tour operators, travel agencies and transportation providers are also bound by certification programmes. They might even be qualified to attest to the natural resources and environmental soundness of tourist locations. The business sector uses a variety of ISO (International Organization for Standardisation) standards to improve the calibre of services. These represent ISO 14000 for environmental management and ISO 9000 for quality management.

It is necessary to implement clear policies in order to counteract the effects of seasonality, as stated by Griffin and De Lacey. These policies should include, among other things, the identification of markets that are most likely to travel during the off-season, the provision of discounted seasonal offers, the scheduling of events during less busy times of the year, the encouragement of attractions to remain open later, and the consideration of alternative solutions in situations where year-round full-time employment is not feasible. At the same time as working hours are altered to meet seasonal patterns, for instance, benefits might be maintained (Griffin & DeLacey, 2002).

In cases where they are not required by law, national and municipal governments ought to make every effort to offer the greatest pay and benefits. When necessary, high-quality amenities like food, housing and healthcare should be offered. The allocation of service charges should follow a transparent and well-documented process.

9. Visitor Exports and Investment

9.1 Visitor Exports

One significant component of travel's direct impact may be the money that foreign visitors to a rural area (or traveller exports) spend there. The world's visitor exports brought in USD 1,643.2 billion in 2022. This is anticipated to increase by 4.0% in 2023 and 1,484,910,000 foreign visitors are anticipated to visit the world.

International visitor arrivals are expected to reach 2,196,090,000 by 2029, resulting in spending of USD 2,483.9 billion, or 3.8% annually.

10. Different Components of Travel and Tourism

In 2022, the value of travel (including domestic and international) for leisure purposes accounted for 78.5% of the total travel value, or USD 4,475.3 billion, while business travel accounted for 21.5% (USD 1,228.0 billion) (Fig. 15.4A).

Spending on leisure travel is projected to increase by 3.8% to USD 4,646.6 billion in 2023 and by 3.9% annually to USD 6,780.7 billion in 2029.

Spending on business travel is anticipated to increase by 3% to USD 1,265.4 billion in 2023 and by 3.2% annually to USD 1,735.1 billion in 2029.

In 2022, the amount spent on domestic travel accounted for 71.2% of the GDP directly related to travel, while visitor exports accounted for 28.8% (Fig. 15.4B).

It is anticipated that domestic travel expenditure will increase by 3.5% to USD 4,203.8 billion in 2023 and by 3.7% per annum to USD 6,031.9 billion in 2029.

It is anticipated that visitor exports will increase by 4.0% to USD 1,708.2 billion in 2023 and by 3.8% annually to USD 2,483.9 billion in 2029.

Both employment and the GDP are impacted by the travel and tourism sector.

Travel and tourism make up three times more of the GDP overall than they do directly.

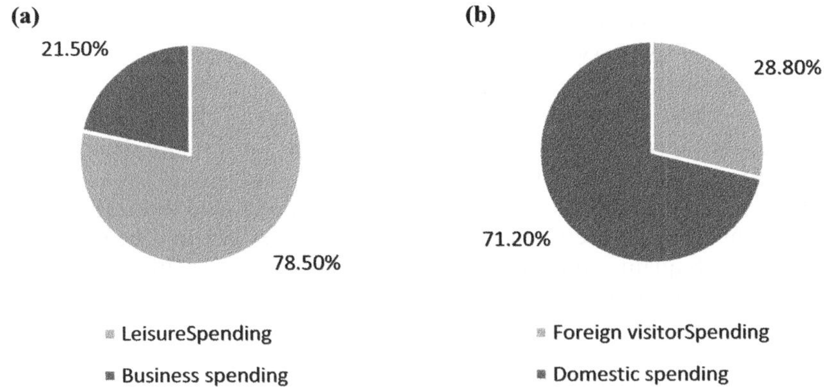

Fig. 15.4. Different Components of Travel and Tourism. *Source:* Produced by author.

11. Conclusion

The business and hospitality industries currently comprise the largest segments of the global economy. It enhances work prospects, revenue creation and the balance of payments. According to recent information, in 2014, travel and business created over 276 million jobs and contributed 9.8% of the global GDP. Additionally, by generating positive employment opportunities, tourism may support the socio-cultural amplification of the populace and empower communities of women and persons with disabilities. Because of this, both established and emerging nations give quality labour to high-quality businesses and encourage industries to contribute to the growth of their respective nations.

According to the review, the tourism and hospitality industries have been responsible for the creation of millions of jobs across the globe in a wide range of industries. These industries include lodging, food and beverage, land transportation and natural and cultural attractions. Adequate climate, political stability, shorter workweeks, technical innovation, higher profit margins and a pleasant atmosphere are the most supportive elements for the development of the explicit company, which allows businesses to create millions of jobs.

The tourism and hospitality industries face a number of challenges when it comes to employment, some of which include unequal treatment of male and female employees; a workforce that is poorly trained; poor working environments; seasonality; sexual provocation; anxiety and excessive labour hours without compensation for overtime. Not only does the sector provide poor salaries, but there are also other sociocultural aspects that must be taken into consideration.

References

Aguayo, E. (2011). Impact of tourism on employment: An econometric model of 50 CEEB regions. *Regional and Sectoral Economic Studies*, *11*(1), 37–46. https://www.usc.es/economet/reviews/eers1114.pdf

Ajala, O. A. (2008). Employment and income potentiality of tourism development in Amhara Region Ethiopia. *Ethiopian Journal of Environmental Studies and Management*, *1*(2), 74–82. https://doi.org/10.4314/ejesm.v1i2.41584

Aynalem, S., Birhanu, K., & Tesefay, S. (2016). Employment opportunities and challenges in tourism and hospitality sectors. *Journal of Tourism & Hospitality*, *5*(6), 1–5. https://doi.org/10.4172/2167-0269.1000257

Bhatia, A. K. (2010). *Tourism development principles and practice* (Vol. 85).

Caribbean Hotel & Tourism Association. (2011). *Travel and tourism's economic impact*.

Dayananda, K. C. (2014). Tourism and employment: Opportunities and challenges in Karnataka – Special reference to Kodagu District. *IOSR Journal of Humanities and Social Science*, *19*(11), 1–11.

Griffin, T., & DeLacey, T. (2002). *Green globe: Sustainability accreditation for tourism*. https://doi.org/10.1016/B978-0-7506-8946-5.50009-5

International Labour Office (ILO). (2013). *Sectoral Activities Department*.

International Labour Organization. (2010). *Development and challenges in tourism and hospitality sector.*

Page, S. J., & Connell, J. (2006). *Tourism a modern synthesis.* Thomson Publication.

Page, S. J., & Connell, J. (2020). Introduction to tourism: Themes, concepts and issues. In *Tourism* (pp. 2–21). Routledge.

Thomas, B. (2013). *International perspectives on women and work in hotels, catering and tourism.* Bureau for Gender Equality, International Labour Office. https://hdl.handle.net/1813/87660

Tosun. (2001). *Study on challenges of sustainability of tourism development in developing world.*

Travel and Tourism Economic Impact 2019–2023 by World Travel and Tourism Council.

United Nations World Tourism Organization (UNWTO). (2011). *A report on the opportunities and challenges which face women with respect to employment in tourism (UNWTO): Global report on women in tourism 2010.*

UNWTO. (2011). *Global report on women in tourism 2010.*

World Travel and Tourism Council (WTTC). (2015). *Travel and tourism economic impact of Ethioipa.*

WTTC. (2014). *Travel and tourism economic impact.*

Chapter 16

Revenge or Renewal? The Nexus of Revenge Tourism, Boomerang Effect and Sustainable Development Goals

Parag Shukla[a], *Jahanvi Bansal*[b] *and Pankaj Kumar Tripathi*[a]

[a]The Maharaja Sayajirao University of Baroda, India
[b]Gujarat State Fertilizers & Chemicals University (GSFCU), India

Abstract

Revenge travel has become a prominent touristic behaviour on a global scale. The study delves into the phenomenon of 'revenge tourism' and its far-reaching implications, emphasising the boomerang effects it triggers. This study proposes conceptual framework that examines the boomerang effects of revenge tourism (economic, environmental and social) and explores their relationship with the SDG8 (Decent work and economic growth), SDG12 (Responsible consumption and production) and SDG14 (Life below water). The study finds that the boomerang effects of revenge tourism positively impact all three Sustainable Development Goals (SDGs). It establishes a positive relationship between 'tourism demand' (RT1) and its 'economic impact' (BE1), along with a positive connection between 'travel behaviour' (RT3) and both its 'environmental impact' (BE2) and 'social impact' (BE3). Additionally, a positive effect of 'destination image' (RT2) on 'travel behaviours' (RT3) is observed. This underscores the tourism industry's inherent capacity to navigate and mitigate economic repercussions arising from the COVID-19 pandemic. The study concludes by emphasising that while COVID-19 is temporary, sustainability implies permanence through optimal resource utilization and maximising benefits for conservation and local communities. Practical implications are discussed as final thoughts, providing actionable insights for the tourism industry.

Keywords: COVID-19; tourism; revenge tourism; boomerang effect; sustainable; SDGs

The Emerald Handbook of Tourism Economics and Sustainable Development, 307–318

doi:10.1108/978-1-83753-708-220241016

1. Introduction

While pandemic has retreated into the shadows across the globe, its lingering repercussions have become apparent. Much like the world witnessed a resurgence in Covid-19 cases following the initial decline, this boomerang or rebound effect could be attributed to the swift relaxation of pandemic control and social restriction measures. Concurrently, a comparable phenomenon unfolded within the tourism industry, where individuals, having endured prolonged confinement within the confines of their homes, embarked on a fervent and unrestrained spree of travel. This phenomenon, aptly named 'revenge travel', epitomized the pent-up desires and suppressed impulse that had accumulated during the pandemic-induced break.

Prior to the pandemic, the tourism industry constituted a significant 'chunk' of the global gross domestic product (GDP) and trade, representing approximately 10% (Spenceley et al., 2021) and 7% (UNTWO, 2023) respectively. In the aftermath of the anticipated downturn brought about by the pandemic, the industry has experienced a remarkable resurgence (Arora & Sharma, 2021). Notably, certain cities, such as Warsaw in Europe and Orlando in the United States, have demonstrated an extraordinary rebound in the realm of travel and tourism, with their respective contributions to GDP surpassing 14% and 10%, exceeding the figures from 2019 (Euronews, 2022).

The issue at hand revolves around the balance between 'revenge travel' and 'responsible travel', stemming from the prioritization and necessity of achieving sustainable growth. Revenge tourism, while it may sound silly, but has a darker side that carries negative connotation that could surge the already unsustainable tourism practices worldwide (Antariksa et al., 2022). This phenomenon runs the risk of causing excessive tourism and other allied issues.

Sustainable development is a comprehensive, multilayered and multifaceted concept aimed at generating value through the integration and harmonization of environmental, social and economic dimensions (Hunjra et al., 2022). The concept of 'sustainable development' has found voluminous mention in policy planning documents, most notably in the 2030 Agenda for Sustainable Development, which was embraced by all United Nations Member States in 2015. This agenda comprises 17 Sustainable Development Goals (SDGs) (Hunjra, Azam, et al., 2023). Notably, the realm of tourism is prominently featured within SDGs 8, 12 and 14 (Table 16.1). With this context in mind, the present study proposes a conceptual model by integrating 'boomerang effect' in the nexus between 'revenge travel' and the three SDGs.

The discourse surrounding revenge travel and its implications on the tourism industry has been initiated, as evidenced by Girish (2021). However, there exists a conspicuous dearth of comprehensive studies that thoroughly delve into this phenomenon, providing valuable and enduring insights for navigating crises akin to those posed by the challenges of the COVID-19 pandemic (Zaman et al., 2021). In the same vein, research focussing on establishing the relationship between revenge travel and its impacts on SDGs is conspicuously absent. Consequently, this study, to the best of the authors' knowledge, marks the inaugural endeavour to scrutinize the repercussions of revenge travel on sustainability. The primary

Table 16.1. Sustainable Development Goals (SDGs) and Their Respective Targets With 'Tourism' in Agenda.

SDG	Target	Particulars
SDG8	8.9	'By 2030, devise and implement policies to promote sustainable tourism that creates jobs and promotes local culture and products'
SDG12	12.b	'Develop and implement tools to monitor sustainable development impacts for sustainable tourism that creates jobs and promotes local culture and products'
SDG14	14.7	'By 2030, increase the economic benefits to small island developing States and least developed countries from the sustainable use of marine resources, including through sustainable management of fisheries, aquaculture and tourism'

Source: https://sdgs.un.org/goals

objective is to illuminate the concept of responsible tourism, thereby furnishing invaluable insights that contribute to the advancement of sustainable and responsible travel practices. The emphasis here lies in the motivation and originality of this chapter within the existing literature. By acknowledging the scarcity of comprehensive studies on revenge travel and its multifaceted impacts, particularly in relation to SDGs, the authors underscore the pioneering nature of their research. The study not only addresses a gap in the current scholarship but also seeks to make a meaningful contribution by shedding light on the unexplored territory of the intersection between revenge travel and sustainability. The emphasis on responsible tourism further underscores the chapter's potential to inform and guide future practices towards a more sustainable and ethical trajectory in the tourism industry.

2. Theoretical Background and Conceptual Framework

This study delves into the multifaceted implications of revenge tourism, highlighting the boomerang effects it sets in motion. It investigates the interplay between destination image and travel behaviour, tourism demand and its economic impact, environmental and social impact and the boomerang effects of these variables with the SDGs related to tourism (Fig. 16.1).

2.1 Revenge Travel

Revenge travel has emerged as a novel trend among post-pandemic travelers (Antariksa et al., 2022). While the term may appear contradictory to the fundamental purpose of travel, which is typically associated with leisure and enjoyment,

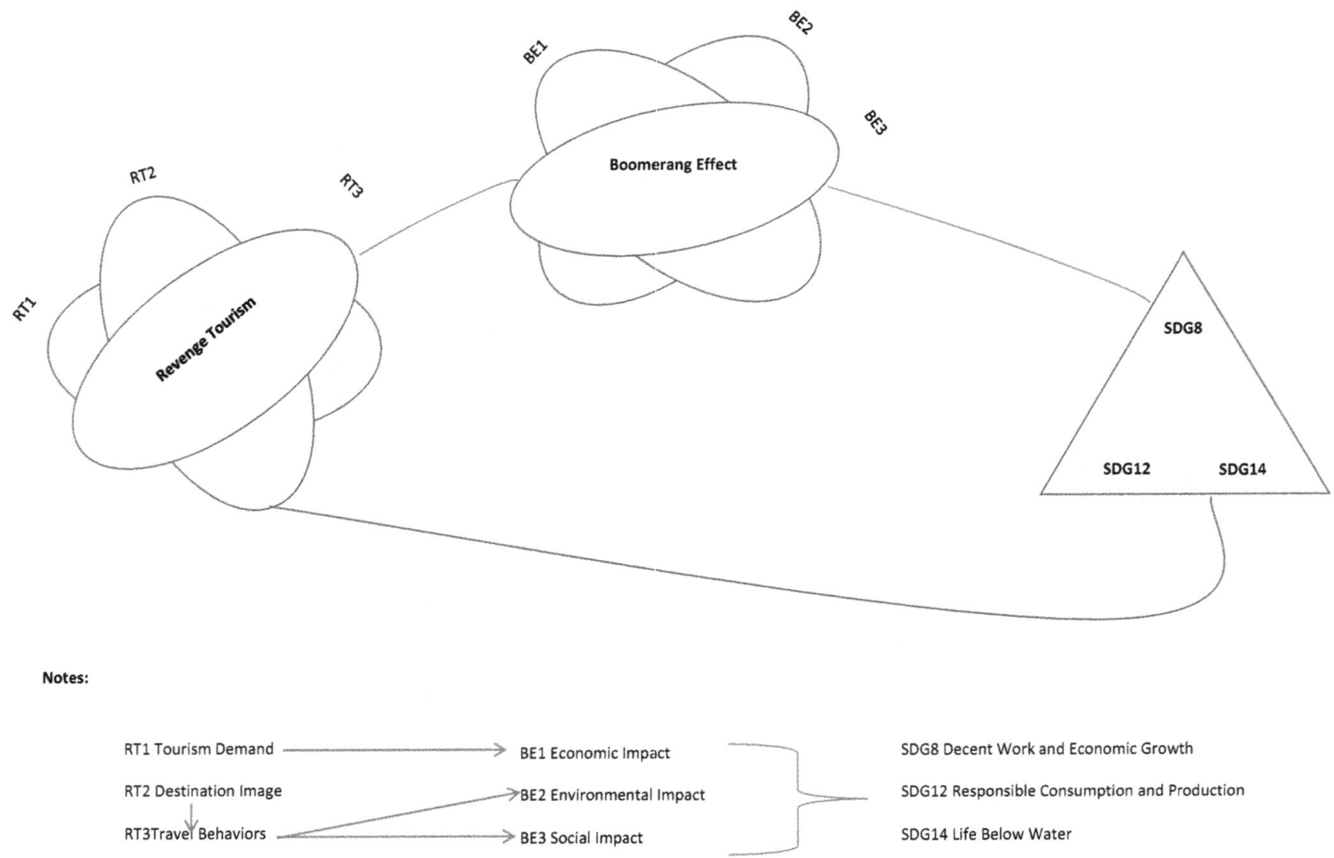

Fig. 16.1. A Conceptual Framework of Boomerang Effects of Revenge Tourism on SDGs [8, 12, & 14].

it has gained traction as a fashionable and trending concept. To comprehend revenge travel, we examined three key variables: tourism demand (RT1), destination image (RT2) and travel behaviours (RT3) (refer Fig. 16.1). The range of emotions, such as anger, lifelong pain from losing loved ones, frustration, strained relationships, a sense of urgency and impatience, and a 'now or never' mindset has emerged among people (Joshi & Sadhale, 2022). This surge in tourism demand, as a form of retaliation or revenge against the impact of COVID-19, has given rise to a boomerang effect. The phenomenon that arises when a threat to one's freedom of choice is perceived and followed by a heightened sense of emotional arousal, as outlined by Byrne and Hart (2009). The heightened demand for travel has once again brought forth questions regarding destination image and attractiveness.

The perception of a destination image relies on a combination of tangible factors, such as the imagery of tourist attractions and landscapes, along with intangible elements like the rich tapestries and mysteries of local stories and the sense safety that characterize these locations (Azeez, 2021). Together, these factors contribute positively to the destination image that forms in the mind of the tourist. Researchers have established that destination attractiveness is determined by the relative importance that a traveller places on the benefits they expect from a destination and the destination's perceived ability to provide those benefits (Orden-Mejía et al., 2022). Nevertheless, it is essential to acknowledge that the actual experiences often diverge from these initial perceptions. The case against travel asserts that travelers can sometimes have a detrimental rather than a beneficial impact on their travel destinations (Callard, 2023). The uncontrolled, impatient, overcrowded and poorly planned influx of tourists resulting from revenge tourism has had significant adverse effects on the environmental quality of tourist destinations (Baloch et al., 2023). These effects include issues like soil erosion, depletion of natural resources, waste accumulation, air pollution and the endangerment of biodiversity, as well as the disruption of socio-cultural ecosystems and the pristine nature of land and sea (Andlib & Salcedo-Castro, 2021). While these problems were present in the past, their severity has notably intensified during the post-pandemic period. Needless to say, the direct of tourism is the climate change giving repeated warning. According to the World Travel & Tourism Council, 8–10% of global CO_2 emissions are caused by the travel and tourism sector (Stüve, 2022). This significant environmental impact poses a threat to the attainment of sustainable development in the 21st century (Hunjra, Hassan, et al., 2023). In light of this, the following proposition is.

P1. There exit a positive relationship between 'destination image' (RT2) and 'travel behaviour' (RT3)

2.2 Revenge Travel and Boomerang Effects

When confronted with accusations of environmental damage, the tourism industry often counters with a compelling economic argument that one in 10 jobs globally depends on it (UNTWO, 2023). While this assertion holds truth, the

COVID-19 pandemic delivered a harsh blow to tourism service operators, leading to layoffs and furloughs, significantly impacting the jobs associated with the industry. Concurrently, travelers' expectations for a surreal vacation experience were affected, creating a boomerang effect that had economic impacts (BE1) and repercussions on tourism (refer Fig. 16.1). As the tourism industry has experienced rapid growth in many cities worldwide, it remains apparent that these impacts will eventually diminish over time.

The contrast between pro-environmental and anti-environmental behaviours among tourists plays a pivotal role in shaping the environmental (BE2) and social (BE3) impacts that will be left on a traveller's destination (refer Fig. 16.1). Speaking of tourist behaviour, prior research has demonstrated that a destination's pro-environmental image tends to foster a general willingness among tourists to invest extra effort in learning and engaging in pro-environmental actions (Bilynets & Cvelbar, 2022). The study further provide evidence that travelers who exhibit above-average dedication to environmental sustainability in their local communities are more likely to demonstrate environmentally conscious behaviour while travelling. While trading off between emotional and cognitive skills, it becomes evident that attitude often takes precedence over awareness or knowledge when it comes to promoting pro-environmental behaviours among travelers (Nilsson & Küller, 2000).

There are certain social effects that are challenging to project and directly link to tourism, yet they remain intertwined with it. This encompasses the anticipated cultural exchange between the host, regional, or local community and the tourists. However, tourists are often less inclined to embrace local customs and habits, except during their temporary stay. Conversely, travelers are more likely to instill enduring changes in the local community through the influence of their modern lifestyle, tastes and preferences. Confirming to this view, a study by Ramkissoon (2020) has purported that host residents' raise concerns about certain negative impacts of travelers' behaviour, including issues such as crime, alcohol abuse and sexual misconduct, among others. These issues can significantly affect the overall tranquility of the community in the region, and it is worth noting that many of these challenges emerged as outcomes of the COVID-19 pandemic in urban areas. The alternative perspective offers several favourable consequences of travelers' conduct. Tourism not only safeguards secluded social values and cultural practices but also involves the processes of recognising, nurturing and rediscovering them (Shahzalal, 2016). Travelers who exhibit respect and thoughtfulness can facilitate constructive cultural interactions while concurrently promoting a harmonious atmosphere, devoid of conflicts and tensions and thus rendering the destination more pleasurable for both visitors and locals. In a study conducted by Alamineh et al. (2023), examining the perceptions of local residents regarding the social impact of tourism, it was observed that residents tend to be more inclined towards the favourable social repercussions of tourism as opposed to the adverse effects. So, the subsequent propositions are as follows.

P2. There exist a positive relationship between 'tourism demand' (RT1) and 'economic impact' (BE1)

P3. There exist a positive relationship between 'travel behaviour' (RT3) and 'environmental impact' (BE2)

P4. There exist a positive relationship between 'travel behaviour' (RT3) and 'social impact' (BE3).

2.3 Revenge Travel, Boomerang Effects and Sustainable Development Goals (SDGs)

During pandemic crises, both the formal and informal sectors experienced significant job losses, but they found a resurgence in the form of revenge travel. Tourism industry, primarily characterized by seasonality and a lack of organization was particularly at risk, with women, youth and the informal workforce being especially most vunerable. Revenge travel has played a pivotal role in rejuvenating the tourism industry, by escalating the frequency and desire of travel within one's own country. Post-pandemic travelers exhibit a strong inclination towards exploring their own homeland rather than venturing to foreign destinations (BBC, 2022). This shift has emerged as a revenue-generating mechanism, with the money spent on travel circulating within the domestic economy. The surge in revenge travel has also given rise to new travel trends, including micro-holidays (short but frequent breaks), workcations (blending work with vacations) and experiential travel (immersing in local culture and traversing less-trodden paths). Consequently, it has fostered the discovery and development of new travel destinations.

The pandemic had a profound impact on small businesses, particularly in the Asia-Pacific region, causing significant hardships. One way to support sustainability is by altering consumer behaviour. Travelers can contribute to the sustainability of local communities by favouring local products, ensuring that money circulates within the community and strengthens livelihoods. The pandemic forced a halt in both domestic and international travel, resulting in a reduction in greenhouse gas emissions and a significant improvement in air quality. This crisis emphasized the importance of maintaining a clean and sustainable environment. In the context of revenge travel, it is notable that people were willing to pay a premium of at least 10% for sustainable travel features, even in the midst of a cost of living crisis, as indicated in the by the London-based market research firm Euromonitor International (2023a). This demonstrates a growing commitment to environmentally responsible travel, suggesting that sustainability has become a significant factor in decision-making for travelers.

Tourism stands out as a pivotal driver of development in Small Island Developing States (SIDS). However, SIDS are characterized by a set of environmental, economic and socio-cultural vulnerabilities primarily stemming from their petite landmass, delicate natural environments, susceptibility to climate change, remote geographical locations and coastal settings. On the one side of the spectrum, tourism can inflict harm on these fragile island ecosystems and erode the authenticity of their cultures. On the other hand, it has the potential to generate crucial local employment opportunities and stimulate foreign investments. The sudden and dramatic reduction in tourist numbers during the COVID-19 pandemic plunged many SIDS into financial crises. The adverse impacts of revenge tourism were particularly pronounced in locations that lacked favourable geographic positioning and connectivity. These factors played a significant role in determining how severely these islands were affected by the sudden influx of tourists during the post-pandemic period. Thus, the ensuing postulations is.

P5. There exist a positive relationship between 'boomerang effects of revenge tourism (BE1, BE2, & BE3)' and SDG8, 12 & 14 with respect to tourism.

3. Discussions and Conclusions

Despite the severe blow dealt by the pandemic, the tourism industry has experienced a resurgence thanks to a phenomenon commonly referred to in literature as 'revenge tourism'. At the outset, study aimed to comprehend the boomerang effect brought about by the revenge tourism. The underlying premise is that revenge travel, in itself, is not necessarily negative; rather, it reflects a fundamental aspect of human nature, the inclination to engage in what was once restricted. However, a lingering question remains: is revenge travel an indicator of the tourism industry's post-pandemic revival, or does it potentially raise issues that could hinder the implementation of sustainable tourism practices? Moreover, the study introduces three key variables concerning revenge tourism, including, tourism demand, destination image and travel behaviour to scrutinize the ins and outs effects of revenge tourism. The study puts forward the idea that a positive destination image and a compelling allure have a central role in moulding preconceived notions about a place and exert a substantial impact on tourist behaviour.

The study firmly establishes that the impactful, albeit brief, life cycle of revenge travel has set in motion a boomerang effect with three variables encompassing the economic, environmental and social aspects. Nevertheless, it is notable that priory is given to environmental sustainability, economic sustainability is often taken for granted, while social sustainability is put at the backseat. Hence, revenge tourism following the pandemic has provided governments and policymakers a distinctive window of opportunity to re-evaluate and re-structure the tourism sector's interactions with economies, the environment and societies. This becomes particularly pertinent in the context of developing countries, where structural issues such as inadequate governance impede progress towards achieving SDGs (Hunjra, Azam, et al., 2023).

The surge in travel associated with revenge tourism has yielded positive outcomes across multiple facets of SDGs. Notably, SDG8, which centres on stimulating decent work and economic growth by harnessing the economic upswing generated by revenge tourism, thereby fostering job creation and bolstering economic stability. In addition, SDG12 is influenced by the behaviour of tourists and their interaction with the environment during their journeys, highlighting the significance of sustainable practices and responsible consumption. Meanwhile, the environmental repercussions of revenge tourism on coastal regions and marine ecosystems potentially intersect with SDG14, reflecting the need to address issues such as resource depletion and environmental impact to safeguard life below water. So, the study opens a window of opportunities to craft a transformative and sustainable roadmap for the future of tourism industry.

4. Practical Implications, Limitations and Future Research Directions

While the hype surrounding revenge tourism may be waning due to escalating inflation and tighter budgets, the lessons learnt are enduring. It is imperative to explore new tourism trends, with a particular focus on diversifying offerings and customer bases to enhance resilience against economic vulnerabilities during crises. One such trend is virtual tourism, which provides viewers with an immersive experience of activities, locations, or destinations through technology. Another promising trend is sustainable and experiential tourism, which forms the crux of this study. In this context, the digital marketing of smaller and medium-scale events plays a crucial role, often being overlooked, especially in the case of religious events where attention tends to be disproportionately concentrated on major religious gatherings (Arora & Sharma, 2020). The repercussions of revenge tourism have underscored the need for more inclusive planning, benefiting society at large, which would naturally promote tourism, rather than creating tourism development primarily aimed at catering to elite tourists and lining the pockets of large hotel chains. The concept of 'public–private partnership' remains highly pertinent for involving all relevant stakeholders in the pursuit of a promising and sustainable world. Such partnerships endure through crises, fostering mutual cooperation among individuals for collective and individual benefits.

To attain a sustainable profile post pandemic, it is evidently crucial to pick the travel destinations wisely. Sweden is identified as the most sustainable destination as per the Sustainable Travel Index 2023, followed by Finland, Austria, Estonia and Norway (Euromonitor International, 2023b). These rankings are based on a set of sustainable criteria, including energy consumption, the impact of tourism on the environment, preservation of local cultures and the social well-being of the local population. However, it's worth noting that flying to these destinations can paradoxically counteract sustainability efforts due to the high carbon emissions associated with passenger jets.

Conceptual studies often involve the synthesis and interpretation of diverse literature. This may lead to challenges in generalising findings, as the scope and nature of revenge tourism can vary widely across different regions, cultures and tourism contexts. Future researchers might investigate cross-country data to comprehend the adverse economic, social and cultural repercussions of this phenomenon. The nascent nature of the research topic, given its relative novelty, poses a significant constraint. Thus, the limited body of literature on the specific intersection of revenge tourism and sustainability, may constrain the depth and breadth of the study. Despite these challenges, the novelty of the topic also presents an opportunity for the conceptual study to contribute to the nascent literature by proposing innovative frameworks, hypotheses, or conceptual models. Within this context, it is crucial to note that the study's scope is confined to empirical investigations of the proposed hypotheses. To fortify support within the tourism sector and propel progress towards the SDGs, additional interventions are imperative.

References

Alamineh, G. A., Hussein, J. W., Endaweke, Y., & Taddesse, B. (2023). The local communities' perceptions on the social impact of tourism and its implication for sustainable development in Amhara regional state. *Heliyon*, *9*(6). https://doi.org/10.3390/su132212834

Andlib, Z., & Salcedo-Castro, J. (2021). The impacts of tourism and governance on CO_2 emissions in selected south Asian countries. *Etikonomi*, *20*(2), 385–396. https://doi.org/10.15408/etk.v20i2.17499

Antariksa, B., Rusata, T., Andriani, D., & Priyatmoko, R. (2022, December). Quality tourism in the emergence of revenge travel: A post-pandemic policy framework. In *International academic conference on tourism (INTACT). "Post pandemic tourism: Trends and future directions" (INTACT 2022)* (pp. 189–201). Atlantis Press. https://doi.org/10.2991/978-2-494069-73-2_14

Arora, S., & Sharma, A. (2020). Digital marketing for religious event of India for tourism sustainability and promotion. In *The Emerald handbook of ICT in tourism and hospitality* (pp. 453–465). Emerald Publishing Limited. https://doi.org/10.1108/978-1-83982-688-720201029

Arora, S., & Sharma, A. (2021). Covid-19 impact on overtourism: Diversion from mass tourism to alternative tourism. In *Overtourism as destination risk* (pp. 275–283). Emerald Publishing Limited. https://doi.org/10.1108/978-1-83909-706-520211018

Azeez, Z. A. (2021). The impact of destination image on tourist behavior: Karbala as a case study. *International Journal of Sustainable Development and Planning*, *16*(7), 1287–1298. https://doi.org/10.18280/ijsdp.160709

Baloch, Q. B., Shah, S. N., Iqbal, N., Sheeraz, M., Asadullah, M., Mahar, S., & Khan, A. U. (2023). Impact of tourism development upon environmental sustainability: A suggested framework for sustainable ecotourism. *Environmental Science and Pollution Research*, *30*(3), 5917–5930. https://doi.org/10.1007/s11356-022-22496-w

BBC. (2022). *India tourism: How 'revenge travel' saved the industry*. https://www.bbc.com/news/world-asia-india-62978198. Accessed on November 7, 2023.

Bilynets, I., & Cvelbar, L. K. (2022). Tourist pro-environmental behaviour: The role of environmental image of destination and daily behaviour. *Annals of Tourism Research Empirical Insights, 3*(2), 100070. https://doi.org/10.1016/j.annale.2022.100070

Byrne, S., & Hart, P. S. (2009). The boomerang effect a synthesis of findings and a preliminary theoretical framework. *Annals of the International Communication Association, 33*(1), 3–37. https://doi.org/10.1080/23808985.2009.11679083

Callard, A. (2023). The case against travel. *The New Yoker*. https://www.newyorker.com/culture/the-weekend-essay/the-case-against-travel. Accessed on October 10, 2023.

Euromonitor International. (2023a). *Sustainable travel index: Ramping up action for positive change*. https://www.euromonitor.com/sustainable-travel-index-ramping-up-action-for-positive-change/report. Accessed on August 9, 2023.

Euromonitor International. (2023b). *Travellers will pay 10% extra for sustainable travel despite cost of living crisis: Euromonitor Report*. https://www.euromonitor.com/press/press-releases/august-2023/travellers-will-pay-10-extra-for-sustainable-travel-despite-cost-of-living-crisis-euromonitor-report. Accessed on September 9, 2023.

Euronews. (2022). *Global tourism rebounds strongly after COVID-19*. https://euronews.com/travel/2022/11/30/global-tourism-rebounds-strongly-after-covid-19. Accessed on January 10, 2023.

Girish, V. G. (2021). COVID-19 pandemic and the emergence of revenge travel. *International Journal of Tourism and Hospitality Research, 35*(10), 125–129. https://doi.org/10.21298/IJTHR.2021.10.35.10.125

Hunjra, A. I., Azam, M., Bruna, M. G., & Bouri, E. (2023). A cross-regional investigation of institutional quality and sustainable development. *Journal of International Financial Markets, Institutions and Money, 84*, 101758. https://doi.org/10.1016/j.intfin.2023.101758

Hunjra, A. I., Azam, M., Bruna, M. G., Verhoeven, P., & Al-Faryan, M. A. S. (2022). Sustainable development: The impact of political risk, macroeconomic policy uncertainty and ethnic conflict. *International Review of Financial Analysis, 84*, 102370. https://doi.org/10.1016/j.irfa.2022.102370

Hunjra, A. I., Hassan, M. K., Zaied, Y. B., & Managi, S. (2023). Nexus between green finance, environmental degradation, and sustainable development: Evidence from developing countries. *Resources Policy, 81*, 103371. https://doi.org/10.1016/j.resourpol.2023.103371

Joshi, A., & Sadhale, M. (2022). Revenge tourism: An overview of the phenomenon in India. *Mahratta-Multi-Disciplinary Journal, 1*(2), 1–5. http://localhost:8080/xmlui/handle/123456789/10099

Nilsson, M., & Küller, R. (2000). Travel behaviour and environmental concern. *Transportation Research Part D: Transport and Environment, 5*(3), 211–234. https://doi.org/10.1016/S1361-9209(99)00034-6

Orden-Mejía, M., Carvache-Franco, M., Huertas, A., Carvache-Franco, W., Landeta-Bejarano, N., & Carvache-Franco, O. (2022). Post-COVID-19 tourists' preferences, attitudes and travel expectations: A study in Guayaquil, Ecuador. *International Journal of Environmental Research and Public Health, 19*(8), 4822. https://doi.org/10.3390%2Fijerph19084822

Ramkissoon, H. (2020). Perceived social impacts of tourism and quality-of-life: A new conceptual model. *Journal of Sustainable Tourism*, *31*(2), 442–459. https://doi.org/10.1080/09669582.2020.1858091

Shahzalal, M. (2016). Positive and negative impacts of tourism on culture: A critical review of examples from the contemporary literature. *Journal of Tourism, Hospitality and Sports*, *20*, 30–35. https://doi.org/10.1080/09669582.2020.1858091

Spenceley, A., McCool, S., Newsome, D., Báez, A., Barborak, J. R., Blye, C. J., Bricker, K., Cahyadi, H. S., Corrigan, K., Halpenny, E., Hvenegaard, G., King, D. M., Leung, Y. F., Mandić, A., Naidoo, R., Rüede, D., Sano, J., Sarhan, M., Santamaria, V., Sousa, T. B., & Zschiegner, A. K. (2021). Tourism in protected and conserved areas amid the COVID-19 pandemic. *Parks*, 27, 103–118. https://doi.org/10.2305/IUCN.CH.2021.PARKS-27-SIAS.en

Stüve, LL. (2022). *How bad for the environment is traveling?* https://www.dw.com/en/how-bad-for-the-environment-is-traveling/a-64092681. Accessed on October 10, 2023.

UNTWO. (2023). *Tourism and Covid-19 – Unprecedented economic impacts.* https://www.newyorker.com/culture/the-weekend-essay/the-case-against-travel. Accessed on September 10, 2023.

Zaman, U., Raza, S. H., Abbasi, S., Aktan, M., & Farías, P. (2021). Sustainable or a butterfly effect in global tourism? Nexus of pandemic fatigue, covid-19-branded destination safety, travel stimulus incentives, and post-pandemic revenge travel. *Sustainability*, *13*(22), 12834. https://doi.org/10.3390/su132212834

Chapter 17

Revitalisation and the Adaptive Re-use of Heritage Buildings for Event Venue in Rajasthan: Opportunities and Challenges

Anukrati Sharma and Shruti Arora

University of Kota, India

Abstract

Venue comes from the Latin word meaning 'come' (venire). Not only the event's location or venue affect attendance but it can also have an impact on the event's personality. Additionally, the location or venue affects the event's atmosphere and the visitor experience. The most significant benefit of revitalisation or adaptive reuse is its positive impact on the environment as razing buildings is a major source of waste and carbon emissions, by contrast, adaptive reuse reduces waste, saves energy and conserves resources. The persistence of this chapter is to investigate the challenges and opportunities in making adaptive re-use or revitalisation of heritage buildings that includes forts/palace/museums as a venue for any event. For the purpose of achieving the goal, secondary data from 2000 to 2023 have been compiled from more than 50 research articles that have been published in peer-reviewed and University Grant Commission (UGC) journals, books written by various authors, including the authors themselves, other researchers working in the related field of event management, conference proceedings and websites. The authors discovered from their examination of the literature that the significance of the event venue cannot be unheeded. The event's tone sets the stage for possible guest experience influences. One must carefully consider the kind of venues or settings that are available in the destination before conducting any event.

Keywords: Venue; heritage; event management; MICE (meeting; incentives; conferences; exhibition); historic/heritage buildings; revitalisation; adaptive re-use

The Emerald Handbook of Tourism Economics and Sustainable Development, 319–333

Published under exclusive licence by Emerald Publishing Limited
doi:10.1108/978-1-83753-708-220241017

1. Introduction

Human nature has always included events; yet, it has only lately turned into a legitimate business. Throughout history, celebrations have been a vital part of human culture, with people coming together for a variety of reasons in civilisations all over the world. In the modern period, the event sector has experienced rapid growth. The Events & Entertainment Industry has expanded to the point where it now includes a variety of events, like press conferences, product launches, wedding celebrations, birthday parties, gala dinners, road shows, music concerts, Bollywood shows, reality shows, company conferences, tourist attractions, sports and cultural programmes, the list is endless. This industry is rapidly growing and has a big impact on both business and tourism. The effective execution of the event depends on choosing the finest location and venue. The term 'event venue' refers to the physical site of an event. Tourism is greatly influenced by events, and the events sector serves as a major engine of economic activity in host locations (Maguire & Hanrahan, 2017) and aids governments in achieving their socioeconomic objectives (Maguire, 2019). There are many choices to be made when planning an event, but one of the key choices that will have the most influence on any event is selecting the appropriate location or venue. Depending on the nature and intent of an event, there are several venue kinds. Like:

- Resorts for weddings, reunions, corporate conferences.
- Hotels for business meetings, product launch, conferences, merrymakings, marriages and so much more.
- Breweries and wineries for hosting private parties, showers, or any other private events.
- Restaurants for business dinners, rehearsal dinners and other important occasions.
- Community centres for business meeting, seminar, or small party would be wise to contact local community centres.
- Museums, aquariums and zoos: In addition to showcasing amazing artwork and historical relics, many museums also organize activities.
- Art galleries may be used to hold parties, fundraisers, networking meetings and other private gatherings.
- Stadiums and stadiums for live music events, conferences for whole brands, major athletic events and touring shows.
- Convention centres offer adaptable interior space for holding public gatherings, rallies, trade exhibits and more.
- Historical location for pre-wedding shoot or wedding.
- Academic venues like universities, colleges for stage shows, comedy night, a play or music night.

There may be several tempting options for venues, in which case the organizers must evaluate each location to ensure that it meets the needs of the event, including best pricing, available dates, environment, guest accommodations, restrooms, reputation or image, accessibility, parking options, appearance and aesthetics, guest capacity, acoustics, professionally managed, medical facility, sound quality, seating configurations, water supply, accessible Wi-Fi, smell and cleanliness. The majority of events nowadays, especially in the digital age, need technical and audio-visual (A/V) infrastructure to enable presentations, live streaming and interactive components. A/V equipment of the highest calibre, including microphones, projectors, screens and sound systems, is crucial before choosing a location. As a result, choosing the appropriate location for the event is crucial and time-consuming. According to Forst (2022), organising a faultless event requires a lot of moving parts to work together flawlessly. Choosing the appropriate site is perhaps the most crucial and significant component. Locations are the starting point for creating incredible experiences. A flawless delivery of excellent experiences and material to an enthusiastic, involved audience is made possible by the appropriate location. In present times, *mixed-style heritage buildings are rising and finding a new lease of life as galleries, restaurants and event venues.* Steeped in tradition and rich architecture, timeless heritage venues are the canvas for iconic events that tell the deeper story of longstanding corporate values and milestones. Think luxury events and exhibitions, which have used heritage venues as unique showpieces to amplify particular themes and ideas. These landmarks lend events their sense of gravitas and help to bind the event's central message and mood in a way that will stick in the minds of guests. *Heritage venues give planners the chance to incorporate an experiential tour of the venue's unique design and history (M&C, Asia, December 2019).* Development and history have a crucial interaction that illustrates the link between sustainability, revival and conservation. Maintaining and caring for existing things while making that the demands of current and future owners and users are fairly balanced is essential to the long-term sustainability of heritage structures. When the state government designated tourism as an economic sector in 1989, Rajasthan was one of the states to introduce novel ideas like historic hotels and cultural excursions. This is in line with a plan to transition from pilgrimages to affluent tourists by heavily promoting its distinctive ancient cities, palaces, fairs and festivals, as well as offering financial incentives for hotels that are mostly geared towards foreign tourists. While some of Rajasthan's monuments, historic structures and forts – like Mehrangarh Fort and Amber Fort – are well-maintained, the majority – like Taragarh Fort in Bundi and Badalgarh Fort in Khetri, among many others – are poorly managed and require substantial attention.

All revitalisation efforts ultimately need funding. Getting the money needed to take projects through to the end is a major challenge. Adaptive reuse is one way to raise these kinds of revenues. In accordance with this method, historic buildings may be leased by private organisations and/or non-governmental organisations for profitable purposes that preserve their historical significance.

2. Review of Literature

Events in the context of tourism are essentially gatherings of people for ceremonial, private, or public celebration, commemoration, or ritual. One of the most crucial factors when planning an event is the venue. A venue should be able to accommodate the number of people expected to attend. Otherwise, it will feel cramped and lack ambience. For example, a too huge venue might drown out the attendees and create a boring mood. However, if it is too little, everyone will be crammed too close, making it appear busy and uncomfortable for visitors. The quality of a space, technology and facilities are other important considerations, requiring to step into the shoes of attendees. The most crucial site selection factors for special events are the staff's attitude, the venue's size, atmosphere and initial impression. Superstructure, or services or amenities provided for tourists, such as hotels, restaurants and other lodging options and infrastructure, or services set up for residents to use, can help or hurt a venue during the site selection process because of the requirement for various amenities as suggested by Rompf et al. (2008), Tsai and Ho (2008). Planners chose one site over another due to its accessibility and plenty of leisure activities (Rompf et al., 2008; Tsai & Ho, 2008). Cities have changed significantly over time, corresponding with all advancements. The physical transformation of the city has been directly impacted not only by historical political, social, demographic and economic events but also by technical advancements, particularly in the last century. Its requirements and demands have drastically changed as a result of this predicament. Numerous structures have outlived their primary use and might be modified to serve new purposes. Adaptive reuse is a new kind of city regeneration that may be retained, according to Yung and Chan (2012). It prolongs the life of buildings and prevents waste from destruction, encourages dynamic embodied recycling and offers significant social and economic advantages to the global community. According to the Australian Department of the Environment and Heritage (2004), adaptive re-use is the process of transforming an object that is no longer needed or useless into a new one that may be utilized for a different purpose. It is also the process of maintaining the majority of the old building while modernising and improving the performance of buildings to meet evolving user requirements and standards. Demolition of existing structures is a material and energy waste. By adding value to underutilized assets, adaptive reuse may revitalize a neighbourhood and significantly lower the cost of new development and land acquisition. Historic structures serve as a reminder of past civilisations and are crucial for passing along cultural identity to future generations. Proposing a new use for these buildings is essential to preserve their preservation, as they can no longer be utilized for their intended purpose (Mısırlısoy & Günçe, 2016), despite the fact that adaptive reuse (AR) is a complex process from beginning to conclusion. Rajasthan's rich cultural legacy is well-known around the world. This encompasses both the visible and intangible cultural legacy, such as traditional water features, religious buildings, palaces, mansions and art forms, as well as fairs, festivals, music, dance and food. Heritage buildings have particular cultural value and are distinctive landmarks located across the city. Using 10 steps – 'initiative', 'analysis of heritage buildings', 'value assessment', 'significance level mapping', 'definition of adaptive reuse potential', 'design strategy definition', 'final decision-making', 'implementation', 'maintenance', and 'assessment after years' – Arfa et al. (2022) developed a conceptual model that depicts the adaptive reuse process of heritage buildings. Unfortunately, a lot of historic structures have declined to disrepair due to improper maintenance management (Adegoriola et al., 2021). The most challenging and effective strategy for preserving cultural assets' worth and revitalisation is an adaptive reuse programme. The goal of revitalisation is to maintain historically significant structures while repurposing new areas in ways that do not undermine the community's cultural values or the preservationist tenets. The goal of the rehabilitation procedure is to raise money for and construct the heritage building by giving it new, acceptable uses. As a result, the building effectively has its own expenses because it is the commercial function that is included. Not asking for drastic changes but only minor changes. It is well acknowledged that cultural heritage preservation may have a positive social impact on the community's economy and feeling of identity. Historic sites draw tourists, commercial ventures, residential construction and ultimately real estate investment. It is apparent that the property market in historic regions commands greater prices than in other places following redevelopment (English Heritage, 2005). A complicated and quickly expanding segment of the events business is event management in historic structures (Connell et al., 2015; Tkaczynski & Rundle-Thiele, 2011). Event managers who work in heritage buildings seem to be at odds with the owners of those buildings, who want to maintain the building's historical significance while event managers want to provide the greatest possible event experience (Weidenfeld & Leask, 2013). In order to balance the amount of revenue generated by foot traffic with the amount of wear and tear brought on by heavy foot traffic, managers of heritage buildings must take these factors into account (Fyall et al., 2008; Morandotti et al., 2013, pp. 1–10). The preservation of legacy buildings aids in the understanding of the property identity by future generations (Shen & Langston, 2010). This is because heritage architecture embodies unique historical, artistic, social and political features of its era. Around the world, there are several instances of adaptive reuse of historic structures that highlight the versatility and inventiveness of this tactic. For instance, the Tate Modern in London is an earlier power station that was converted into a museum of contemporary art, fascinating millions of guests every year. An additional instance is the almost 113-year-old oldest Alembic Industrial building in Vadodara, India, which underwent renovations in 2018 by Karan Grover and Associates. Originally designed to produce penicillin, this iconic building is now a museum including areas set aside for art studios, displays and exhibitions. The Jai Vilas Palace, located in Gwalior, India, was built in the 19th century by the Scindia Dynasty in the European architectural style. While some of the palace is still occupied by the royal family's descendants, the majority of it has been transformed into the 'Jivajirao Scindia Museum', which showcases the rich culture of Gwalior under the Scindias. The museum also features opulent and tastefully decorated drawing rooms, bedrooms, banquet halls and dining rooms. The adaptive reuse strategy has faced criticism recently because of the willingness of local authorities and owners, which has negatively impacted heritage sites (Haidar & Talib, 2015). Although they maintain the historical features, they may have a direct or indirect impact on the site's validity (Tikhonova & Alho, 2015). The history, worth and continued existence of the priceless historic sites may be at jeopardy if an appropriate adaptive reuse strategy is not implemented. Therefore, using adaptive reuse correctly is the greatest approach to preserve cultural buildings' originality while also making a major contribution to the reduction of environmental burdens. With the owner and local government's desire to embrace the adaptive reuse technique in historic (Adams, 1983; Alivizatou, 2012; Bertolini et al., 2011; Burke, 2007; Foster, 2020) of the historical buildings, this plan has lately come under fire for allegedly compromising authenticity. By avoiding demolition and rebuilding operations, adaptive reuse of historic structures—that is, changing a historical building's original function to suit new conditions or demands—is essential to the sustainable growth of communities (Aplin, 2002). Reusing properties has the potential to revitalize existing towns, limit sprawl, drastically reduce the cost of developing and purchasing property and offer vital community resources (Godwin, 2011). Snyder (2005) contends that while a building's history and structure must be respected when a new use is added, a repurposed building does not have to be a significant piece of architecture in order for the process to be effective. Effective adaptive reuse preserves a building's historical qualities while working to give it a contemporary look that preserves its individuality (Department of Environment and Heritage, DEH). It is crucial to do this in order to preserve and transmit the true legacy to next generations (Mısırlısoy & Günçe, 2016). It not only prolongs the building's lifespan (Othman & Heba, 2018), but it also serves as a catalyst for the development and maintenance of environmental, social and economic values (Othman & Heba, 2018), all of which strengthen the building's sense of self (Djebbour & Biara, 2020). According to some research, adaptive reuse improves social cohesiveness and tolerance in the neighbouring communities (Chen et al., 2018). When adaptive reuse techniques are used on historic structures, they aim to preserve the original builders' labour, expertise and commitment in addition to protecting the building itself (Bullen & Love, 2009). According to Md Ali et al. (2019), a building that meets contemporary demands and activities is preferable to one that is empty. According to Bosone et al. (2021), adaptive reuse of underutilized and abandoned sites of various forms of cultural heritage can be a tactic for enhancing heritage conservation, promoting sustainable development processes through the co-creation of new meanings, repurposing neglected heritage resources and turning them into the new living cultural heritage of the area. The building's historic characteristics are given new life through adaptive reuse (Conejos et al., 2011). Furthermore, according to PPP Knowledge Lab (2019), legacy revitalisation is the restructuring of an existing city structure as a result of social and economic decline. It includes concepts for quality enhancement and the urban environment. Additionally, revitalising historic structures gives locals more employment options. Numerous roles and stakeholders, including specialists, employees and consulting departments, are needed to complete the revitalisation work and any associated monitoring in a single revitalisation project (Jacobus & Hickey, 2006). Adaptive reuse is becoming a potent tool for architects and builders in this day and age, when sustainability is at the forefront of architectural innovation. In the meanwhile, there can be some detrimental effects from heritage revival. One of the drawbacks of historic rehabilitation is the rising real estate values of nearby residential buildings. Conejos et al. (2016) attested to the difficulties adaptive reuse projects face, especially with relation to modern legal constraints, regulatory compliance and design specifications. Maintaining the original character and value of historic buildings while transforming them to meet modern norms and purposes is another difficulty. This calls for a collaborative, interdisciplinary approach including a range of stakeholders, including the public, architects, planners, users, owners and conservationists. Avoiding the detrimental consequences of adaptive reuse, such as gentrification, displacement, commercialisation and loss of authenticity, which might diminish the social and cultural significance of heritage buildings, is the second problem.

3. Methodology

The core objective of this research is to investigate the re-use and revitalisation of heritage buildings for event venue in Rajasthan with its challenges and opportunities in today's event industry. The research is descriptive in nature. For the purpose of achieving this objective, secondary data from 2000 to 2023 has been collected from more than 50 research articles that have been published in various journals, books written by various authors, including the authors themselves, conference proceedings and websites.

3.1 TheDevelopment and Re-use of Heritage Buildings for Event Venues in Rajasthan

Historic structures are valuable assets to any country. The economies, cultures, histories and sustainability of the countries, organisations and individuals all benefit from the reuse of these structures. According to Farjami and Turker (2019), heritage tourism is made possible by the preservation of historic structures, and preservation is financed in part by tourism. This leads to socio-economic sustainability. Reuse is a restoration strategy that helps to maintain the identity and legacy of an area. Hosting any personal or professional event is no longer the preserve of hotels or in-built conference and exhibition centres. Today, open theatres, museums, forts, palaces, universities and sporting complex also compete for their share in events sector. Because of the disruption of the work-life balance brought on by high-paying corporate positions, people nowadays scarcely have time for one another. However, they have the money to spend on event management services or venue planners to provide their loved ones with well-planned events. Particularly, the increasingly popular nuclear family structure raises the bar for this service in the events sector. The transformation of the consumer mindset plays an essential part in the growth and development of the event industry. The choice of location for an event can greatly influence its success since it sets the tone for the experience that attendees will have. The days of holding successful events only requiring a stage and sound equipment are long gone. Event venues must continue to push the boundaries in order to provide the ideal atmosphere and setting that complement the type of event being arranged since they have developed from outdoor concert venues to underground basements to nature trails and more. Innovative ideas and outside-the-box thinking assist to create a special experience for the visitors that makes them eagerly anticipate the upcoming event on the calendar. When it comes to selecting the ideal locations, today's event management companies push the boundaries of inventiveness to unprecedented heights. Choosing a versatile venue can suggestively impact the success of the event. With its flexibility, it can provide a personalized attendee experience, handle last-minute changes and offer cost-effective solutions. But one cannot ignore the fact, that with the change in time still there are few iconic venues around the world that provide an unforgettable environment or framework like The Brewery (London), Natural History Museum (London), Tijara Fort Palace (India), 1AQ (India), Yale Bowl (US), Caesars Palace (US), Hannover Messe (Germany), Feria Valencia (Spain), Mount Smart Stadium (New Zealand) or Dalhalla (Sweden) and the list goes on. Some of the examples in India that involved the palaces and forts of Rajasthan:

- Six Senses Fort Barwara: A stunning fort from the 14th century that was painstakingly transformed into a Six Senses wellness haven. Facing the Chauth ka Barwara Mandir temple, it was once possessed by the Rajasthani Royal Family. The 700-year-old courteous and royal atmosphere of a bygone period is recreated in the design. The lavish Six Senses Fort Barwara castle hosted Katrina Kaif and Vick Kaushal's (a Bollywood celebrity) wedding in 2021. It is a three-hour drive from Jaipur (https://www.sixsenses.com/en/resorts/fort-barwara/).
- Jaisalmer's Suryagarh Palace, Jaisalmer: Situated on Meru Hill and known as Trikoot Garh, Jaisalmer Fort was constructed in 1156 by the Bhati Rajput monarch Jaisal and has seen several conflicts. Based on this fort, the renowned Indian filmmaker Satyajit Ray created a detective book that he subsequently adapted into a movie called Sonar Kella (The Golden Fortress). Bollywood celebrities Kiara Advani and Sidharth Malhotra were married in 2023 at the castle (https://www.suryagarh.com/about).
- Umaid Bhawan Palace, Jodhpur: Maharaja Umaid Singh laid the foundation stone for this majestic palace in 1929 and construction was completed by 1943. In December 2018, Priyanka Chopra, a celebrity from Bollywood and Hollywood, wed Nick Jonas at Umaid Bhawan Palace in Jodhpur (https://www.tajhotels.com/en-in/taj/umaid-bhawan-palace-jodhpur/).
- Alila fort, Bishangarh: Built in the 18th century on a single granite hill. In a quaint amalgamation of their Punjabi and Tamil traditions, Sanjeevani Lamba (Business Operations Manager in a Pharmaceutical company) and Kelvin Kistnen's (Chartered Accountant) beguiling weddinging 2023 at Alila Fort Bishangarh was set against the picturesque beauty and rich heritage of Rajasthan. The vibrant multicultural celebration was attended by guests from all over the world and included unforgettable moments of love and beauty (https://www.hyatt.com/en-US/hotel/india/alila-fort-bishangarh/jaial?src=agn_pfxin_all_sem_JAIAL_BR_Exact_SubBrand-Property_IN_ALL_Jaipur-Feeder_google_sbp_e+p_x_x&src=pfxswain_sem___alila+fort+bishangarh&mckv=s-dc_pcrid_672110519421_mtid_529dkt13792&gad_source=1&gclid=CjwKCAiAivGuBhBEEiwAWiFmYV6JsziJXQoIer_aJi68IgI0gmstXQak8F7dP4urDLT1BsQNCcHRWRoCRcAQAvD_BwE&gclsrc=aw.ds).

- Chomu Palace: It is the epitome of elegance and upholds the vibrant culture of Rajasthan. This 300-year-old magnificence embodies the gracious lifestyle of the royals who previously lived there and traits its guests to the finest things life has to offer. The venue has done a splendid job in blending modernity and ancient heritage. The historic monument is immersed in grandeur and makes any celebration unforgettable (https://www.chomupalacehotel.com/).
- Hotel Castle, Mandawa: If deserts blossom with a multitude of hues, it is in Rajasthan's Shekhawati area. Every house in Shekhawati is embellished with brightly painted murals that provide light on the parched surroundings. This location is well-known for hosting conferences and royal weddings (http://castlemandawa.com/).
- Chittorgarh Fort, Chittorgarh: Ministry of Culture under the aegis of Azadi Ka Amrit Mahotsav celebrated the launch event of Qila Aur Kahaniya Campaign at Chittorgarh Fort, Rajasthan from 28 to 30 April 2023. Various artists from different parts of Rajasthan came and registered for Phad Art and Water Colour Workshops. In the evening, a light and sound show on Chittorgarh fort was organised which narrated the vast history of the fort and its various battles. An exhibition on the forts of Rajasthan as well as the Forts falling under Qila and Kahaniya campaign was put up at the venue. Also, a special live performance on the Bhajans of Meerabai enchanted the audience (https://amritmahotsav.nic.in/event-detail.htm?236625).
- Hawa Mahal, Jaipur: One of the most well-known tourist destinations in Jaipur is Hawa Mahal, popularly referred to as the 'Palace of Winds'. Maharaja Sawai Pratap Singh constructed it in 1799 to resemble the Hindu deity Krishna's crown. Basically, a theatre show, it is a fabulous mix of history, stories, music, dance and embellishments. The perfectly clear acoustics and the depiction of each subtlety of Rajasthani culture is the feature of each act. Traditional acrobats, young fellows and ladies performing incredible tricks, Ghoomar, fire dance and the Kalbeliya tribe displaying their moves are a portion of the performances that one gets the opportunity to experience with the Hawa Mahal Music Show (https://www.tourism.rajasthan.gov.in/content/rajasthan-tourism/en/shows.html).
- Laxmi Niwas Palace, Bikaner: The Indo-Saracenic style of architecture was created between 1898 and 1902 by Col. Samuel Swinton Jacob on a contract from Maharajah Sir Ganga Singh to serve as his royal home. It was created by Jacob in a scenic style and cost one lakh rupees to build. The exquisite stone detailing was a testament to the skill of the region's designers and craftspeople. Possessing a remarkable sense of hospitality, the Maharajah entertained a multitude of notable figures and dignitaries, such as King George V and Queen Mary, the Greek King, viceroys, other Maharajahs and well-known statesmen. Laxmi Niwas Palace was transformed into an opulent luxury hotel at the start of the 21st century, honouring its rich history with well conserved historic details. It is a stunning location for weddings with a touch of true royalty (https://laxminiwaspalace.com/).

- Tijara Fort, Alwar: Constructed during the 1800s, the Tijara Fort-Palace complex is an incomplete masterpiece consisting of three buildings designed in the Rajput-Afghan style with early colonial influences: the Mardana, which was built for the royal men, the Rani Mahal, which was built for the maharanis and the Hawa Mahal, or palace of winds, which is a pleasure palace perched on a plateau and overlooks a body of water. Tijara Fort-Palace, with its expansive and secluded surroundings and commanding views of the neighbouring mustard fields and countryside, is a charming place for weddings in Rajasthan (https://www.neemranahotels.com/tijara-fort-palace-alwar-rajasthan/).

Rajasthan of today is the ideal fusion of tradition and cutting-edge technology advancement. According to Nandwani (July, 2019), Indian hoteliers are converting historic havelis, mansions and palaces into historical hotels that appeal mostly to foreign tourists and give a nostalgic and romantic experience; almost 40% of all affluent global travellers. In Rajasthan, the heritage hospitality movement was pioneered by Maharana Bhagwat Singh, by renovating their summer palace Jag Niwas into the Taj Lake Palace Hotel in the initial 1960s. Of crucial importance is revitalising and restoring a heritage structure to welcome guests. Neemrana Fort Palace, formerly a desolate ruin that blended into the dusty Aravalli hills, was one of the first fort palaces to be renovated into a hotel, which Aman Nath, founder and chairman of Neemrana Hotels and his late partner, Francis Wacziarg discovered and restored to pristine condition (Nandwani, July, 2019). The wrecks were preserved and reestablished using traditional materials where necessary, as well as employing modern technologies and materials. The fort-palace was restored with the help of local artisans.

3.1.1 Challenges in Making Reuse of Historical Building

There are unique design issues when renovating an older structure, particularly when it has to adhere to modern user expectations and requirements. From infrastructure to furniture, the daunting task of keeping intact the historical relevance is sometimes overwhelming. According to the authors, the encounters accompanying with holding events in historic buildings are preservation, management and sustainability.

- The commercial real estate markets and building owners have provided little assistance in making buildings more sustainable.
- Disagreement on the new applications of the heritage with the neighbourhood.
- Technical installations, creative solutions and refurbishment methods are needed.
- Expensive upkeep and repairs because of flaws and physical degradation.
- Protracted and challenging rehabilitation or reuse, which frequently results in lower profitability.

- The building's artistic exterior look and finishing.
- Balancing the demands of the present and the future with the requirement to maintain the historic asset's distinctive character and value is the biggest problem when planning for historic structures.
- Not every structure is suitable for refurbishment; chances for inventive reuse may be limited by site constraints, structural issues and condition of repair.
- Some buildings have no alcohol at all, while others prohibit wine in specific locations.
- People are not permitted to wear heels with blades in some buildings.
- Because these places lack power, running water, or kitchens, food must be 'hot-boxed' into these spaces. It was said that storage in these structures was difficult.

3.2 Opportunities in Making Adaptive Re-use of Heritage Buildings for Event Venue

As a commercial activity in and of itself, as a location factor for other economic enterprises, as a contributor to environmental amenity and local identity, as a component of place image promotion and branding, and as a frequently catalytic element in neighbourhood regeneration, heritage is frequently called upon to perform many instrumental roles in largely local economic development strategies (Rizzo & Mignosa, 2013). According to Sharma (Federation of Rajasthan Event Management: FOREM, April 2022), events is a Rs 5,000-crore industry in Rajasthan and trying to take it to Rs 10,000 crore in the next 2–3 years. Sharma also states that weddings are a big segment but there should be some more events. In today's world, where everyone wants the finest event venue, Rajasthan has emerged as the preferred location for organising a grand event. To put it simply, it provides the royal experience like no other, not only with its multitude of forts and palaces, the majority of which have been turned into hotels, but also with the vibrancy that one connects with a celebration, whether its gastronomy or culture. Throughout the state, luxury venues such as historic palaces, forts and havelis that combine history and magnificence are accessible.

4. Conclusion and Future Directions

To arrange the ideal event, try to choose a location that has won awards, exclusive event spaces and environmentally friendly amenities. The setting determines whether a party or event will be successful. A carefully picked event site will undoubtedly complement the event's goals. Consider selecting the location with the most 'wow' factor. Choosing the ideal location is therefore a complex procedure. However, due to the amount of difficulty in adding elements to create the mood, certain locations might not be appropriate for hosting the event. One of the most important and difficult parts of organising an event is choosing the location. A badly picked location may cause attendees to have a bad time, which might lower attendance and cause them to stop doing business with you. For this reason, choosing the ideal location is crucial to producing an enjoyable and unforgettable event experience.

According to authors of the chapter, experienced venue managers may be employed to make improvements to the event venue management and planning in order to improve the attendee experience and move the prospective attendee dial from intrigued to registered. The hiring, scheduling, training and supervision of internal staff members, such as bartenders, ticket takers, maintenance personnel and security, is done by these venue managers. They also ensure that the area is tidy and safe and that the tools and equipment housed on site are properly maintained. Try to have a clear notion of the kind of location or venues that would work best for the event as the planner and make a pitch to it. Customers enjoy unusual or interesting locations, and they are frequently impressed by some 'new' discoveries. Give the customer a virtual tour of the event's layout in relation to the location. This can be important since a location that is architecturally fascinating might compliment the event, or be 'hidden' by decorations if it is a simple room. Heritage buildings can contribute significantly to a modern urban environment and meet the functional requirements associated with it while maintaining their original facades, historical patterns, layouts, architectural elements and historical associations through compatible reuse and appropriate assessment methods. Like in 2020, the Ministry of tourism, Government of India has developed a project for the growth/erection/upgradation of facilities and amenities along with their complete operation and maintenance for the heritage sites/monuments/tourist sites which was 'Adopt a Heritage' – 'Apni Dharohar, Apni Pehchaan'. This research presents few limitations, namely, it is limited to Rajasthan and the study is based on secondary data base. India, is a country with many forts, palaces, museums and many older and heritage building, there is still a lack of researches or studies, especially the research on the basis of primary data base the adaptive reuse and revitalisation of heritage building. Hence, it is recommended to study the authentic value of buildings. It is further motivated to evaluate the existing building conditions and to find the connection between adaptive reuse and authenticity of the building for event venues.

References

Adams, K. B. (1983). *Maryland heritage: A family history*. KB Adams.

Adegoriola, M., Lai, J. H. K., Chan, E. H., & Darko, A. (2021).Heritage building maintenance management (HBMM): A bibliometric-qualitative analysis of literature. *Journal of Building Engineering*, *42*. 102416

Alivizatou, M. (2012). The paradoxes of intangible heritage. *Safeguarding Intangible Cultural Heritage*, 9–21.

Aplin, G. (2002). *Heritage: Identification, conservation, and management*. Oxford University Press.

Arfa, F. H., Zijlstra, H., Lubelli, B., & Quist, W. (2022). Adaptive reuse of heritage buildings: From a literature review to a model of practice. *The Historic Environment: Policy & Practice*, *13*(2), 148–170. https://doi.org/10.1080/17567505.2022.2058551

Bertolini, L., Carsana, M., Gastaldi, M., Lollini, F., & Redaelli, E. (2011). Corrosion assessment and restoration strategies of reinforced concrete buildings of the cultural heritage. *Material and Corrosion*, *62*(2), 146–154. https://doi.org/10.1002/maco.201005773

Bosone, M., De Toro, P., Fusco Girard, L., Gravagnuolo, A., & Iodice, S. (2021). Indicators for ex-post evaluation of cultural heritage adaptive reuse impacts in the perspective of the circular economy. *Sustainability*, *13*(9), 4759. https://doi.org/10.3390/su13094759

Bullen, P., & Love, P. (2009). Residential regeneration and adaptive reuse: Learning from the experiences of Los Angeles. *Structural Survey*, *27*(5), 351–360. https://doi.org/10.1108/02630800911002611

Burke, S. (2007). ICOMOS: Twentieth century heritage International Scientific Committee. *Journal of Architectural Conservation*, *13*(2), 143–150. https://doi.org/10.1080/13556207.2007.10785001

Chen, C. S., Chiu, Y. H., & Tsai, L. (2018). Evaluating the adaptive reuse of historic buildings through multicriteria decision-making. *Habitat International*, *81*, 12–23. https://doi.org/10.1016/j.habitatint.2018.09.003

Conejos, S., Langston, C., Chan, E. H., & Chew, M. Y. (2016). Governance of heritage buildings: Australian regulatory barriers to adaptive reuse. *Building Research and Information*, *44*(5–6). https://doi.org/10.1080/09613218.2016.1156951

Conejos, S., Langston, C. A., & Smith, J. (2011). Improving the implementation of adaptive reuse strategies for historic buildings. In *International forum of studies titled SAVE heritage: Safeguard of architectural, visual, environmental heritage* (pp. 1–10). Institute of Sustainable Development and Architecture.

Connell, J., Page, S. J., & Meyer, D. (2015). Visitor attractions and events: Responding to seasonality. *Tourism Management*, *46*, 283–298. https://doi.org/10.1016/j.tourman.2014.06.013

Department of the Environment and Heritage. (2004). *Adaptive reuse: Preserving our past building our future*. Commonwealth of Australia. https://www.dcceew.gov.au/sites/default/files/documents/adaptive-reuse.pdf

Department of Environment and Heritage (DEH) *Adaptive reuse: Preserving our past, building our future*. https://www.environment.gov.au/system/files/resources/3845f27a-ad2c-4d40-8827-18c643c7adcd/files/adaptive-reuse.pdf

Djebbour, I., & Biara, R. W. (2020). The challenge of adaptive reuse towards the sustainability of heritage buildings. *International Journal of Conservation Science*, *11*(2), 519–530. https://www.proquest.com/openview/2561d7343150de0baca02a9fe7463a3f/1?pq-origsite=gscholar&cbl=5327637

English Heritage. (2005). *Regeneration and the historic environment heritage as a catalyst for better social and economic regeneration*. English Heritage.

Farjami, E., & Turker, Ö. O. (2019). Integrating adaptive reuse of heritage buildings with environmental rating systems. In D. Vanneste & W. Gruijthuijsen (Eds.), *Value of heritage for tourism: Proceedings of the 6th UNESCO UNITWIN conference* (pp. 75–85). University of Leuven. https://ees.kuleuven.be/eng/unitwin2019/proceedings/ProceedingsUNITWIN2019-Farjami.pdf

Forst, J. (2022, August). The art and science of event venue selection. https://www.eventindustrynews.com/featured/the-art-and-science-of-event-venue-selection. Accessed on 15 August 2023.

Foster, G. (2020). Circular economy strategies for adaptive reuse of cultural heritage buildings to reduce environmental impacts. *Resources, Conservation and Recycling*, *152*, 104507. https://doi.org/10.1016/j.resconrec.2019.104507

Fyall, A., Garrod, B., Leask, A., & Wanhill, S. (2008). *Managing visitor attractions: New directions* (2nd ed.). Butterworth Hienemann.

Godwin, P. J. (2011). Building conservation and sustainability in the United Kingdom. *Procedia Engineering*, *20*, 12–21. https://doi.org/10.1016/j.proeng.2011.11.135

Haidar, L., & Talib, A. (2015). Adaptive reuse practice in tower houses of old city Sana'a Yemen. *Procedia-Social and Behavioral Sciences*, *202*, 351–360. https://doi.org/10.1016/j.sbspro.2015.08.239

Jacobus, R., & Hickey, M. (2006). *Commercial revitalization planning guide: A toolkit for community based organizations*. The Local Initiatives Support Corporation Center for Commercial Revitalization.

M and C, Asia (2019, December). *Harnessing the power of history with heritage venues*. https://www.meetings-conventions-asia.com/News/Venues/Harnessing-the-power-of-history-with-heritage-venues. Accessed on 25 October 2023.

Maguire, K. (2019). Examining the power role of Local Authorities in planning for socio-economic event impacts. *Local Economy: The Journal of the Local Economy Policy Unit*, *34*(7), 657–679. https://doi.org/10.1177/0269094219889603

Maguire, K., & Hanrahan, J. (2017). Assessing the economic impact of event management in Ireland: A local authority planning perspective. *Event Management*, *21*(3), 333–346. https://doi.org/10.3727/152599517X14942648527545

Md Ali, Z., Zawawi, R., Myeda, N. E., & Mohamad, N. (2019). Adaptive reuse of historical buildings: Service quality measurement of Kuala Lumpur museums. *International Journal of Building Pathology and Adaptation*, *37*(1), 54–68. https://doi.org/10.1108/IJBPA-04-2018-0029

Mısırlısoy, D., & Günçe, K. (2016). Adaptive reuse strategies for heritage buildings: A holistic approach. *Sustainable Cities and Society*, *26*, 91–98. https://doi.org/10.1016/j.scs.2016.05.017

Morandotti, M., Besana, D., & Cinieri, V. (2013). *The sustainable management of historic buildings; Central Europe towards sustainable building*. Sustainable Refurbishment of Existing Building Stock

Nandwani, D. (2019, July). *Hotelier India*. https://www.hotelierindia.com/business/8005-indian-hoteliers-are-restoring-ancient-palaces-mansions-and-havelis-into-heritage-hotels-that-offer-a-blend-of. Accessed on 28 October 2023.

Othman, A. A. E., & Heba, E. (2018). Adaptive reuse: An innovative approach for generating sustainable values for historic buildings in developing countries. *Organization, Technology & Management in Construction: An International Journal*, *10*(1), 1704–1718. https://doi.org/10.2478/otmcj-2018-0002

PPP Knowledge Lab. (2019). *Urban revitalization*. https://pppknowledgelab.org/sectors/urban-revitalization

Rizzo, I., & Mignosa, A. (Eds.). (2013). *Handbook on the economics of cultural heritage*. Edward Elgar Publishing.

Rompf, P. D., Breiter, D., & Severt, K. (2008). Destination selection criteria: Key success factors evolve and dominate. *Event Management*, *12*(1), 27–38. https://doi.org/10.3727/152599509787992599

Shen, L. Y., & Langston, C. (2010). Adaptive reuse potential: An examination of differences between urban and non-urban projects. *Facilities*, *28*(1/2), 6–16. https://doi.org/10.1108/02632771011011369

Snyder, G. H. (2005). *Sustainability through adaptive reuse: The conversion of industrial buildings*. Doctoral dissertation. University of Cincinnati. http://rave.ohiolink.edu/etdc/view?acc_num=ucin1115644591

Tikhonova, O., & Alho, C. (2015). Authenticity criteria in castles of Ivano-Frankivsk Region. *European Scientific Journal.*

Tkaczynski, A., & Rundle-Thiele, S. R. (2011). Event segmentation: A review and research agenda. *Tourism Management*, *32*(2), 426–434. https://doi.org/10.1016/j.tourman.2010.03.010

Tsai, H. Y., & Ho, P. T. (2008). Assessing site selection criteria for Taiwan Lantern Festival by analytic hierarchy process. *Event Management*, *12*(3–4), 187–197. https://doi.org/10.3727/152599509789659713

Weidenfeld, A., & Leask, A. (2013). Exploring the relationship between visitor attractions and events: Definitions and management factors. *Current Issues in Tourism*, *16*(6), 552–569. https://doi.org/10.1080/13683500.2012.702736

Yung, E. H., & Chan, E. H. (2012). Implementation challenges to the adaptive reuse of heritage buildings: Towards the goals of sustainable, low carbon cities. *Habitat International*, *36*(3), 352–361. https://doi.org/10.1016/j.habitatint.2011.11.001

Printed and bound by CPI Group (UK) Ltd, Croydon, CR0 4YY

15/09/2026

14962490-0002